Analysis and Design of FRP Reinforced Concrete Structures

Analysis and Design of FRP Reinforced Concrete Structures

Shamsher Bahadur Singh

Ph.D. (Stru. Engg.), PDF (LTU, USA), PE (Michigan, USA)
Professor of Civil Engineering Department
BITS Pilani, Pilani-333 031, Rajasthan, India

Mc Graw Hill Education

New York Chicago San Francisco
Athens London Madrid
Mexico City Milan New Delhi
Singapore Sydney Toronto

1 2 3 4 5 6 7 8 9 0 QVS/QVS 19 18 17 16 15 14

ISBN 978-0-07-184789-6
MHID 0-07-184789-8

The sponsoring editor for this book was Lauren Poplawski and the production supervisor was Pamela Pelton. The art director for the cover was Jeff Weeks.

This book was previously published as *Analysis and Design of FRP Reinforced Concrete Structures* by McGraw-Hill Education (India) Private Limited, New Delhi, copyright © 2014.

This book is printed on acid-free paper.

To

My beloved son (Late) Mr. Sneh Vardhan,
My wife Kiran Singh and daughter Aparajita Singh,
for their great sacrifices made during the preparation of this book

Contents

Preface *xv*

Notation *xvii*

Chapter 1. Introduction **1**

1.1. Evolution of FRP Reinforcement *2*

1.2. Review of FRP Composites *3*

1.3. The Importance of the Polymer Matrix *4*

1.3.1. Matrix polymers *4*

1.3.2. Polyester resins *6*

1.3.3. Structural considerations in processing polymer matrix resins *8*

1.3.4. Reinforcing fibers for structural composites *8*

1.3.5. Effects of fiber length on laminate properties *9*

1.3.6. Bonding interphase *9*

1.3.7. Design considerations *10*

1.4. Description of Fibers *11*

1.4.1. Forms of glass fiber reinforcements *12*

1.4.2. Behavior of glass fibers under load *13*

1.4.3. Carbon fibers *13*

1.4.4. Aramid fibers *15*

1.4.5. Other organic fibers *16*

1.4.6. Hybrid reinforcements *17*

1.5. Manufacturing and Processing of Composites *17*

1.5.1. Steps of fabrication scheme *17*

1.5.2. Manufacturing methods *18*

1.6. Sandwich Construction *23*

1.7. Compression Molding *24*

1.8. Multi-Axial Fabric for Structural Components *24*

1.9. Fabrication of Stirrups *24*

1.10. FRP Composites *25*

1.11. FRP Composite Applications *26*

1.12. Composite Mechanics *26*

1.12.1. Laminate terminology *27*

1.12.2. Composite product forms *27*

1.13. Laminates Types and Stacking Sequence *28*

Chapter 2. Material Characteristics of FRP Bars **39**

2.1. Physical and Mechanical Properties *39*

2.2. Physical Properties *39*

2.3. Mechanical Properties and Behavior *41*

2.3.1. Tensile behavior *41*

2.3.2. Compressive behavior *42*

2.3.3. Shear behavior *43*

2.3.4. Bond behavior *43*

2.4. Time-Dependent Behavior *43*

2.4.1. Creep rupture *43*

2.4.2. Fatigue *44*

2.5. Durability *45*

2.6. Recommended Materials and Construction Practices *46*

2.6.1. Strength and modulus grades of FRP bars *46*

2.6.2. Surface geometry, bar sizes, and bar identification *46*

2.7. Construction Practices *48*

2.7.1. Handling and storage of materials *48*

2.7.2. Placement and assembly of materials *49*

2.8. Quality Control and Inspection *49*

Chapter 3. History and Uses of FRP Technology **53**

3.1. FRP Composites in Japan *53*

3.1.1. Development of FRP materials *53*

3.1.2. Development of design methods in Japan *55*

3.1.3. Typical FRP reinforced concrete structures in Japan *55*

3.1.4. FRP for retrofitting and repair *57*

3.1.5. Future uses of FRP *58*

3.1.6. FRP construction activities in Europe *58*

3.2. Reinforced and Prestressed Concrete: Some Applications *59*

3.2.1. Rehabilitation and strengthening *60*

3.2.2. Design guidelines *61*

3.3. FRP Prestressing in the USA *61*

3.3.1. Historical development of FRP tendons *62*

3.3.2. Research and demonstration projects *62*

3.3.3. Future prospects *65*

Chapter 4. Design of RC Structures Reinforced with FRP Bars **67**

4.1. Design Philosophy *67*

4.1.1. Design material properties *67*

4.1.2. Flexural design philosophy *69*

4.1.3. Nominal flexural capacity *71*

4.1.4. Strength reduction factor for flexure (ϕ) *72*

4.1.5. Check for minimum FRP reinforcement *73*

4.1.6. Serviceability *73*

4.2. Shear *79*

4.2.1. Shear design philosophy *79*

4.2.2. Shear failure modes *80*

4.2.3. Minimum shear reinforcement *80*

4.2.4. Shear failure due to crushing of the web *80*

4.2.5. Detailing of shear stirrups *80*

4.2.6. Punching shear strength of FRP reinforced, two-way concrete slab *81*

4.3. ISIS Canada Design Approach for Flexure *81*

4.3.1. Flexural strength *83*

4.3.2. Serviceability *88*

4.4. Design Approach for CFRP Prestressed Concrete Bridge Beams *90*

4.4.1. Theoretical development of design equations *91*

4.4.2 Deflection and stesses under service load condition *102*

4.4.3. Nonlinear response *103*

E4.1. Design Example 1 *104*

E4.2. Design Example 2 *110*

E4.3. Design Example 3 *114*

E4.4. Design Example 4: A Case Study Problem *116*

E4.5. Design Example 5: Case Study of CFRP Prestressed Concrete Double-T Beam *120*

E4.6. Design Example 6: Case Study of CFRP Prestressed Concrete Box-Beam *132*

E4.7. Design Example *137*

Chapter 5. Design Philosophy for FRP External Strengthening Systems 143

5.1. Introduction *143*

5.1.1. Non-prestressed soffit plates *143*

5.1.2. End anchorage for unstressed (non-prestressed) plates *144*

5.1.3. Prestressed soffit plates *144*

5.2. Flexural Failure Modes and Typical Behavior *144*

5.2.1. Flexural failure *144*

5.2.2. Shear failure *144*

5.2.3. Plate-end debonding failures *145*

5.2.4. Plate-end interfacial debonding *146*

5.2.5. Intermediate crack-induced interfacial debonding *146*

5.2.6. Other debonding failures *147*

5.2.7. Some additional aspects of debonding *147*

5.3. Flexural Design Considerations *148*

5.3.1. Flexural design philosophy (ACI 440-2R-02) *148*

5.3.2. Strengthening limits *148*

5.4. Design Material Properties *149*

5.5. General Considerations for Flexural Strengthening *151*

5.5.1. Assumptions *151*

5.5.2. Section shear strength *151*

5.5.3. Existing substrate strain *152*

5.6. Nominal Strength, M_n *152*

5.6.1. Controlling failure modes *152*

5.6.2. Strain level in FRP reinforcement *153*

5.6.3. Stress level in the FRP reinforcement *153*

5.7. Ductility *153*

5.8. Serviceability *154*

5.9. Creep-Rupture and Fatigue Stress Limits *154*

5.10. Applications of Flexural Design Considerations to a Singly Reinforced Rectangular Section *155*

5.10.1. Ultimate flexural strength *155*

5.10.2. Stress in steel under service loads *156*

5.10.3. Stress in FRP under service loads *157*

5.11. Shear Strengthening *158*

5.11.1. Nominal shear strength (ACI 440.2R-02) *159*

5.11.2. Shear strength contribution of FRP system *159*

5.12. Spacing of FRP Strips *161*

5.13. Reinforcement Limits *161*

5.14. Design Procedure for Strengthening of RC Beam Using NSM Bars *161*

5.14.1. Flexural strengthening *162*

5.14.2. Design procedure for flexural strengthening using NSM FRP rebars *164*

5.14.3. Shear strengthening *166*

5.14.4. Anchorage length requirement *168*

5.15. ACI 440.2R-02 Design Approach for NSM FRP Strengthening *169*

5.15.1. Flexural design approach *169*

5.15.2. Nominal flexural strength *170*

5.16. Design for Shear Strength *171*

5.16.1. Ultimate shear strength *171*

5.17. Serviceability *174*

5.18. Detailing *175*

5.19. Development Length of NSM FRP Bars *175*

5.20. ISIS Canada Design Approach for External FRP Strengthening *176*

5.20.1. Flexural strengthening of beam and one-way slab *177*

5.20.2. Flexural design approach *178*

5.21. ISIS Canada Design Guidelines for Shear Strengthening *182*

5.21.1. Design principles *182*

5.22. External Strengthening of Columns *185*

5.22.1. Slenderness limits of circular columns *186*

5.22.2. Confinement *186*

5.23. Fundamentals of Seismic Retrofit of Columns *188*

5.23.1. Potential failure modes *188*

5.23.2. Flexural ductility of retrofitted columns *189*

5.23.3. Shear strength contributions *191*

5.23.4. Flexural plastic hinge confinement *194*

5.23.5. Lap splice clamping *194*

E5.1. Design Example 1 *195*

E5.2. Design Example 2 *203*

E5.3. Design Example 3: Shear Strengthening Using CFRP Laminates—A Case Study *206*

E5.4. Design Example 4: A Case Study Problem *213*

E5.5. Design Example 5 *219*

E5.6. Design Example 6 *222*

E5.7. Design Example 7 *227*

E5.8. Design Example 8 *231*

E5.9. Design Example 9 *233*

E5.10. Design Example 10 *236*

E5.11. Design Example 11 *238*

E5.12. Design Example 12 *243*

E5.13. Design Example 13 *251*

E5.14. Design Example 14 *253*

E5.15. Design Example 15: Shear Strengthening as per ISIS Canada Design Approach *255*

Chapter 6. Durability-Based Design Approach for External FRP Strengthening of RC Beams **263**

6.1. Designing of Reinforced Concrete Beams *263*

6.1.1. Compute design material properties *263*

6.1.2. Compute the existing substrate strain *264*

6.1.3. Compute the balanced plate ratio ($\rho_{f,b}$) *264*

6.1.4. Compute the maximum allowable plate ratio ($\rho_{f,\max}$) *266*

6.1.5. Proportion of the FRP plate *266*

6.1.6. Compute the balanced plate ratio ($\rho_{f,bb}$) to determine failure modes *266*

6.1.7. Determine the critical plate ratio ($\rho_{f,c}$) *267*

6.1.8. Determine the mode of failure *267*

6.1.9. Nominal moment capacity of strengthened beams *268*

6.1.10. Compute design moment capacity *270*

6.1.11. Allowable services stresses *270*

E6.1. Design Example 1: A Case Study Problem *271*

E6.2. Design Example 2: A Case Study Problem *280*

E6.3. Design Example 3: A Case Study *300*

Bibliography 317

Index 319

PREFACE

With the advent of advanced composite materials in the form of fiber reinforced polymer (FRP), which has high strength-to-weight ratio, high stiffness-to-weight ratio, and most importantly non-corrodible characteristics, these innovative FRP materials have been utilized in many demonstration projects across the world as internal reinforcements, external reinforcements, and prestressing tendons/strands. Outcomes of the research carried out on the demonstration projects using FRP materials have shown the potential of FRP materials as efficient constructions materials requiring least maintenance and hence, providing minimal life cycle cost of structures reinforced and/or prestressed with FRP materials. In addition to normal reinforcements and prestressing applications, FRP has been utilized as external strengthening material to retrofit the deficient structures and to upgrade the strength of structure to meet new code specifications and loading situations.

In spite of many advantages, the use of FRP materials for reinforcement, prestressing, and strengthening of RC structures has been limited, due to lack of unified design procedures which can be readily referred to design the structures using FRP reinforcements. There are many research works available on the use of FRP. However, the research data are in a scattered form. Also the research findings cannot be readily utilized by designers, practitioners and academicians. Recently, ACI 440 committee and ISIS Canada have come out with design guidelines based on the research works available in literature. However, these draft documents lack the variety of design problems and are also not unified in nature. Today, there is no text book across the world giving detailed design examples related to analysis and design of FRP reinforced and prestreesed concrete structures with and without discussion of case study problems. The goal of this book is to present various aspects of FRP composite materials characteristics, manufacturing techniques, real life projects, various forms of FRP products, and most importantly, detailed design procedures for design of new structures using FRP as internal reinforcements, external strengthening materials, and prestressing materials.

The main topics covered in this book consist of introduction of FRP composites (Chapter 1), material characteristics (Chapter 2), history and uses of FRP technology (Chapter 3), design of RC structures using FRP bars (Chapter 4), design philosophy for FRP exteranal strengthening systems (Chapter 5), and durability based design approach for external FRP strengthening of RC beams (Chapter 6). Furthermore, other topics of interest such as case study problems on FRP prestressed concrete bridges have also been included along with other case study problems. In the whole book, SI units has been used in tables as well as for design examples. However, few emprical design equations have been presented with both the SI and English units for better understanding of the empirical equations. This book will be of prime interest to wide range of readers such as researchers, academicians in general, consultants, practitioners, designers, writers of design codes, structural engineers, senior under-graduate and graduate students as a consolidated source of design guidelines and design examples on FRP reinforced and/or prestressed concrete structures.

This book has been grown out of the many research and teaching works carried by the author during his post-Doctoral fellowship at Lawrence Technological University, Southfield, Michigan, USA, and current teaching and research works leading to Indian Lead Partner of prestigious UK–India Collaborative Research Project between Indian Universities (BITS Pilani, IIT Delhi, NIT Jalandhar, NIT Surathkal, NIT Surat, MNIT Jaipur, and SRM University Chennai), University of Dundee, UK and University of Bath, UK.

The author is indebted to Prof. Nabil F. Grace, Lawrence Technological University, USA for providing opportunity to work on many projects of international repute using FRP materials and Prof. R.K. Dhir, University of Dundee for his continuous support and guidance during the execution UKIERI project leading to many constructive inputs for this book. Moreover, the author is also thankful to other UKIERI partners from India and UK for their constructive suggestions made during UKIERI technical research paper presntations which has helped me to improve the quality of the book. Furthermore, the author expresses appreciation to his graduate research scholars especially S.Madappa, Dinesh Kumar, Bhavin, R., and K. Santosh for their continuous support during the preparation of this book.Finally, the author wishes to express his profound gratitude to his beloved son (Late) Mr. Sneh Vardhan, daughter (Ms. Aparajita Singh), and wife (Mrs. Kiran Singh) for their sufferings, sacrifices, understanding, and support during the preparation of this book.

Shamsher Bahadur Singh

NOTATION

The notations of the various parameters used in this paper are given below:

a = depth of equivalent rectangular compression block, mm

A_c = cross-sectional area of composite DT-beam, mm^2

A_p = cross-sectional area of precast DT-beam, mm^2

A_{fb} = cross-sectional area of bottom bonded prestressing tendons in each row, mm^2

A_{fi} = cross-sectional area of tension reinforcement of a particular material, mm^2

A_{fn} = cross-sectional area of non-prestressing tendons in each row, mm^2

A_{fnt} = cross-sectional area of non-prestressing tendons at flange top, mm^2

A_{fnb} = cross-sectional area of non-prestressing tendons at flange bottom, mm^2

A_{fu} = total cross-sectional area of unbonded post-tensioning tendons, mm^2

b = width of compression face of member, mm

b_w = width of the web, mm

C = resultant compressive force, kN

C_t = resultant compression force in concrete topping, kN

C_f = resultant compression force in flange of precast DT-beam, kN

c.g.c = axis passing through the centroid of concrete cross-section of the DT-beam

c.g.p = axis passing through the center of gravity of the resultant pretensioned force

d = distance of center of gravity of the resultant tension force from the extreme compression fiber, mm

d_f = total flange thickness of the beam, mm

$\overline{d}$ = distance of center of gravity of the resultant compression force from the extreme $\overline{d}$ compression fiber, mm

d_j = distance of centroid of bonded prestressing tendons of an individual row from the extreme compression fiber, mm

d_m = distance of centroid of bottom bonded prestressing tendons (*m*th row) from the extreme compression fiber, mm

d_b = distance of centroid of non-prestressing tendons at flange bottom from the extreme compression fiber, mm

d_t = distance of centroid of non-prestressing tendons at flange top from the extreme compression fiber, mm

d_u = distance of centroid of unbonded post-tensioning tendons from the extreme compression fiber, mm

D = overall depth of the beam, mm

e_b = eccentricity of resultant pretensioning force from the centroid of the precast concrete cross-section, mm

e_{up} = eccentricity of unbonded post-tensioning tendons from centroid of the precast concrete cross-section, mm

e_{uc} = eccentricity of unbonded post-tensioning tendons from the centroid of composite section, mm

E_c = modulus of elasticity of precast concrete, GPa

E_{ct} = modulus of elasticity of concrete topping, GPa

E_f = modulus of elasticity of bonded tendon, GPa

E_{fn} = modulus of elasticity of non-prestressing tendons in webs, MPa

E_{fp} = modulus of elasticity of unbonded tendon, MPa

f_c = stress in the concrete at extreme compression fiber in very-under-reinforced beam, MPa

f_{ct} = stress in the concrete at extreme compression fiber in over-reinforced beam, MPa

f'_c = specified compressive strength of precast concrete, MPa

f_{fu} = guaranteed ultimate tensile strength of bonded prestressing tendons, MPa

f_{fui} = guaranteed ultimate strength of tension reinforcement of a particular material, MPa

$(f_{fu})_d$ = guaranteed ultimate strength of reinforcement governing the minimum balanced ratio

f_{fun} = guaranteed ultimate tensile strength of non-prestressing tendons in webs, MPa

f_{fup} = ultimate tensile strength of unbonded post-tensioning tendons, MPa

f_{pbj} = total stress in bonded prestressing tendons of an individual row, MPa

f_{pbm} = total stress in bottom prestressing tendons (*m*th row), MPa

f_{pbmi} = initial effective prestress in bottom bonded prestressing tendons (*m*th row), MPa

f_{pnb} = total stress in non-prestressing tendons at flange bottom, MPa

f_{pnj} = total stress in non-prestressing tendons of an individual row in webs, MPa

f_{pnk} = total stress in non-prestressing tendons of bottom row (*k*th row), MPa

f_{pnt} = total stress in non-prestressing tendons at flange top, MPa

f_{pu} = total stress in unbonded post-tensioning tendons, MPa

f_r = modulus of rupture of concrete, MPa

F_{pi} = total initial effective pretensioning and post-tensioning forces, kN

F_{pre} = resultant effective pretensioning force in bonded tendons, kN

F_{post} = resultant effective post-tensioning force in unbonded tendons, kN

F_{pbj} = resultant tensile force in bonded prestressing tendons of an individual row, kN

F_{pbm} = resultant tensile force in bonded bottom (*m*th row) prestressing tendons, kN

F_{pnb} = resultant compression force in non-prestressing tendons at flange bottom, kN

F_{pnj} = resultant tensile force in non-prestressing tendons of individual row in webs, kN

F_{pnk} = resultant tensile force in non-prestressing tendons of bottom row (*k*th row) in webs, kN

F_{pnt} = resultant compression force in non-prestressing tendons at flange top, kN

F_{pu} = resultant tensile force in unbonded post-tensioning tendons, kN

F_R = resultant of the tensile forces in bonded and unbonded tendons, kN

F_u, F_i = ultimate and initial forces in unbonded post-tensioning strands, kN

h_f = flange thickness of DT-beam, mm

h_j = distance of centroid of non-prestressing tendons of an individual row in webs, mm

h_k = distance of centroid of bottom non-prestressing tendons (*k*th row), mm

h_t = thickness of concrete topping, mm

I_c = moment of inertia of composite concrete cross-section, mm^4

I_{cr} = gross transformed moment of inertia of cracked section, mm^4

I_{tr} = gross transformed moment of inertia of gross cross-section, mm^4

k = number of rows of non-prestressing tendons in webs of DT-beam

k_u = neutral axis depth coefficient

k_{up} = neutral axis depth coefficient for a specific reinforcement ratio

L = effective span of the beam, m

m = number of rows of bonded tendons

M = applied maximum moment due to service loads, kN-m

M_{cr} = cracking moment capacity of section, kN-m

M_u = design moment capacity of section, kN-m

M_D = maximum bending moment due to dead load, kN-m

M_L = maximum bending moment due to live load, kN-m

M_n = nominal moment capacity of section, kN-m

M_{required} = required moment capacity of the section, kN-m

n = depth to the neutral axis from the extreme compression fiber

$N.A.$ = neutral axis of the DT-beam section

p = number of materials used for tension reinforcement

P_{cr} = midspan load causing cracking of DT-beam, kN

q = total number of bonded prestressing and non-prestressing tendon layers

S_b = section modulus corresponding to the bottom extreme fiber of composite section, mm^3

W = total midspan load, kN

W_d = self weight of beam/unit length, kN/m

$\bar{y}$ = distance of centroid of tension reinforcement from the extreme compression fiber, mm

y_{tc} = distance of centroid of composite cross-section from the top fiber, mm

β_1 = factor defined as the ratio of the equivalent rectangular stress block depth to the distance from the extreme compression fiber to the neutral axis

ε_{cr} = strain at extreme tension fiber at first cracking of beam

ε_{cu} = ultimate compression strain in concrete (0.003)

ε_{fbj} = flexural strain in the bonded prestressing tendons of an individual row

ε_{fbm} = flexural strain in the bonded prestressing tendons of bottom row (mth row)

ε_{fu} = ultimate tensile strain capacity of bonded prestressing tendons

ε_{fun} = ultimate tensile strain capacity of non-prestressing tendons in webs

ε_{fup} = ultimate tensile strain capacity of unbonded tendons

ε_{pbj} = total strain in bonded prestressing tendons of an individual row

ε_{pbji} = initial strain in bonded prestressing tendons of an individual row

ε_{pbm} = total strain in bonded prestressing tendons of mth row

ε_{pbmi} = initial strain in bonded prestressing tendons of mth row

ε_{pnb} = total strain in non-prestressing tendons at flange bottom

ε_{pnj} = total strain in non-prestressing tendons of an individual row in webs

ε_{pnk} = total strain in non-prestressing tendons of bottom row (kth row)

ε_{pnt} = total strain in non-prestressing tendons at flange top

ε_{pu} = total strain in unbonded post-tensioning tendons

ε_{pui} = initial strain in unbonded post-tensioning tendons

$\Delta\varepsilon_{pu}$ = flexural strain in unbonded post-tensioning tendons

ε_t = strain at top of the beam at specific load stage

ε_b = strain at bottom of the beam at specific load stage

α, β = stress block factors

α_1, β_1 = stress block factors for rectangular section

α_2, β_2 = stress block factors for flanged section

α_m = moment capacity factor

α_i = ratio of guaranteed ultimate tensile strength of a particular tensile reinforcement to the guaranteed ultimate strength of reinforcement governing the minimum balanced ratio

ϕ = strength reduction factor

ϕ_{cr} = curvature of the beam at first cracking

σ_{bp} = resultant prestress at the extreme tension fiber of the beam due to effective pretensioning and post-tensioning forces, MPa

ρ = tensile reinforcement ratio

ρ_b = balanced reinforcement ratio

Ω = bond reduction coefficient for uncracked section

Ω_c = bond reduction coefficient for cracked section

Ω_u = bond reduction coefficient at ultimate

δ = maximum midspan deflection of the beam under service loads, mm

δ_a = maximum midspan deflection of the beam due to applied load, mm

δ_d = maximum midspan deflection of the beam due to dead load, mm

δ_p = maximum midspan deflection of the beam due to prestressing forces, mm

CHAPTER 1

Introduction

Both non-prestressed and prestressed steel have been used as reinforcing bars as well as prestressing tendons in the construction of conventional concrete structures. The steel is initially protected against corrosion by the alkalinity of the concrete, which results in durable construction. When Structures subjected to aggressive environments, such as, marine structures, bridges, parking garages exposed to deicing salts, combination of moisture, temperature and chlorides, which reduce the alkalinity of the concrete and to undergo/experience corrosion of reinforcing and prestressing steel. Ulitmately, the corrosion of steel causes concrete deterioration and loss of serviceability. This corrosion problem was initially addressed by professionals, researchers and construction industry by using epoxy-coated steel bars (ACI 440.1R-03). Such remedies are although found to be suitable in some situations, they are however unable to completely eliminate the problem of steel corrosion (Keesler and Powers, 1988).

Recently, innovative fibrous composite materials, such as, fiber-reinforced polymers (FRPs) are being used as an alternative to conventional steel material for reinforcement or prestressing applications. These composite materials are made of fibers embedded in polymeric resin. FRP materials exhibit several properties such as:

- High longitudinal tensile strength, which varies with sign and direction of loading relative to fibers;
- Corrosion resistant which does not dependent on coating;
- Insensitive to magnetic field;
- High fatigue endurance which varies with type of reinforcing fibers;
- Lightweight (about $\frac{1}{5}$ to $\frac{1}{4}$ the density of steel);
- High specific strength (strength to weight ratio);
- High specific stiffness (stiffness to weight ratio);
- High impact and fatigue resistance; and
- Low thermal and electric conductivity.

Since, FRP materials are non-magnetic and non-corrosive, the problems of electromagnetic interference and steel corrosion can be avoided with FRP reinforcement. In spite of many advantages, FRP materials have some disadvantages which require special attention in their use as structural materials. Most prominent disadvantages of FRP composites include:

- Unlike steel they are brittle and exhibit no yielding before brittle rupture;
- Low transverse strength which varies with sign and direction of loading relative to fibers;
- Low modulus of elasticity, which varies with type of reinforcing fibers;
- Susceptibility of damage to polymeric resins and fibers under ultraviolet radiation exposure rupture;

- Low modules of elasticity, which varies with type of reinforcing fibers; and
- Susceptibility of damage to polymeric resins and fibers under ultraviolet radiation exposure.

The difference in the mechanical performance of FRP reinforcement with that of steel necessiates changes in the design philosophy of concrete structures. Because FRP materials are anisotropic and are characterized by high tensile strength only in the fiber directions, the anisotropic behavior affects the shear strength and dowel action of FRP bars, as well as bond performance of FRP bars to concrete. The linear elastic behavior up to failure of these materials results in lack of ductility in concrete reinforced with these materials. Several countries such as Japan (JSCE, 1997), Canada (Canadian Standards Association, 1996) and the United States of America (ACI 440.1R-06) have established design procedures especially for the use of FRP reinforcements.

Fiber-reinforced polymer products were first used to reinforce concrete in the mid 1950s (ACI 440R-96). These FRP products are available in various forms—such as, bars, cables, 2-D and 3-D grids, fabric sheets, strips, plates, etc. FRP products may achieve the same or better reinforcement objective of commonly used metallic products—such as, steel reinforcing bars, prestressing tendons and bonded plates. Current widespread application and product development efforts are able to address many opportunities for reinforcing concrete members. Some of these efforts are: (i) high volume production techniques to reduce manufacturing costs; (ii) modified construction techniques to better utilize the strength properties of FRP and reduce construction costs; and (iii) optimization of the combination of fiber and resin matrix to ensure optimum compatibility with Portland cement.

It must be noted that the use of continuous fibers—glass, aramid and carbon etc.—embedded in a polymeric resin matrix—the glue that allows the fibers to work together as a single filament—forms the common linkage among all FRP products. Organic thermoset and thermoplastic matrices are usually used in FRP products. Most notable thermoset matrices are epoxy, polyester, and vinylester, whereas most notable thermoplastic matrices are nylon, polyethylene, and terephalate. Characteristics of these matrices will be discussed later.

FRP composites are differentiated from the short fibers used widely to reinforce non-structural cementitious products known as fiber reinforced concrete (FRC). The use of continuous fibers together with resin matrix allows FRP materials to be tailored, such that optimized reinforcement of the concrete structure is achieved. The pultrusion process is one such manufacturing method widely practiced today. This method is used to produce consumer and construction products—such as, fishing rods, bike flags, shovel handles, structural shapes of constant profile etc. In this process, continuous forms of reinforcement are combined with resin to produce high-fiber volume, and directionally oriented FRP products. The primary interest of using FRP reinforcements in the concrete industry is that these do not pose durability problems as those associated with steel reinforcement. Japan, Europe and North America are the leading countries in the production of FRP products.

1.1. Evolution of FRP Reinforcement

The corrosion problem began to surface with the steel reinforced concrete used in highway bridges and structures. Use of road salts (in colder climates) and marine salts (in coastal areas), accelerated corrosion of the reinforcing steel. Since the corrosion products would expand and cause the concrete to fracture, the galvanized coating to the reinforcing bars was seen as a first solution to the corrosion problem. In the late 1960s, several companies developed an electrostatic-spray fusion-bonded (powdered resin) coating for steel *oil and gas* pipelines. In the early 1970s, the Federal Highway Administration funded research to evaluate over 50 types of coatings for steel reinforcing bars, which led to the current use of epoxy-coated steel reinforcing bars.

In the late 1960s, research on the use of resins in concrete started with a program at the Bureau of Reclamation on polymer-impregnated concrete. However, steel reinforcement could not be done with polymer concrete because of incompatible thermal properties. This fact led Marshall-Vega Corporation to manufacture a glass fiber reinforced polymer (GFRP) reinforcing bar. Thus the experiment was successful and the resultant composite reinforcing bar became a reinforcement of choice for polymer concrete.

In spite of earlier research on the use of FRP reinforcement in concrete, commercial application of this product in conventional concrete which was not recognized until the late 1970s. Hence, at this juncture, research started in earnest, to determine if companies were a significant improvement over epoxy coated steel. During the early 1980s, International Grating Inc. (a pultrusion company) recognized the product potential and entered the FRP reinforcing bar industry. Thus there was increased use of FRP reinforcing bars in applications with special performance requirements or where reinforcing bars were subjected to several chemical attacks. This material finds special market for the reinforced concrete to support or surround magnetic resonance imaging (MRI) medical equipment, since conventional steel reinforcement cannot be used. GFRP reinforcing bars have continued to be selected by structural designers over non-magnetic (nitronic) stainless steel.

The world's first highway bridge using composite reinforcement was built in Germany in 1986. From then onwards, bridges have been constructed throughout Europe, North America, and Japan. The U.S. and Canadian governments are investing significant sums for FRP product evaluation and further development. By the end of 1993, there were nine companies actively marketing commercial FRP reinforcing bars.

1.2. Review of FRP Composites

Composites are a material system, where the term 'composites' can be applied to, any combination of two or more separate materials having an identifiable interface between them. The FRP composites consist of polymer matrix reinforced with fibers (aramid, carbon, glass). The polymeric matrix are whether thermosetting (polyester, vinylester, epoxy, phenolic) or thermoplastic (e.g., nylon, PET). Thermoset matrices are resins that are formed by cross-linking polymer chains. A thermoset cannot be melted and recycled because the polymer chains form a three-dimensional network. Thermoplastic resin is not cross-linked and can be remelted and recycled. Specific definitions used within this text book include glass-fiber reinforced plastic (GFRP), aramid fiber reinforced plastic (AFRP), and carbon-fiber reinforced plastic (CFRP). For the definitions of various difficult terms see glossary at the end of the book. Although these composites are defined as a polymer matrix that is reinforced with fibers, this definition must be further refined when describing composites for use in structural applications. In the case of structural applications, such as FRP composite reinforced concrete, at least one of the constituent materials must be a continuous reinforcement phase supported by stabilizing matrix material. For the special class of matrix material (i.e., thermosetting polymers), the continuous fibers will usually be stiffer and stronger than the matrix. However, if the fibers are discontinuous in form, the fiber volume fraction should be 10% or more, in order to provide a significant reinforcement function. The performance of any composite depends on the materials of which the composite is made, the arrangement of the primary load-bearing portion of the composite (reinforcing fibers), and the interaction between the materials (fibers and matrix). The major factors affecting the physical performance of the FRP matrix composite include: fiber mechanical properties, fiber orientation, length, shape, and composition of the fibers, the mechanical properties of resin matrix, and the adhesion of bond between the fibers and the matrix.

1.3. The Importance of the Polymer Matrix

The reinforcing fibers are the primary structural constituents in composites. However, it is essential to consider and understand the important role that the matrix polymer plays. The polymer matrix transfer stresses between the reinforcing fibers and the surrounding structures and protect the fibers from environmental and mechanical damage. The polymer matrix properties influence interlaminar shear as well as the in-plane shear properties of the composite. Moreover, matrix resin also provides lateral support against fiber buckling under compression loading.

1.3.1. Matrix Polymers

A polymer is defined as a long-chain molecule having one or more repeating units of atoms joined together by strong covalent bonds. A polymeric material (i.e., a plastic) is a collection of a large number of polymer molecules of similar chemical structure. If, in a solid phase, the molecules are in random order, the plastic is said to be amorphous. If the molecules are in combinations of random and ordered arrangements, the polymer is said to be semi-crystalline. Moreover, portions of polymer molecule may be in state of random excitation. These segments of random excitation increase with temperature, giving rise to the temperature-dependent properties of polymeric solids.

Polymer matrix materials differ from metals in several aspects that can affect their behavior in critical structural applications. The mechanical properties of composites depend strongly on ambient temperature and loading rate. In the glass transition temperature (T_g) range, polymeric materials change from a hard, often brittle solid to a soft, tough solid. The tensile modulus of the matrix polymer can be reduced by as much as five orders of magnitude. The polymer matrix material is also highly viscoelastic. When an external load is applied, it exhibits an instantaneous elastic deformation followed by slow viscous deformation. As the temperature is increased, the polymer changes into a rubber like solid, capable of large, elastic deformations under external loads. As the temperature is increased further, both amorphous and semi-crystalline thermoplastic reach highly viscous liquid states, with the latter showing a sharp transition at the crystalline melting point.

The glass transition temperature of a thermoset is controlled by varying the amount of cross-linking between the molecules. For a very highly cross-linked polymer, the transition temperature and softening may not be observed. For a thermosetting matrix polymer such as a polyester, vinylester or epoxy, no "melting" occurs. In comparison to most common engineering thermoplastics, thermosetting polymers exhibits greatly increased high-temperature and load bearing performance. Normally, the thermosetting polymers char and eventually burn at very high temperatures. The effect of loading rate on the mechanical properties of a polymer is opposite to that due to temperature. At high loading rates, i.e., in the case of short duration of loading), the polymeric solid behaves in a rigid, brittle manner. At low loading rates (i.e., long duration of loading), the same material may behave in a ductile manner and exhibit improved toughness values. A comparison of thermoset and thermoplastic matrix materials is provided in Table 1.1.

In structural applications, fillers are used selectively to improve load transfer and also to reduce cracking in unreinforced areas. Fillers are available in variety of forms and are normally treated with organo-functional silanes to improve performance and reduce resin saturation. Although minor in terms of the composition of the matrix polymer, a range of important additives, including UV inhibitors, initiators (catalysts), wetting agents, pigments and mold release materials are frequently used. A more detailed explanation of the commercial thermosetting matrix polymers used to produce composite concrete reinforcing products including dowel bars, reinforcing rods, tendons, and cable stays are given in the following sections.

Table 1.1. Thermoset versus thermoplastic matrix materials.

S. No.	Thermosetting matrix polymer	Thermoplastic matrix polymer
1.	Thermosetting polymers are always processed in a low viscosity and liquid state. Therefore, it is possible to get good fiber wet-out without resorting to high temperature or pressure. To date, thermosetting matrix polymers (polyesters, vinylesters, and epoxies) have been the materials of choice for the great majority of structural composite applications.	The progress of commercial structural uses of thermoplastic matrix polymers has been slow. A major obstacle is that thermoplastic matrix polymers are much more viscous and are difficult to combine with continuous fibers in a viable production process. Recently, however, a number of new promising process operations, especially filament winding and pultrusion have been developed.
2.	Thermosetting matrix polymers are low molecular weight liquids with very low viscosities.	Thermoplastic matrix materials have high molecular weight and high viscosity.
3.	Thermosetting matrix polymers provide good thermal stability and chemical resistance. They also exhibit reduced creep and stress relaxation.	They provide poor thermal stability and have high creep and stress relaxation.
4.	Thermosetting matrix polymers generally have a short shelf-life (usable storage life) after mixing with curing agents (catalysts), low strain-to-failure, and low impact strength.	Thermoplastic matrix polymers have high shelf-life, high strain-to-failure, and high impact strength, as well as high fracture resistance. Other advantages are shorter molding cycles, secondary formability, and ease of handling and damage tolerance.
5.	Fillers can be added to reduce resin cost, control shrinkage, improve mechanical properties, and impart a degree of fire retardancy.	Fillers can be added to reduce resin cost, control shrinkage, improve mechanical properties, and impart a degree of fire retardancy.

1.3.2. Polyester Resins

Unsaturated polyester (UP) is the polymer resin most commonly used to produce large composite structural parts. Approximately 85% of United States (US) composite production is based on unsaturated polyester resins. These resins are typically in the form of low viscosity liquids during processing or until cured. However, partially processed materials containing fibers can also be used under specific conditions of temperature and pressure. This class of material has its own terminology with the most common preproduction forms of partially reacted or chemically –thickened materials being preimpregnation (prepreg).

Unsaturated polyesters are produced by the polycondensation of dihydroxyl derivatives and dibasic organic acids or anhydrides, yielding resins that can be compounded with styrol monomers to form highly cross-linked thermosetting resins. The resulting polymer is then dissolved in a reactive vinyl monomer such as styrene. The viscosity of solutions typically range from 200 to 2000 centipoises (cps), but depends on the ingredients. Addition of heat and/or a free-radical initiator such as organic peroxide causes a chemical reaction that result in non-reversible cross-linking between the unsaturated polyester polymer and the monomer. Room temperature cross-linking can also be accomplished by using peroxides and suitable additives (typically promoters). Common commercial types of unsaturated polyester resin are briefly explained in the following:

Orthophthalic polyester (ortho polyester): This was the original form of unsaturated polyester. This resin includes phthalic anhydride, and maleic anhydride or fumaric acid. Ortho polyesters do not have the mechanical strength, moisture resistance, thermal stability or chemical resistance of the higher-performing isophthalic resin polyesters or vinylesters described in the following section. Thus, it is unlikely that ortho polyesters will be used for demanding structural applications such as composite-reinforced concrete.

Isophthalic polyester (iso polyester): These resins include isophthalic acid and maleic anhydride or fumaric acid. Iso polyesters demonstrate superior thermal resistance, improved mechanical properties, greater moisture resistance and improved chemical resistance compared to ortho polyesters. Iso polyester resins are more costly than ortho polyester resins, but are highly processable in convention oriented-fiber fabricating processes such as pultrusion.

Vinylesters (VE): Vinylester resins are produced by reacting a monofunctional unsaturated acid (methacrylic or acrylic acid), with a bisphenol diepoxide. The polymer has unsaturation sites only at the terminal positions, and is mixed with an unsaturated monomer such as styrene. Vinylesters process and cure are essentially like polyesters and are used in many of the same applications. Although vinylesters are higher in cost than ortho or iso polyesters, they provide increased mechanical and chemical performance. Vinylesters are also known for their toughness, flexibility and improved retention of properties in aggressive environments including high pH alkali environments associated with concrete. For these reasons, vinylesters should be considered for composite-reinforced concrete applications.

Bisphenol A fumerates (BPA): Bisphenol A fumerates offer high rigidity, improved thermal and chemical performance compared to ortho or iso polyesters.

Chlorendics: These resins are based on a blend of chlorendic (HET) acid and fumaric acid. They have excellent chemical resistance and provide a degree of fire retardancy due to the presence of chlorine. There are also brominated polyesters having similar properties and performance advantages.

Epoxy resins: Epoxy resins are used in advanced applications including aircraft, aerospace, and defense, as well as many of the first generation composite reinforcing concrete products currently available in the market. Epoxy resins are available in a range of viscosities, and will work with a number of curing agents or hardeners. The nature of epoxy allows it to be manipulated into a partially cured or advanced cure state commonly known as a prepreg'. Epoxy resins are more expensive than commercial polyesters and vinylesters.

Although some epoxies harden at temperature as low as 30°C (80°F), all epoxies require some degree of heated postcure to achieve satisfactory high temperature performance. Currently, specially formulated epoxies which when heated, have low viscosities in order to be compatible with the, process parameters of a new generation of resin-infusion processes. Large parts fabricated with epoxy resins exhibit good fidelity to the mold shape and dimensions of the molded part. Epoxy resins can be formulated to achieve very high mechanical properties. It may be noted that no styrene or other monomer released during molding. However, certain hardeners—particularly amines as well as the epoxy resins—can be skin sensitizing, and hence appropriate personal protective procedures must always be followed. Some epoxies are also more sensitive to moisture and alkali. This behavior must be taken into account in determining the long-term durability and suitability for any given application.

The raw materials for most epoxy resins are low-molecular-weight organic liquid resins containing epoxide groups. The epoxide group comprises rings of one oxygen atom and two carbon atoms. The most common starting material used to produce epoxy resin is diglycidyl ether of bisphenol-A (DGEBA), which contains two epoxide groups, one at each end of the molecule. Other materials that can be mixed with the starting liquid include dilutents to reduce viscosity flexibilizers to improve impact strength of the cured epoxy resin.

Cross-linking of epoxy: Cross-linking of epoxies is initiated by use of a hardener or reactive curing agent. One common commercial curing agent is diethylenetriamine (DETA). Hydrogen atoms in the amine groups of the DETA molecule react with the epoxide groups of DGEBA molecules. As this reaction continues, DGEBA molecules cross-link with each other and a three dimensional network is formed, creating the solid cured matrix of epoxy resins. Curing time and increased temperature required to complete cross-linking (polymerization) depend on the type and amount of hardener used. Additives also known as accelerators are sometimes added to the liquid epoxy resin, to speed up reactions and decrease curing cycle times.

The continuous use temperature limit for DGEBA epoxy is 150°C (300°F). However, higher heat resistance can be obtained with epoxies based on novalacs and cycloaliphatics. The epoxy based on cycloaliphatics will have continuous use temperature of 250°C (489°F). The heat resistance of an epoxy is improved if it contains more aromatic rings in its basic molecular chain. The epoxy resin would be in B-staged form, if its curing reaction is slowed by external means—i.e. by lowering the reaction temperature—before all molecules are cross-linked. In this form, the resin has formed cross-links at widely spaced positions in the reactive mass, but is uncured. Hardness, tackiness and the solvent reactivity of these B-staged resins depend on the degree of curing. However, curing can be completed at a later time usually by application of external heat. In this way, a prepreg, which in this case of epoxy matrix polymer is a B-staged epoxy resin containing structural fibers or suitable fiber array, can be handled as a tacky two-dimensional combined reinforcement and placed on the mold for manual or vaccum/pressure compaction, followed by the application of external heat to complete the cure (cross-linking).

Hardeners for epoxies: Epoxy resins can be cured at different temperatures ranging from room temperature to elevated temperatures as high as 175°C (347°F). Post-curing is usually done. Epoxy polymer matrix resins are approximately twice as expensive as polyester matrix materials. Compared to polyester resins, epoxy resins provide the following general performance characteristics:

1. A range of mechanical and physical properties can be obtained due to diversity of input materials.
2. No volatile monomers are emitted during curing and processing.
3. There is low shrinkage during cure.
4. Excellent resistance to chemicals and solvents is seen.
5. Good adhesion to a number of fillers, fibers and substrates.

However, in addition to the aforementioned advantages, epoxy matrix polymers have some drawbacks which are as follows:

1. Matrix cost is generally higher than that for iso-polyester or vinylester resins.
2. Epoxies must be carefully processed to maintain moisture resistance.
3. Cure time can be lengthy.
4. Some hardeners require special precautions in handling and resin and some hardeners cause skin sensitivity reactions in production operations.

1.3.3. Structural Considerations in Processing Polymer Matrix Resins

The matrix resin must have significant levels of fibers within it, at all important load-bearing locations. In the absence of sufficient fiber reinforcement, the resin matrix may shrink excessively, can crack, or may not carry the load imposed upon it. Fillers, especially those with a high aspect ratio, can be added to the polymer matrix resin to obtain some measure of reinforcement. However, it is difficult to selectively place fillers. Hence, use of fillers can reduce the volume fraction available for the load-bearing fibers. This forces compromises to made on the designer and processor.

The other controlling factor is the matrix polymer viscosity. Reinforcing fibers must be fully wetted by the polymer matrix to insure effective coupling and load transfer. Thermoset polymers of major commercial utility either have suitably low viscosity, or this can be easily managed with processing methods utilized. Processing methods for selected thermoplastic polymers having inherently higher viscosity, are currently being developed to a state of prototype practically.

1.3.4. Reinforcing Fibers for Structural Composites

Glass, carbon and aramid are the principal commercially used fibers for civil engineering applications, including composite reinforced concrete. The most common form of fiber-reinforced composite used in structural applications is called a laminate. Laminates are made by stacking a number of thin layers (laminae) of fibers and matrix, and consolidating them into the desired thickness. Fiber orientation in each layer as well as the stacking sequence of the various layers can be controlled to generate a range of physical and mechanical properties.

A composite can be of any combination of two or more materials, so long as they are distinct, and recognizable regions of each material. The performance of composite depends upon the materials, of which the composite is constructed, the arrangement of the primary load-bearing reinforcing fiber portion of the composite, and the interaction between these materials. The major factors affecting the performance of the fiber matrix composite are: fiber orientation, length, shape, and composition of the fibers, the mechanical properties of the resin matrix, and adhesion or bond between the fibers and the matrix. A uni-directional or one-dimensional fiber arrangement is anisotropic. This fiber orientation results in maximum strength and modulus in the direction of the fiber axis. A planar arrangement of fibers is two-dimensional and has different strengths at all angles of fiber orientation. A three-dimensional array is isotropic but has sufficiently reduced strength over the one-dimensional arrangement. Mechanical properties in any one direction are proportional to the amount of fiber by volume oriented in that direction. Many fiber-reinforced composites exhibit high internal damping properties, which lead to better vibrational energy absorption within the material and reduce transmission to adjacent structures. This aspect of composite behavior may be relevant in civil engineering structures, such as bridges, highways, etc., that are subject to loads that are more transitory and of shorter duration than sustained excessive loading.

Functional relationship of polymer matrix to reinforcing fibers: The matrix gives form and protection to the fibers from the external environment. Chemical, thermal, and electrical performance can be affected by the choice of matrix resin. In addition, matrix maintains the position of the fibers. Under loading, the matrix resin deforms and distributes the stress to the higher modulus fiber constituents. The matrix should have an elongation at break greater than that of the fiber. It should not shrink excessively during curing to avoid placing internal strains on the reinforcing fibers. If the designer wishes to have anisotropic properties, he will use appropriate fiber orientation and forms of uni-axial fiber placement. However, in the case of a complex part, it may be necessary to resort to shorter fibers to reinforce the molding effectively in three dimensions. In this way, quasi-isotropic properties can be achieved in the composite.

1.3.5. Effects of Fiber Length on Laminate Properties

In structural part fabrication, fiber placement can be affected with both continuous and short fibers. There may be a part or process constraints, which impose choice limitations on designers. The alternative in these cases may require changes in composite cross section area or shape. Variables in continuous-fiber manufacture as well as in considerations in part fabrication, make it impossible to obtain equally stressed fibers throughout their length.

1.3.6. Bonding Interphase

Fiber reinforced composites withstand higher stresses than their constituent materials because matrix and fibers interact to redistribute the stresses of external loads. The distribution of stresses within the composite structure depends on the nature and efficiency of bonding. Both the mechanical and chemical processes are operational in any given structural situation. Coupling agents are used to improve the chemical bond between reinforcement and matrix since the fiber-matrix interface is frequently in a state of shear when the composite is under load.

1.3.7. Design Considerations

The FRP materials are generally more expensive on unit weight basis, however, they are quite cost effective on a specific strength basis, i.e., cost per unit of load carried. With the exception of carbon fibers, the modulus of fiber-reinforced composite is significantly lower than conventional materials. Hence, innovative design with respect to shape, fiber choice, fiber placement, or hybridization with fibers must be utilized by designers to take this factor into account. The following considerations are representative of the choices which are commonly made:

1. Composites are anisotropic and can be oriented in the direction (s) of the load (s) required.
2. There is a high degree of design freedom. Variations in thickness and compound part geometry can be molded into the part.
3. FRP composites have very high tensile strength in comparison to the conventional design materials such as steel, but do not have comparable stiffness except in the case of CFRP. However, designers may have to be concerned about impact and brittleness. Table 1.2 gives a comparative idea of thickness and weight for equal strength materials.
4. Provision of designing with the maximum stiffness with the minimum materials should be made.
5. Advantage of anisotropic nature of materials and oriented fibers should be taken by making sure that the manufacturing process is compatible with selections.
6. Maximum strain limit of the laminate should be optimized. Also, for a large structural part, elongation of resin is an important factor in choosing the matrix resin. However, the effect of stress corrosion in chemical or environmentally stressful conditions may reduce the long term performance and hence a more conservative design may be required. This conservative design allows for effects of creep, cracking, aging, deleterious solutions, etc.
7. Creep and fatigue properties of the laminate under constant and intermittent loads should be understood.
8. Properties of the matrix should be such that the matrix should be able to sustain higher strain than the fibers.
9. The composite should be tough. The toughness properties of the composite can be indicated by the energy stored at the failure, i.e., the area under the stress–strain curve. The energy stored at failure should be as large as possible.

Table 1.2. Comparative thickness and weight for equal strength materials.

Materials	Equal tensile strength		Equal tensile stiffness		Equal bending stiffness	
	Thickness	Weight	Thickness	Weight	Thickness	Weight
Mild steel	1.0	1.0	1.0	1.0	1.0	1.0
Aluminum	1.8	0.3	3.0	1.1	1.5	0.5
GFRP*	2.4	0.07	25	5.0	3.0	0.6
GFRP†	0.3	0.1	6.8	1.5	1.9	0.5

* Based on random fiber orientation.
† Based on the uni-directional fiber orientation.
Source: Parklyn (1971)

1.4. Description of Fibers

Different types of fibers, such as, glass, carbon and aramid fibers, as used in FRP composites are described in the following sections

Glass Fibers: Glass fiber has been a predominant fiber for many civil engineering applications because of its economical balance of cost and specific strength properties.These fibers are commercially available in E-Glass formulation (for electrical grade), the most widely used general purpose form of composite reinforcement; high strength S-2 glass and ECR glass—a modified E-glass which provides improved acid resistance. Other glass fiber compositions include AR, R, and Te. Although considerably more expensive than glass, carbon and aramid fibers are used for their strength or modulus properties or in special situations as hybrids with glass. Properties of common high performance reinforcing fibers are shown in Table 1.3. Glass fibers are made with different compositions as shown in Table 1.4, utilizing glass chemistry to achieve the required chemical and physical properties

A brief description of various types of glass fibers are given in the following:

E-Glass: This is the family of calcium-alumina-silicate glasses, which is used for general-purpose molding and virtually all electrical applications. E-glass comprises approximately about 80%–90% of the glass fiber commercial production. The nomenclature "ECR-glass" is used for boron-free modified E-glass compositions. This formulation offers improved resistance to corrosion by acids.

S-Glass: This is a proprietary magnesium alumino-silicate formulation that achieves high strength and higher temperature performance. S-Glass and S-2 Glass have the same composition, but use different surface treatments. S-Glass is the most expensive form of glass fiber reinforcement and is produced under specific quality control and sampling procedures to meet military specifications.

C-Glass: This glass has a soda-lime-borosilicate composition and is used for its chemical stability in corrosive environments. It is often used in composites that contact or contain acidic materials, for corrosion resistant service in the chemical processing industry.

Table 1.3. Comparison of inherent properties of fibers (impregnated strand per ASTM D 2345, ACI 440R-96).

Fiber type	Specific gravity	Tensile strength MPa	Tensile modulus GPa
E-glass	2.58	2689	72.4
S-2 glass®	2.48	4280	86.0
ECR-glass*	2.62	3625	72.5
K-49 Aramid	1.44	3620	131.0
AS4 Carbon	1.80	3790	234.0

Table 1.4. Compositional ranges for commercial glass fibers.

Ingredients	E-glass range	S-glass range	C-glass range
Silicon dioxide	52–56	65	64–68
Aluminum oxide	12–16	25	3–5
Boric oxide	5–10	—	4–6
Sodium oxide and potassium oxide	0–2	—	7–10
Magnesium oxide	0–5	10	2–4
Calcium oxide	16–25	—	11–25
Barium oxide	—	—	0–1
Zinc oxide	—	—	—
Titanium oxide	0–1.5	—	—
Zirconium oxide	—	—	—
Iron oxide	0–0.8	—	0–0.8
Iron	0–1	—	—

Units = percent by weight, ACI 440R-96.

1.4.1. Forms of Glass Fiber Reinforcements

Glass fiber reinforced composites contain fibers having lengths far greater than their cross sectional dimension (aspect ratios >10:1). The largest commercially produced glass fiber is 'T' fiber filament, having a nominal diameter of 22.86–24.12 μ. A number of fiber forms available are discussed in the following sections:

Rovings: This is the basic form of commercial continuous fiber. Rovings are a grouping of a number of strands and are available in the form of bundle. In the case of a 'direct pull' rovings, the entire roving is formed at one time, which results in a more uniform product and eliminates catenary associated with roving groups of strands under unequal tension.

Woven Roving: The fiber bundle roving is also used as input to woven roving reinforcement. The product is defined by weave type, which can be at 0° and 90°; at 0°, +45°, and other orientations depending on the manufacturing process. These materials are sold in weight per square yard. Common weights are 18 oz/yd^2 [(610.3 g/m^2) and 24 oz/yd^2 (813.7 g/m^2)].

Mats: These are two-dimensional random arrays of chopped strands. The fiber strands are deposited onto a continuous conveyor and pass through a region where thermosetting resin is dusted on them. This resin is heat set and holds the mat together. The binder resin dissolves in the polyester or vinylester matrix thereby allowing the mat to conform to the shape of the mold.

Cloth: Cloth glass fiber reinforcement is made in several weights as measured in ounces-per-square-yard. Cloth is made from continuous strand filaments that are twisted and plied and then woven in conventional textile processes. It may be noted that all composite reinforcing fibers, including glass, are anisotropic with respect to their length. There are fiber placement techniques and textile–type operation that can further arrange fibers to approach a significant degree of quasi-isotropic composite performance. Glass fibers and virtually all other composite fibers are also available in a range of fabric like forms—including braided, needle punched, stitched, knitted, bonded, multi-axial, and multi-ply materials.

1.4.2. Behavior of Glass Fibers under Load

Glass fibers are elastic until failure and exhibit negligible creep under controlled dry conditions. The modulus of elasticity of mono-filament E-glass is approximately 73 GPa. The ultimate fracture strain varies from 2.5% to 3.5%. Glass fibers are much stronger than a comparable glass formulation in bulk form such as window glass or bottle glass. The strength of glass fibers is well-retained, if they are protected from moisture and air-borne or contact contamination. When glass fibers are held under a constant load at stresses below the instantaneous static strength, they will fail at some point as long as the stress is maintained above a minimum value. This is called creep rupture. This phenomenon is affected by atmospheric conditions with water vapor being most deleterious. It is known that the surface of glass contains submicroscopic voids that act as stress concentrations. Moist air can contain weakly acidic carbon dioxide. The corrosive effect of such exposure can affect the stress in the void regions of the glass fiber filaments until failure occurs. This corrosive phenomenon is also known as stress corrosion. In addition, exposure to high pH environments may cause aging or a rupture associated with time. However, to eliminate the problem associated with stress corrosion and other problems associated with defects in glass fibers, a number of special organo-silane functional treatments have been developed for this purpose. Both multi-functional and environmental-specific chemistries have been developed for the classes of matrix materials in current use.

1.4.3. Carbon Fibers

The commercial carbon fibers are produced from three sources, viz., (i) pitch (a by-product of petroleum distillation, (ii) Polyacrylonitrile (PAN), and (iii) rayon. The properties of carbon fiber are controlled by molecular structure and degree of freedom from defects. Formation of carbon fibers requires processing temperatures above 1000°C (1830°F). At this temperature, most synthetic fibers will melt and vaporize. However, acrylic fibers do not melt and its molecular structure is retained during high temperature carbonization. There are two types of carbon fibers: (i) the high modulus (Type I) and (ii) the high strength (Type II). The properties of Type I and Type II are due to difference in fiber microstructure. These properties are derived from the arrangement of the graphene (hexagonal) layer networks present in graphite. If these layers are present in three-dimensional stacks, the material is defined as graphite. If the bonding between layers is weak and two dimensional layers occur, the resulting material is defined as carbon. Carbon fibers have two-dimensional ordering.

High modulus carbon fibers of 200 GPa (30×10^6 psi) require that stiff graphene layers be aligned approximately parallel to the fiber axis. Rayon and pitch precursors are used to produce low modulus carbon fibers (50 GPa or 7×10^6 psi). Both PAN and liquid crystalline pitch precursors are made into higher modulus carbon fibers by carbonizing above 800°C (1400°F). Fiber modulus increases with heat treatment from 1000°C to 3000°C (1830°F to 5430°F). The results vary with the precursor selected. However, fiber strength appears to maximize at a lower temperature 1500°C (2730°F) for PAN and some pitch precursors fibers, but increases for most mesophase (anisotropic) pitch precursor fibers.

The axial-preferred orientation of graphene layers in carbon fibers determines the modulus of the fiber. Both axial and radial textures and flaws affect the fiber strength. Orientation of graphene layers at the fiber surface affects wetting and strength of interfacial bond to the matrix.Carbon fibers are not easily wet by resins; particularly the higher modulus fibers. Surface treatments that increase the number of active chemical groups—and sometimes roughen the fiber surface—have been developed for some resin matrix materials. Carbon fibers are frequently shipped with an epoxy size treatment applied to prevent fiber abrasion, improve handling and provide an epoxy resin matrix compatible interface. Fiber and matrix interfacial bond strength approaches the strength of resin matrix for lower modulus carbon fibers. However, higher modulus PAN based fibers show substantially lower interfacial bond strengths. Failure in high modulus fiber occurs in its surface layers in much the same way as with aramids. The carbon fibers are available as 'tows' or bundles of parallel fibers. The range of individual filaments in tow is normally from 1000 to 200 000 fibers. Carbon fiber is also available as prepreg, as well as in the form of unidirectional tow sheets. In Table 1.5, typical properties of composite reinforcing fibers are shown.

Table 1.5. Typical Properties of commercial reinforcing fibers.

Fiber	Typical diameter (μ)	Specific gravity	Tensile modulus, GPa	Tensile strength, GPa	Strain to failure (%)	Coefficient of thermal expansion, 10^{-6}/°C	Poisson's ratio
E-Glass	10	2.54	72.4	3.45	4.8	5.0	0.2
S-glass	10	2.49	86.9	4.30	5.0	2.9	0.22
Carbon PAN-Carbon T-300[a]	7	1.76	231	3.65	1.4	−0.1 to −0.5 (longitudinal), 7–12 (radial)	−0.2
AS[b]	7	1.77	220	3.1	1.2	−0.5 to −1.2 (longitudinal), 7–12 (radial)	—
t-40[a]	6	1.81	276	5.65	2.0	—	—
HSB[b]	7	1.85	344.5	2.34	0.58	—	—
Fortafil 3™ [c]	7	1.80	227	3.80	1.7	−0.1	—
Fortafil 5™ [c]	7	1.80	345	2.76	0.8	—	—
Pitch-Carbon P-555[a]	10	2.0	380	1.90	0.5	−0.9 (longitudinal)	—
P-100[a]	10	2.16	758	2.41	0.32	−1.6 (longitudinal)	—
Aramid Kevlar™ 49[d]	11.9	1.45	131	3.62	2.8	−2.0 (longitudinal), 59 (radial)	0.35
Twaron™ 1055[e]*	12.0	1.45	127	3.6	2.5	−2.0 (longitudinal), 59 (radial)	0.35

[a] Amoco; [b] Hercules; [c] Akzo-Nobel/Fortafil fibers; [d] DuPont de Nemours and Co. ; [e] Akzo-Nobel Fibers;
* Minimum lot average values.
Source: Akzo-Nable, 1998 and Mallic, 1988b.

1.4.4. Aramid Fibers

Amongst the several organic fibers, aramid fibers are the most popular for structural applications. The fiber is poly-para-phenyleneterephthalamide, known as PPD-T. Aramid fibers are produced commercially by DuPont (Kevlar™) and Akzo Nobel (Twaron™). They belong to liquid crystal polymers and are very rigid and rod-like. The aromatic ring structure contributes to high thermal stability, while the para configuration leads to stiff, rigid molecules that contribute to high strength and high modulus. When PPD-T solutions are extruded through a spinneret and drawn through an air gap during manufacture, the liquid crystal domains can align in the direction of fiber flow. The fiber structure is anisotropic with higher strength in axial direction and lower in the transverse direction. The fiber is also fibrillar and usually tensile failure initiates at fibril ends and propgates via shear failure between the fibrils.

Kevlar 49 and Twaron 1055 are the major forms of aramid fibers used today because they have higher modulus. Kevlar 29 and Twaron 2000 are used for ballistic armor and applications requiring increased toughness. Kevlar 149 is having ultra-high modulus. Aramid fibers are available in tows, yarns, rovings, and various woven cloth products. These can be further processed to intermediate stages, such as prepregs. Representative properties of para-aramid (p-aramid) fibers are given in the following section.

1. Tensile modulus is a function of molecular orientation.
2. P-aramid fiber is 50% stronger than E-glass. High modulus p-aramid yarns show a linear decrease of both tensile strength and modulus, when tested at an elevated temperature. More than 80%of strength is retained after temperature conditioning.
3. At room temperature, the effect of moisture on tensile strength is <5%.
4. Para-aramid is resistant to fatigue and creep rupture. Creep rate is low and similar to that of fiberglass. It is less susceptible to creep rupture.
5. P-aramid exhibits nonlinear, ductile behavior under compression. Yield is observed at a compression strain of 0.3%–0.5%. This corresponds to the formation of structural defects known as kink bands, which are related to compressive buckling of p-aramid molecules. This non-linear behavior of p-aramid fibers limits their use in applications that are subject to high compressive or flexural loads.
6. Toughness of p-aramid fiber is directly related to the area under stress-strain curve. However, p-aramid fibrillar structure and compressive behavior contributes to composites that are less notch sensitive.
7. The p-structure results in a high degree of thermal stability. Fibers will decompose in air at 425°C (800°F). They are usable over the temperature range of –200°C–200°C, although, not used on long-term basis at temperatures above 150°C (300°F) because of oxidation. These fibers have a slightly negative longitudinal coefficient of thermal expansion of -2×10^{-6}/K.
8. P-aramid is an electrical insulator. Its dielectric constant is 4.0 measured at 106 Hz.
9. P-aramid fiber can be degraded by strong acids and bases. However, it is resistant to most other solvents and chemicals. UV degradation can also occur. In polymeric composites, strength loss has not been observed.

It may be noted that the compressive behavior as noted here results in local crumpling and fibrillation of individual fibers, thus leading to the lower strength under conditions of compression and bending. This is the reason aramids are unsuitable in FRP shell structures unless hybridized with glass or carbon fibers to carry high compressive or bending loads. Such hybridized fiber structures lead to a high vibration

Table 1.6. Properties of ARAMID yarn and reinforcing fibers.

Property	Kevlar 49	Twaron 1055*
Yarn tensile strength, MPa	2896	2774
Tenacity dN/tex	20.4	19.0
Modulus, GPa	117.2	103.4
Elongation at break, percent	2.5	2.5
Density g/cm^3	1.44	1.45
Reinforcing fibers tensile strength, MPa	3620	3599
Modulus, GPa	124.1	127.0
Elongation at break, percent	2.9	2.5
Density g/cm^3	1.44	1.45

* Minimum lot average values. *Source:* DuPont, 1994 and Akzo-Nobel, 1994

damping factor, which may offer advantages in dynamically loaded FRP structures. The properties of Kevlar 49 and Twaron 1055 are given in Table 1.6.

1.4.5. Other Organic Fibers

In addition to aramid fibers, some other important organic fibers are described as follows:

Ultra-high-molecular-weight-polyethylene fibers: Spectra^{TM} is one of the ultra-high-molecular-weight-polyethylene fibers, which is manufactured by Allied Signal Corp. (USA). It was originally developed by Dutch State Mines (DSM) in Netherlands. Mechanical properties of these fibers have been presented in Table 1.7. The major applications of Spectra^{TM} have been in rope, special canvas, woven goods and ballistic armor. Its lightness combined with strength and low tensile elongation makes it attractive for these uses. However, these fibers have the following drawbacks:

1. They breakdown at temperatures above 130°C (266°F).
2. None of the current resin matrix materials bond well to this fiber.
3. Plasma treatment has been used to etch the surface of fibers for a mechanical bond to the resin matrix, which is expensive and is not readily available in commercial production.

Table 1.7. Properties of Spectra^{TM} fibers.

Properties \ Spectra Fibers	Spectra 900	Spectra 1000
Density, g/cm^3	0.97	0.97
Filament length, m	38	27
Tensile modulus, GPa	117	172
Tensile strength, GPa	2.6	2.9–3.3
Tensile Elongation, percent	3.5	2.7
Available yarn count (number of filaments)	60–120	60–120

Source: Pigliacampi (1987).

1.4.6. Hybrid Reinforcements

Since, so-called high performance fibers also have high cost, these fibers can be combined in lamina, and in uni-axial arrangements as hybrids to give appropriate properties at an appreciable cost. Infrastructure applications are natural opportunities for evaluation and utilization of such combinations. Table 1.8 shows the results obtained from carbon-glass-polyester hybrid composites at different carbon/glass fiber ratios. Both polymer matrix resin and reinforcing fibers exercise an interactive effect on the fabrication used to join composite materials used to form finished part.

1.5. Manufacturing and Processing of Composites

The main objectives of the manufacturing and processing of composites are as follows:

- Translate the potential fiber and matrix properties into reality in the composite.
- Insure process consistency, i.e., many processing methods are available and each process has specific attributes that must be considered as part of the design process.
- Minimize processing costs.

1.5.1. Steps of Fabrication Scheme

The fabrication of FRP components should involve the following steps:

- Design-Stress and geometric envelope,
- Material selection,
- Arrangement (Orientation and configuration of reinforcement),
- Assembly of reinforcement and resin system,
- Application of heat and pressure as appropriate to cure the composite,
- Finishing processes,
- Assembly, and
- Quality control and non-destructive inspection.

Table 1.8: Properties of carbon-glass-polyester hybrid composites*.

Carbon/ Glass Ratio	Tensile strength, MPa	Modulus of elasticity (Tension) GPa	Flexural strength, MPa	Flexural modulus, GPa	Interlaminar shear strength, MPa	Density, g/cm^3
0: 100	604.7	40.1	944.6	35.4	65.5	1.91
25: 75	641.2	63.9	1061.8	63.4	74.5	1.85
50: 50	689.5	89.6	1220.4	78.6	75.8	1.80
75: 25	806.7	123.4	1261.7	1261.7	82.7	1.66

*Fiber contents are by volume; Resin is 48% Thermoset Polyester plus 52% continuous unidirectional oriented fiber by volume equivalent to 30% resin and 70% fiber by weight; Properties apply to longitudinal fiber direction only.
Source: Schwarz (1992).

1.5.2. Manufacturing Methods

The manufacturing processes of FRP components are classified into the following categories.

Direct Methods
Those methods that use the reinforcement and resin directly without any pre-processing changes, e.g., Wet layup, Pultrusion, Filament winding, Resin Transfer Moulding and SCRIMP (Seemanns Composite Resin Infusion Moulding Process, which is a form of resin infusion or vacuum assisted resin transfer moulding).

Indirect Methods
Those methods that use the combination of fiber and matrix before the process such as in prepregging or formation of injection moulding pellets or Sheet Moulding Compound (SMC) blanks, such as Compression Moulding, Pultrusion, Winding, Autoclave Curing, etc.

Details of various manufacturing processes are given in the following sections.

1.5.2.1. Hand Layup Process

In this method, the material (usually prepreg) is cut and laid up to produce laminate. This method is used for small quantities of complex and/or high quality parts. However, this method is very labor intensive and thus an expensive process. This method has been automated partially in last few years. Hand lay-up may use woven or non-woven fabric or uni-directional tape, usually in the prepreg form. The lower performance applications may use woven glass fabric impregnated with wet resin rather than prepreg, while the high performance applications usually require prepreg. This method produces laminates with relatively high fiber volume (50%–65%) and low void content (1%–2%). Variation on hand lay-up process consists of roll wrapping, used to make fly rods and most golf club shafts. Schematics of hand layup process are shown in Fig. 1.1.

The Curing of a lay-up is relatively expensive and elevated temperature cure is usually required. Sometimes curing is performed in a laminating press or by vacuum bagging and oven curing the laminate. High-performance hand lay-ups require vacuum bagging and autoclave curing. For thermoset matrix composites autoclave cure requires pressure steps, wherein time at cure temperature and pressure is ≥2 hr. The pressure step is followed by cool-down cycle. Some systems require an oven postcure.

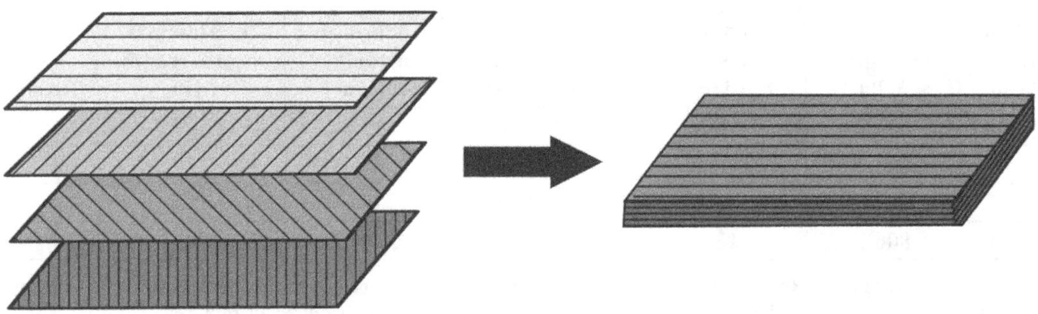

Figure 1.1. Schematics of hand layup process.

1.5.2.2. Semi-Automated and Automated Lay-Up Processes

These processes greatly reduce labor costs. The more highly automated process, the more complex the equipment and higher the capital cost as well as costs of set-up and "debugging". Automation is cost effective if costs can be amortized over a large production volume. Reduction of labor costs is the main goal of processes such as mechanically-assisted lay-up and controlled tape laying. Other benefits include improved part quality, reduced material waste, and decreased processing time. Simple mechanical assistance is used only for flat or moderately controlled parts. More complex parts require much more complex automation. Figure 1.2 shows the hand- held tape laying machine such as semi-automated machine.

1.5.2.3. Filament Winding

In this method, fiber bundles are impregnated with resin and wound upon a rotating mandrel to produce a shape such as a tube or pressure vessel. Alternatively, tape winding is used where prepreg tape is employed in place of fiber bundles. Schematics of filament winding techniques are shown in Figs. 1.3 and 1.4. In this method composite parts have excellent strength-to-weight ratio. Continuous, reinforced filaments, usually glass, in the form of roving are saturated with resin and machine wound onto mandrels having shape of desired finished part. Once winding is completed, part and mandrel are cured. Mandrel can then be removed through pothole at the end of wound part. High strength reinforcements can be oriented precisely in direction where strength is required. This method also results in good uniformity of resin distribution in finished part or mainly circular objects such as pressure bottles, pipes, and rocket cases.

Advantage of filament winding techniques: The followings are the main advantages of Filament winding techniques:

- High speed up to 700 lbs of material per hour and low cost;
- Wet winding avoiding the use of prepreg further reduces the cost;
- Void content is much higher than with hand lay-up, but the part quality can be excellent;

Figure 1.2. Hand-held tape laying machine.

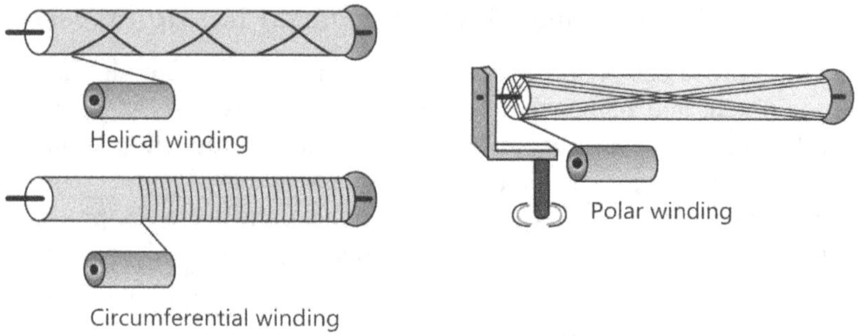

Figure 1.3. Filament winding techniques.

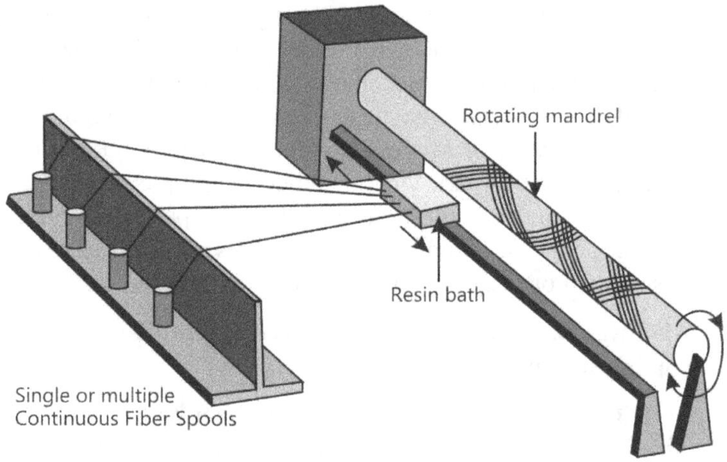

Figure 1.4. Filament winding techniques.

- By adjusting the winding pattern, the fiber volume can be varied to fit the application; and
- Filament wound include rocket motor casings, pressure vessels, specialty piping, aircraft fuselages, aircraft wings, railcar bodies, and wind turbine blades.

1.5.2.4. Fiber Placement

This is an automated method which combines automatic tape laying with filament winding to permit winding concave surfaces, changing dimensions and winding cutouts. Fiber placement equipment and setup is expensive, but offers excellent part quality including low void content as well as tremendous design flexibility. This technique has been applied to parts such as tail and wing spars, flex beams and skins. This process accommodates window openings, etc. very efficiently. Fiber placement is also used to manufacture the pivot shaft of the horizontal stabilizers for the F-22 Raptor. This reduces the weight by 90 lb. per pair of shafts versus the titanium alloy used previously for such applications. Figure 1.5 shows the schematics of Fiber Placement Technique.

1.5.2.5. Pultrusion Process

As shown in Figs. 1.6 and 1.7, in this process, continuous fiber reinforcement is impregnated with resin i.e. by passing through resin bath. These impregnated fibers are pulled through a forming die,

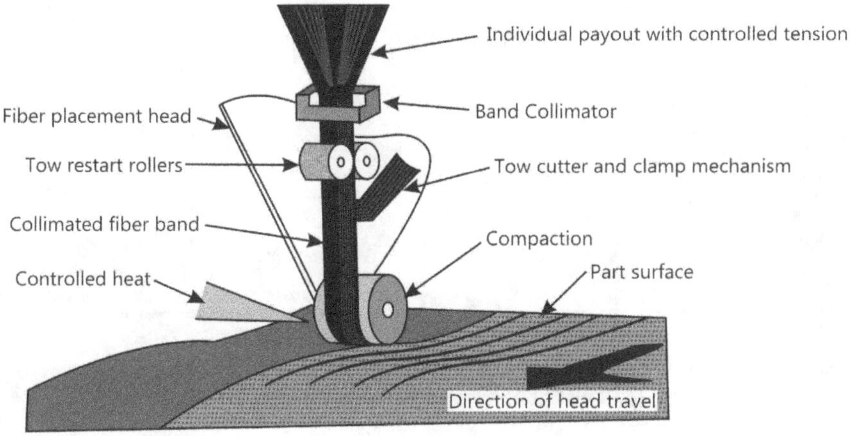

Figure 1.5. Schematics of fiber placement.

consolidated, cured, cooled quickly and cut to length—all as a continuous automated process. In spite of the high initial cost, high production volume of the pultrusion results in low cost/part. Production speeds are usually two to four feet per minute. Quality of the product ranges from good to excellent with a low void content. However, this method does not provide flexibility and uniformity of product control, and automation. This method is used for continuous production of simple shapes, such as rods, tubes and angles, principally incorporating fiberglass or other reinforcement and large output is possible.

Main limitations of pultrusion process

The main limitations of pultrusion process are as follows:

1. The process produces mainly two dimensional shapes.
2. It is not possible to vary cross sectional shape within the product.
3. Main reinforcements are in the axial direction and slow curing resins cannot be used.

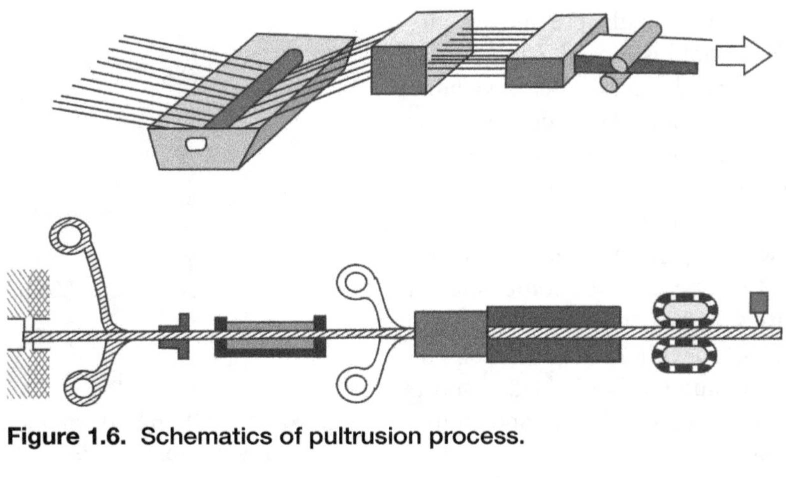

Figure 1.6. Schematics of pultrusion process.

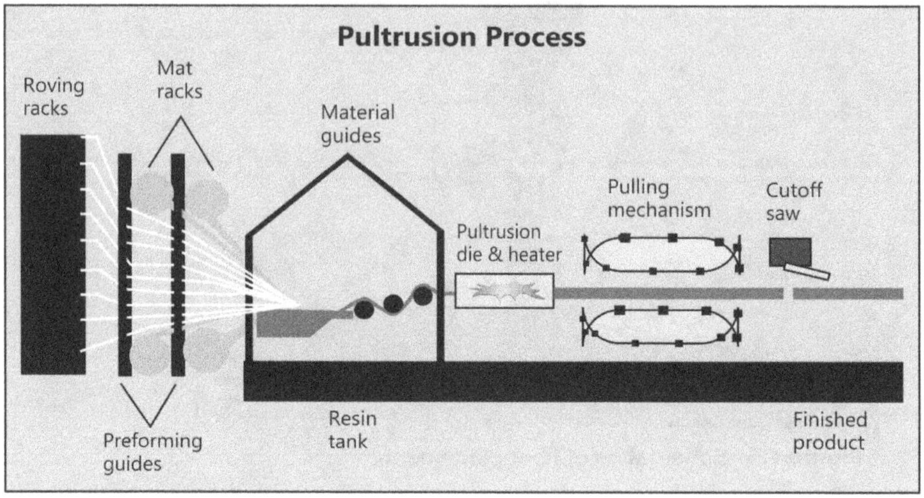

Figure 1.7. Schematics of pultrusion process.

1.5.2.6. Pull Forming Process

This is a sophisticated modification of the pultrusion process, which allows changing cross -sections and fabrication of the curved parts etc. Examples of the such pultruded products include I-beams—for construction applications and oil platforms—rebar, prestressing strands and twisted cables automotive drive shafts, corrosion resistant and high strength structural pipes, floor gratings and hand rails for off-shore oil platforms, fiber optic communication cables, non-conductive bar stock, foam core residential siding, and light weight non-conductive ladder rails.

1.5.2.7. Resin Transfer Molding (RTM)

This process is one of a family of processes known as liquid composite molding. This is a closed-mold low-pressure process. Matched male and female molds are loaded with pre-shaped dry reinforcement and the mold is closed. Liquid resin is pumped into the mold usually with vacuum assistance to impregnate the reinforcement. The composite is cured under ambient or elevated temperature conditions depending upon the resin system and part thickness. The process can produce large, complex low-cost parts at near net shapes. Theoretically, RTM can be used for composites of all sizes, complexities, and performance levels, and can produce composites with 3-D reinforcement. This is widely used in automotive industry, and is being developed for smaller volume, and higher performance applications.

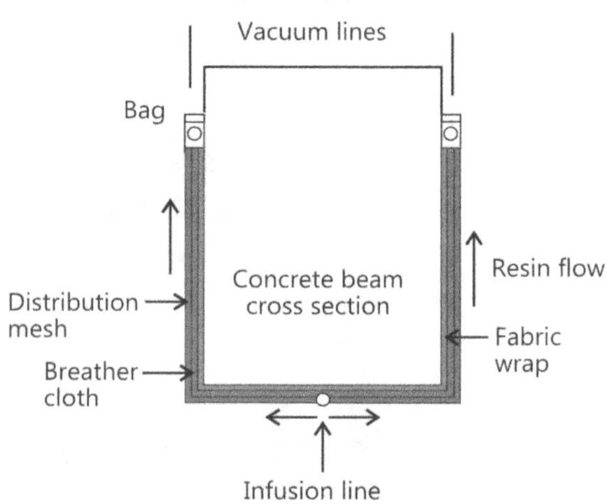

Figure 1.8. Schematics of vacuum-assisted resin transfer molding (VARTM).

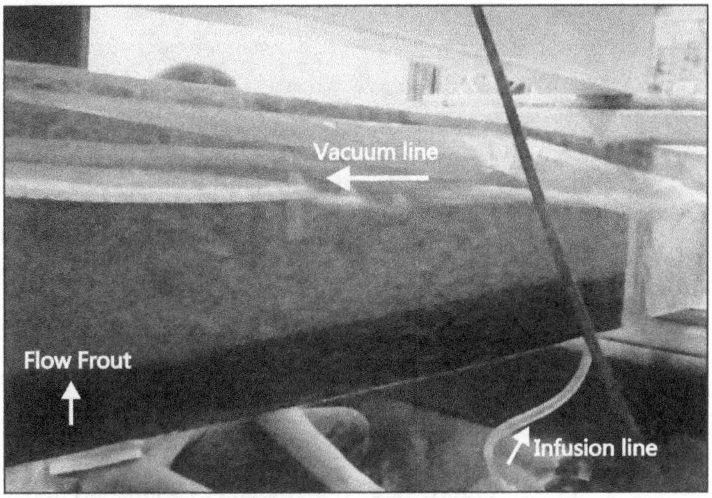

Figure 1.9. Vacuum assisted RTM process for external strengthening of rectangular beam FRP fabrics.

1.5.2.8. Vacuum Assisted RTM Process for Rectangular Beam Strengthening

As shown in Fig. 1.8, a vacuum assisted resin transfer molding process has been used to create vacuum between the concrete surface and FRP wrap for uniform distribution of resin on the concrete surface. After the composite material is entered into the mold, the part is vacuum bagged. In this case, a vacuum of ≥14 psi and cure temperature of less than 350°F is applied. In this situation, the vacuum actually compacts the composite and helps the resin wet out the preformed composite part. Figure 1.9 shows the vacuum assisted RTM process for external strengthening of rectangular beam.

1.6. Sandwich Construction

This is developed for aircraft design to make the structures as light as possible without sacrificing strength and stiffness. In sandwich construction, thin, stiff, strong skins on a core material provide the highest strength-to-weight ratio and rigidity-to-weight ratios obtainable. The key to success of this construction depends on the right choice of core and of adhesives to bond on the face sheets. The core materials used are honeycomb or some types of foam. Figures 1.10 shows components of sandwich construction.

Characteristics of honeycomb core: The followings are the characteristics of Honeycomb Core used in the construction of sandwich panels.

1. It provides the highest performance per unit weight.
2. Almost every aircraft flying today depends upon the honeycomb construction.
3. Honeycomb materials include aluminum and many non-metals—such as paper, fiberglass, aramid fiber and papers, carbon fiber, etc.
4. The first honeycomb was developed in 1940s—canvas with phenolic resin.
5. Other usage of honeycomb construction include—cargo containers, lightweight air-transportable buildings, aircraft seats, electronic blackboards, photocopier housings, air-transportable shelters, and curtain walls for high-rise buildings.

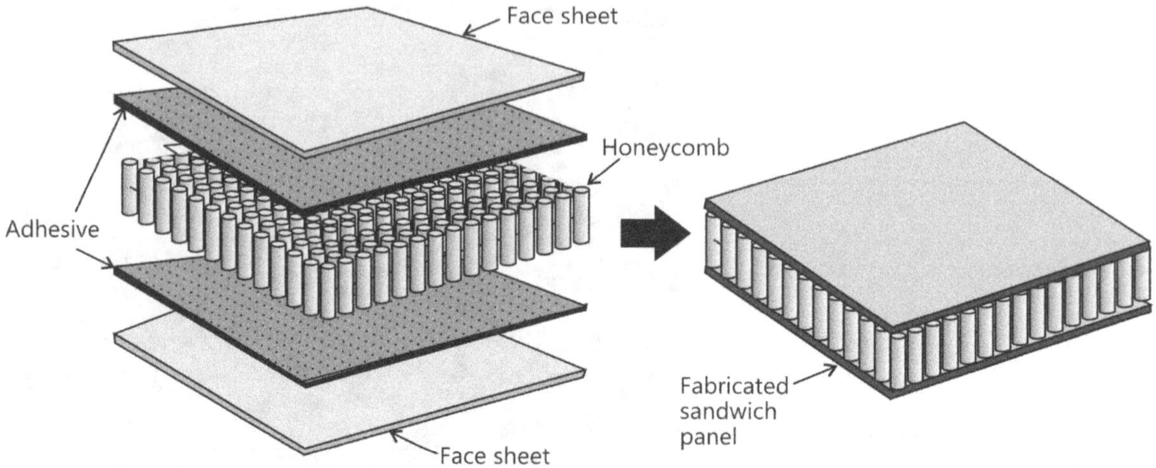

Figure 1.10. Schematics of sandwich panel fabricated with face sheet and honeycomb core material bonded with adhesive.

1.7. Compression Molding

The Compression molding methods includes Sheet Molding Compound, Dough Molding Compound (DMC), Bulk Molding Compound (BMC), and general compression molding processes. A preblended mixture of reinforcement, thermosetting resin, and filler is placed between a pair of heated matched metal dies and squeezed under pressure. Molding cycles of less than 2 minutes are quite common. Excess material (or flash) is machined off after molding of parts into shapes. In some cases, dry preforms are placed between layers of the charge—a mixture of resin, fiber, and filler. The method requires use of a press operable under hydraulic or pneumatic pressure. This method is used for non-load bearing components, but has recently been used with pultrusion for the fabrication of a high stiffness composite reinforcement bar.

1.8. Multi-Axial Fabric for Structural Components

As shown in Fig. 1.11, multi-axial fabrics are used for boats, pier repairs, wind mill blades, RTM processing, and auto frame etc. These are fabricated by stitching together the different layers of fabrics with different fiber orientations.

1.9. Fabrication of Stirrups

As shown in Fig. 1.12, FRP cables such as 7-wires carbon fiber twisted cables (CFTC 1×7) are held between the jigs and then tensioned using torque wrench as shown in Fig. 1.13 to fabricate the box stirrups. The whole assembly of stirrups is then cured at about 90°F for 24 hr for making them ready to use for structural applications.

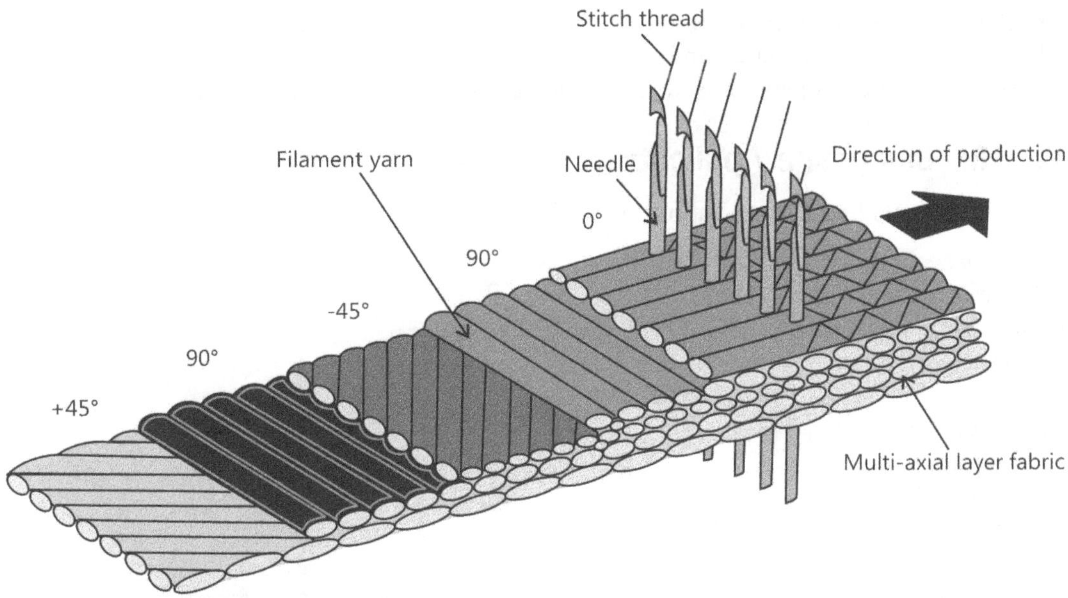

Figure 1.11. Multi-axial fabrics for structural components.

Figure 1.12. Jigs for holding CFTC 1 × 7 box stirrup.

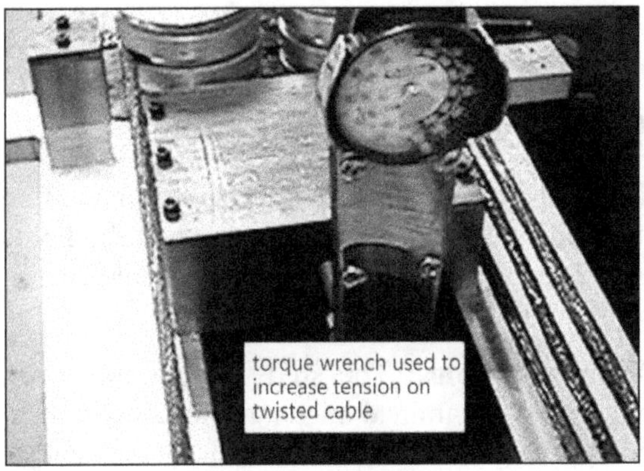

Figure 1.13. Application of torque wrench to increase tension in box-stirrups.

1.10. FRP Composites

Fibre-Reinforced Polymer Composites (FRPCs) are used as primary structural members in many fields due to the following reasons:

- High strength-to-weight ratio,
- High stiffness-to-weight ratio,
- Better environmental resistance,

- Higher fatigue endurance limit,
- Facility to tailor the material, and
- Less noisy while in operation and provide lower vibration transmission than metals.

1.11. FRP Composite Applications

The FRP composite find its applications in the following fields:

- Aerospace,
- Automotive,
- Chemical,
- Electrical/Electronics,
- Medical,
- Semiconductor,
- Oil Refinery,
- Materials Handling,
- Heavy Equipment,
- Offshore Exploration,
- Food Processing,
- Wastewater Treatment,
- Infrastructure, and
- Future's Industry.

1.12. Composite Mechanics

The study of composite material is based on Micromechanics and Macromechanics. Details of composite mechanics can be found elsewhere in the book entitled Mechanics of Composite Materials by Jones (1975). However, for the sake of completeness, a brief description of the mechanics of composite and terminology used are given here.

Micromechanics: The study of composite material behavior where the interaction of constituent material is examined in detail, and used to predict and define the behavior of the heterogeneous composite material is called micromechanics. Approaches to the study of micromechanics consist of the followings:

- Mechanics of Materials,
- Elasticity
  - Bounding Principles
  - Exact Solution
  - Approximate Solutions.

Macromechanics: The study of composite material behavior where the material is presumed homogeneous and the effects of constituent materials are detected only as averaged 'apparent' properties of the composite material.

1.12.1. Laminate Terminology

The following terminology is used for description of various kinds of laminates:
Symmetric Laminate: Laminate composed of plies such that both geometric and material properties are symmetric about the middle surface (mid-plane).
Balanced Laminate: For every + ply there exists a − ply of the same thickness and material property.

Cross-ply Laminate: Laminate is composed of 0° and 90° plies.

Angle-ply Laminate: Laminate is composed of + and − plies.

Generally orthotropic lamina In this lamina, the principal material directions—fiber and transverse to fiber directions—are not parallel to the body coordinate axes of the laminate as shown in Fig. 1.14.

1.12.2. Composite Product Forms

Composites are referred to by type of fibre/type of matrix (AS4/3501-6). Layers can be fibre only for later 'wet' layup (adding resin), or can be 'prepreg' (already containing the resin). When all fibers are aligned in single direction, then it is called tape or unidirectional composite. When fibres are aligned in multiple (usually two) directions, the composite form is known as cloth or woven fabric. When fibers are placed over and under one fiber at a time, it is called plain weave and when they are placed over many and under one at a time, it is called Satin weave. Figure 1.15 shows the schematics of unidirectional and woven fabric.

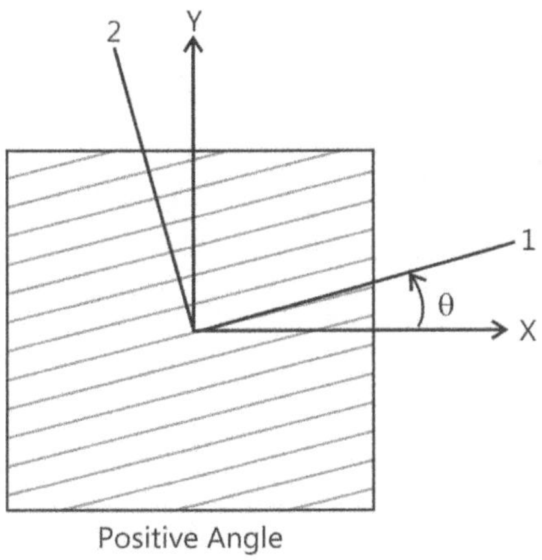

Figure 1.14. Generally orthotropic lamina.

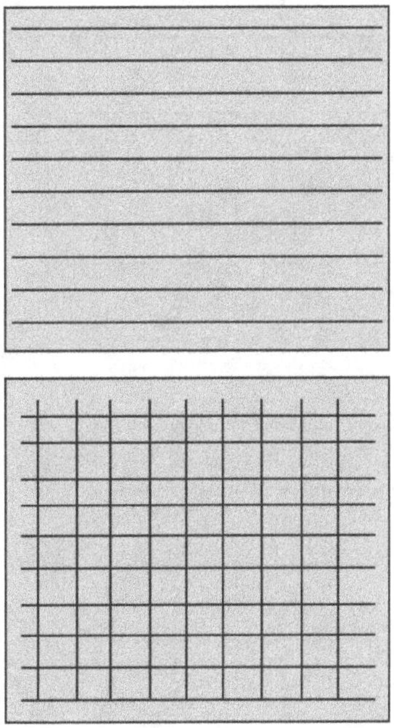

Figure 1.15. Schematics of unidirectional and woven fabric.

1.13. Laminates Types and Stacking Sequence

Depending on the stacking sequence of lamina in structural laminates and lamina fiber orientations, laminates are classified into different categories as described in the following:

Symmetric Laminate

Symmetric laminate implies that the material and orientation of layers above the laminate mid-plane are identical to those below (No bending – extensional coupling). Figure 1.16 shows typical symmetric laminates.

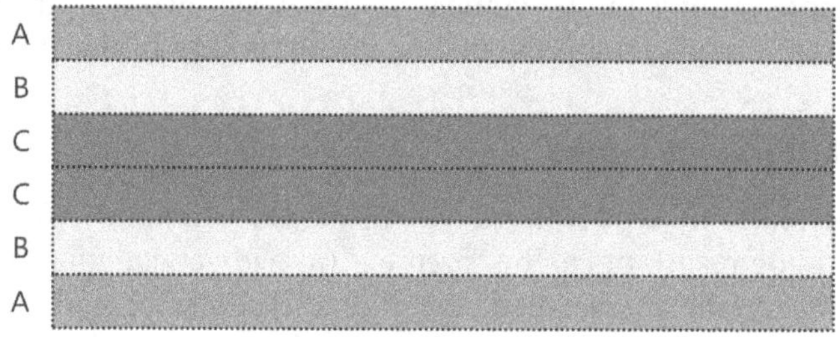

Figure 1.16 (a). Schematics of symmetric laminate.

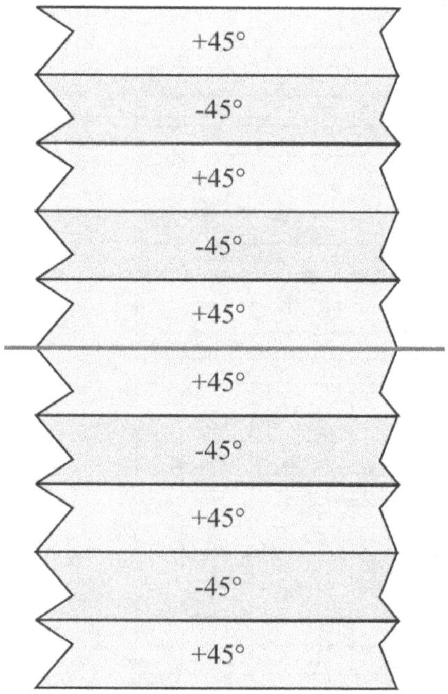

Figure 1.16 (b). Symmetric angle-ply laminate.

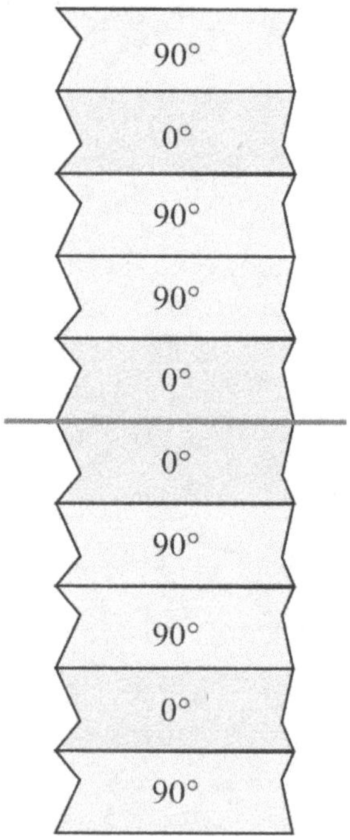

Figure 1.16 (c). Symmetric cross-ply laminate.

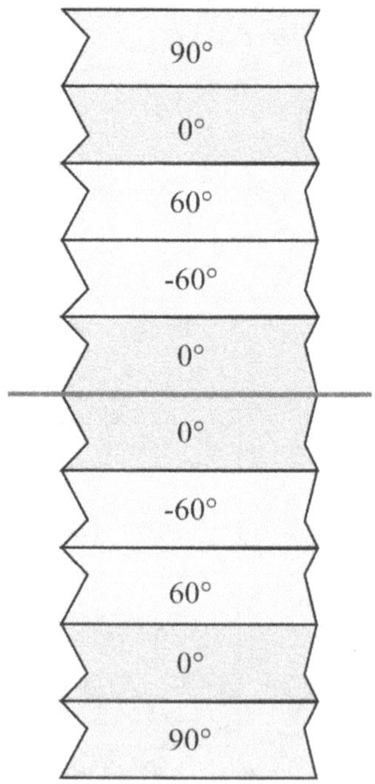

Figure 1.16 (d). General symmetric laminate.

Antisymmetric Laminate

Antisymmetric laminate implies that the material of layers above the laminate mid-plane is identical to those below, but the orientations are of opposite sign. Figure 1.17 shows the schematics of typical antisymmetric laminates.

Figure 1.17 (a). Schematics of an antisymmetric laminate.

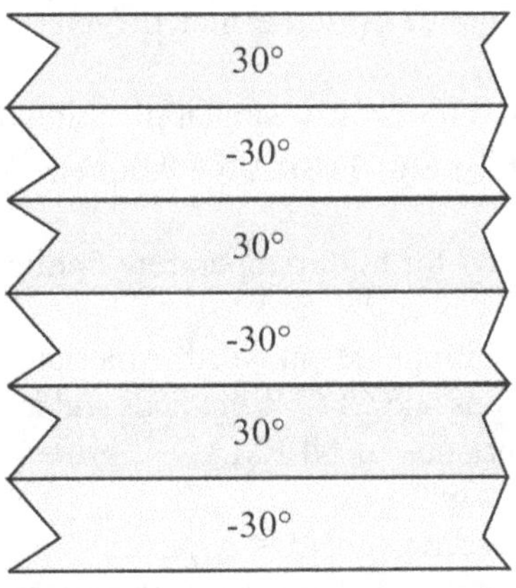

Figure 1.17 (b). Schematics of an antisymmetric angle-ply laminate.

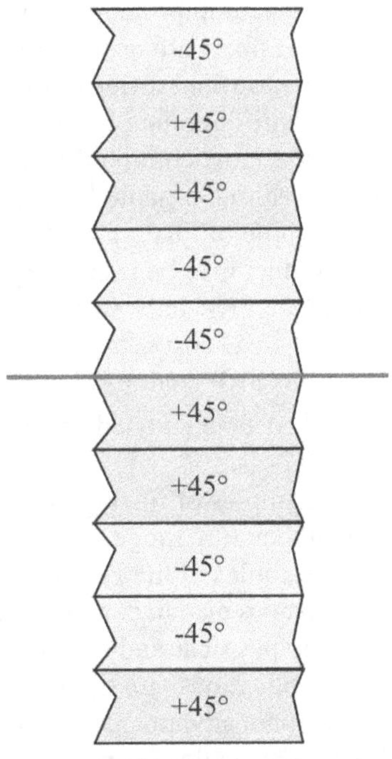

Figure 1.17 (c). Schematics of an antisymmetric angle-ply laminate.

Exercise Problems

A. Answer the following questions briefly:

1. What kinds of matrices are widely used for the construction of civil engineering infrastructures?
2. Why are Fiber Reinforced Polymers (FRP) used as substitutes for steel reinforcement?
3. What are the main functions of fibers in an FRP reinforced material?
4. What is a high performance composite?
5. Give at least two examples of most commonly used conventional reinforcements in civil engineering infrastructure.
6. Give an example of laminar composite.
7. What are the main advantages of general composite materials with respect to conventional materials?
8. What is the major disadvantage of bulk glass? Which property of the bulk glass can be improved using fiberglass?
9. What do you mean by redundancy of structures? How can the FRP reinforced or laminated structures, play an important role in increasing the redundancy of structures?
10. What kinds of fibers are usually referred to as wire instead of fiber?
11. How does the tow/end differ from yarn with regard to filament bundle?
12. How can you differentiate Roving, Mat and Fabric?
13. Which are the most commonly used organic fibers? Do the glass and carbon fiber reinforced polymers constitute organic fibers?
14. Give an example of FRP manufactured rope, which does not use matrix for bonding the fibers.
15. If a FRP product is to be used in aggressive alkaline environmental condition, can you suggest which kind of FRP bars would be most suitable? Assume that off-axis properties are not important whereas cost is an important factor.
16. Which kind of FRP tendon is not recommended for prestressing application and why?
17. Which kind of fiber is used in the manufacturing of FiBRA™, Arapree™ and Technora™ rods/tendons?
18. Give few examples of fibers, which can be used for high temperature applications along with their use temperature range.
19. Give few examples of the FRP manufacturing companies in North America.
20. Which companies manufacture Leadline™ tendons and CFCC™ strands and in which countries?
21. What are the physical and structural characteristics of NEFMAC™ grids? For what kind of structural applications, are these grids mostly used?
22. What is the main advantage of 3-D FRP reinforced panels?
23. Which kind of tendon is recommended for unbonded or external prestressing but not for pretensioning applications?
24. Give two examples of thermosetting and thermo-plastic resins.
25. What is the primary difference between thermo-plastic and thermo-setting resins?
26. How does the matrix of an FRP influence the quality of FRP products?
27. Define glass transition temperature of a matrix? What is its significance in field application of the polymeric matrix?

28. What is the major obstacle for thermoplastic matrix in viable FRP production process?
29. Define a polymer. State different forms of polymers and their characteristics.
30. Define specific gravity and specific weight of FRP materials.
31. Define Shelf-life. Which kinds of resins have a higher shelf-life?
32. What are the major disadvantages of thermosetting matrices?
33. What are the major advantages of thermosetting matrices?
34. In some applications of FRP production, fillers are used? Can you state why fillers are added to the thermosetting and thermoplastic polymers?
35. Give an example of a hardener used with epoxy resin.
36. State major advantages of epoxy in comparison to polyester resin.
37. State major disadvantages of epoxy in comparison to polyester resin.
38. What are the major factors affecting the performance of the fiber matrix composite?
39. Many fiber-reinforced composites exhibit high internal damping properties. This leads to better vibrational energy absorption within the material and reduces transmission to adjacent structures. Give examples of civil infrastructures where this aspect of composite behavior may be relevant.
40. In the case of a complex part, it may be necessary to resort to shorter fibers to reinforce the molding effectively in three dimensions. In this way, quasi-isotropic properties can be achieved in the composite. Can you suggest other ways to obtain the quasi-isotropic laminates/composites?
41. On what factors does the stress distribution within an FRP depend?
42. In the case of CFRP materials, what parameters may be of concern primarily to the designers?
43. How can the toughness of composite predicted and what is its structural significance?
44. How can the manufacturers improve the resistance to alkaline environment of the GFRP?
45. What is the most widely used general-purpose form of glass fiber and in which application is it primarily used?
46. For which application is the ECR glass more suitable?
47. Which kind of glass FRP is most appropriate for structural applications?
48. If GFRP is to be used for its chemical stability against corrosive environments where it may contact or contain acidic material in the chemical processing industry, which specific form of glass can be used in this situation?
49. What is creep rupture? Which FRP is most susceptible to creep-rupture?
50. What is stress corrosion and how does it occur in the case of GFRP bars?
51. What factors control the modulus of elasticity of Carbon fibers?
52. What is a precursor? Give a few examples of most widely used precursors for production of carbon fibers?
53. What is a prepreg and what is its advantage?
54. Which kind of manufacturing process is used for FRP components for civil engineering infrastructures?
55. How can the vacuum assisted resin transfer molding process be a potential repairing and external strengthening technique?
56. Give two examples of commercially available Aramid fibers.
57. Which structural applications limit the use of AFRP and why?

58. Comment on the notch sensitivity of AFRP.
59. Which fibers have the negative coefficient of longitudinal thermal expansion?
60. If Aramid fibers have to be used for FRP shell structures, what property of this fiber will be of most concern to the designers? What could be the solution for this concern?
61. Give an example of commercially available Ultra-high-molecular-weight-polyethelene fiber.
62. What is the major disadvantage of Spectra™ fiber and how can it be eliminated?
63. What kind of structural components can be easily manufactured by Filament winding process?
64. Which matrices are most commonly used in Pultrusion process?
65. How can the ultraviolet rays affect the properties of FRP materials?
66. What do you mean by durability of FRP system? How can the water and freeze-thaw affect the material properties?
67. Do you think fire and temperature can significant affect the strength of FRP materials? If so, how does this effect from the direct exposure of FRP to fire from that used internally in the concrete? Comment on this behavior of FRP.
68. What do you understand by accelerated aging method?
69. Why are gripping mechanisms so important in prestressing and material properties testing of FRP?
70. What are the different ways to improve the bond and shear strength of FRP bars/tendons?
71. Why are natural fibers not recommended for structural applications?
72. Do you think that FRP materials such as AFRP, GFRP, and CFRP show linear elastic relationships up to the failure? If so, definitely structures reinforced with these materials will not exhibit significant ductility. Could you suggest how a designer can take into account the brittleness of FRP in his design?
73. What kind of natural fibers were used for FRC in the olden days?
74. What is implication of the assumption—'Plane section remains plane before and after bending'—in the design and analysis equation?
75. Design moment capacity of an FRP reinforced structure consisting of two strength reduction factors such as ϕ and ψ. Are these factors used for the same purpose? What is the basis and purpose of their use in the design capacity of a section?
76. What design philosophy has been adopted in ACI 440.1R-06 standard?
77. On average what is the fiber volume fraction of FRP composites?
78. What do you understand by guaranteed strength and guaranteed rupture strain of FRP materials? What is its significance?
79. From test results of 30 specimens, the average strength of an FRP material was predicted to 2468 MPa. If standard deviation of test results was 68.95 MPa. What is the guaranteed strength of material? If modulus of Elasticity of this material is 147 GPa, what is its approximate guaranteed rupture strain.
80. What is deformability factor? What is its significance? Is it the same factor as ductility index? Comment on this?
81. Define aging of materials.
82. What do you understand by endurance limit and endurance time? To what phenomena are these parameters related?
83. Define fatigue strength of an FRP material. Which FRP material is most susceptible to fatigue effect?

84. Define notch sensitivity of FRP material. What is its significance with regard to the strength of FRP material?
85. In general, what is the relationship between tensile strength and compressive strength for AFRP, CFRP and GFRP materials?
86. In general, what is relationship between tensile modulus and compression modulus for AFRP, GFRP, and CFRP materials?
87. When anchoring a reinforcing bar in concrete, there are three main mechanisms, which affect the transfer of bond force to the concrete. What are those mechanisms?
88. Do you think that creep-rupture is an issue with steel rebars in the concrete structures? If so, in what situation could creep-rupture may be predominant failure mode for steel reinforcing bars?
89. What is the relationship between creep-rupture strength and endurance time?
90. What test conditions significantly influence the ambient environment fatigue behavior?
91. What do you understand by S-N curve? What is its significance?
92. What factors control the bent strength of FRP bars?
93. Do you think that failure of FRP reinforced concrete structure can be based on formation of plastic hinges? If so, please comment on this.
94. Define serviceability of structures. What parameters control the serviceability of structures?
95. What is the significance of Branson's equation?
96. For computing the long-term deflection of FRP-reinforced concrete structures, what factors are taken into account?

Answers

The answers of the above questions are given below:

1. Thermoset matrices.
2. They are non-corrodible, insensitive to magnetic effects, lightweight, easy to handle, less labor cost and low maintenance cost.
3. These are load carrying elements. Provides longitudinal strength and stiffness to FRP in combination with matrix materials.
4. Composites whose mechanical properties are better than conventional structural materials.
5. Steel and Aluminum.
6. Plywood.
7. High longitudinal strength (varies with sign and direction of loading relative to fibers), corrosion resistant (not dependent on coating), nonmagnetic, high fatigue endurance (varies with type of reinforcing fiber), Lightweight (about ⅕ to ¼ the density of steel), low thermal and electric conductivity (for glass and aramid fibers), fiber composites are inhomogeneous, fiber composites have better properties than any of the materials they are made from. For example, neither straw nor mud has great structural properties, but straw-filled mud bricks aren't too bad.
8. Bulk glass is very stiff and fails at very low stress levels due to rapid growth of cracks. The larger the pre-existing cracks, the lower the stress level at which glass fails. Fracture resistant of bulk glass can be improved using fiberglass.
9. Progressive failure of structures refers to redundancy of structure. FRP reinforced or laminated structures can play an important role in increasing the redundancy of structure by progressive failure of laminates.
10. Boron or some silicon carbide, since their smallest fiber unit is fairly large (100–200 micrometer).
11. Tow/end is fairly large bundle of filament and untwisted. Yarn refers to a small bundle of filaments that are twisted together.

12. Roving consists of large fiber bundle made by combining more than one tow/end, usually with little or no twist. Mat is a felt-like material made by chopped or swirled short filaments combined loosely with a binder material. Fabric is a planar material made by weaving fibers (bundles/wires). Many different weave patterns are possible. Non-wovens are also sometimes called fabrics.
13. Aramid is the most commonly used. Glass and carbon fibers do not constitute organic fibers.
14. Parafil^{TM} tendons (Type G), which consists of a closely packed parallel core of continuous aramid (Kevlar 49^{TM}) fibers contained within a thermoplastic sheath. The sheath maintains the circular profile of rope and protects the core without adding to its structural properties.
15. Polyethylene fiber reinforced polymers would be most appropriate FRP.
16. GFRP, because of creep-rupture at a very low stress level.
17. Aramid fibers.
18. Ceramic fibers (upto 1200°C/2200°F, however electron-beam cured ceramic fibers are usable for short period at 3300°F (1800°C), Quartz fibers [1920°F (1250°C)].
19. Autocon Composites, Corrosion Proof Products, Creative Pultrusions, International Grating, Marshall Industries Composites (no longer in use), Diversified Composites inc., Marshall-Vega Corporation, Polystructures, Polygon and Pultrall.
20. Mitsubishi Functional Products Inc., Tokyo Rope Manufacturing Co.
21. NEFMAC^{TM} is a 2-D grid-type reinforcement consisting of glass and carbon fibers impregnated with resin. It was developed by Shimizu Corporation, one of the largest Japanese general contractors. This is formed into a flat or curved grid shape by a pin-winding process similar to filament winding. It is available in several combinations of fibers (e.g., glass, carbon and glass-carbon) and cross-sectional areas (5 to 400 mm^2). These grids are most commonly used in tunnel lining applications, offshore construction and bridge decks. Applications in buildings include lightweight curtain walls.
22. 3-D FRP reinforced panels have sufficient strength and rigidity to withstand design wind loads and can easily achieve fire resistance for 60 min.
23. Parafil tendons.
24. Epoxy and Vinylester are thermosetting resins, whereas Polyethylene, polyimides, peek and pps are thermoplastic matrices.
25. A thermoset is produced by cure reaction. Once cured, it cannot be non-destructively softened or reformed by the action of heat or solvents. Thermoplastic matrices can be reformed by the use of heat or solvents without destroying its chemical integrity.
26. The matrix binds the fibers together and keep them in proper orientation and position, transfer the load to and in between fibers, protects the fiber from moisture and chemical attack, protects the fibers from mechanical damage during processing and use, enhances toughness and impact resistance of composite system, largely determines the composite's compressive and transverse mechanical properties and usually determines the overall service temperature of the composite.
27. This is the temperature at which polymeric materials change from a hard, often brittle solid to a soft, tough solid. The tensile modulus of the matrix can be reduced by as much as five orders of magnitude.
28. A major obstacle is that thermoplastic matrix polymers are much more viscous and are difficult to combine with continuous fibers in a viable production operation.
29. A long-chain molecule having one or more repeating units of atoms joined together by strong covalent bonds. A polymeric material (i.e., a plastic) is a collection of a large number of polymer molecules of similar chemical structure. If, in a solid phase, the molecules are in random order, the plastic is said to be amorphous. If the molecules are in combinations of random and ordered arrangements, the polymer is said to semi-crystalline.
30. The ratio of density of FRP materials to density of water is specific gravity, while ratio of unit weight of materials to the unit weight of water is known as specific weight.
31. Length of time in which a material can be stored and continue to meet the specifications requirements, remaining suitable for its intended use. Thermoplastic matrices have higher shelf life.

32. Thermosetting matrix polymers have a short shelf-life after mixing with curing agents (catalysts), low strain to failure and low impact strength.
33. Thermosetting matrix polymers are low-molecular weight liquids with very low viscosities. The polymer matrix is converted to solid by using free radicals to effect cross-linking and curing. In comparison to most common engineering thermoplastics, thermosetting polymers exhibit greatly increased high-temperature and load bearing performance. Normally, thermosetting polymers char and eventually burn at very high temperatures.
34. Fillers are added to thermosetting or thermoplastic polymers to reduce resin cost, control shrinkage, improve mechanical properties and impart a degree of fire retardancy. In structural applications, fillers are used selectively to improve load transfer and also to reduce cracking in unreinforced area. Clay, Calcium Carbonate and glass milled fibers are frequently used depending on the requirements of applications
35. Diethylenetriamine (DETA). Hydrogen atoms in the amine groups of the DETA molecule react with the epioxide group of DGEBA molecules. As this reaction continues, DGEBA molecules cross-link with each other and a three dimensional network is formed, creating the solid cured matrix of epoxy resins.
36. A range of mechanical and physical properties can be obtained due to diversity of input materials. No volatile monomers are emitted during curing and processing. Low shrinkage during cure. Excellent resistance to chemicals and solvents. Good adhesion to number of fillers, fibers and substrates.
37. Matrix cost is generally higher than for iso-polyester or vinylester resins. Epoxy must be carefully processed to maintain moisture resistance. Cure time can be lengthy. Some hardeners require special precautions in handling and resin and some hardeners can cause skin sensitivity reactions in production operation.
38. Fiber orientation, length, shape and composition of the fibers, the mechanical properties of the resin matrix and adhesion or bond between the fibers and the matrix.
39. Bridges, highways, etc that are subjected to loads that are more transitory and of shorter duration than sustained excessive loadings.
40. Yes, by (+45, –45, 0 and 90)'s lamination scheme.
41. Nature and efficiency of the bonding.
42. Impact and brittleness.
43. By area under the stress-strain diagram. Higher the toughness, the higher the impact resistance.
44. By adding zirconia. AR glass, R glass, S glass fibers have high alkaline resistance.
45. E-glass used widely in electrical industry.
46. In acidic environment. ECR-glass is used for boron-free modified E-glass composition and offers improved resistance to corrosion by most acids.
47. S-glass and S-2 glass. S-glass and S-2 glass have the same composition, but use different surface treatments. S-glass is the most expensive form of glass fiber reinforcement and is produced under specific quality control and sampling procedures to meet the military specifications.
48. C-Glass
49. When FRP is held under a constant load at stresses below the instantaneous static strength, they will fail at some point as long as stress is maintained above a minimum value. This is called creep-rupture. GFRP.
50. It has been theorized that the surface of glass contains submicroscopic voids that act as stress concentrations. Moist air can contain weakly acidic carbon dioxide. The corrosive effect of such exposure can affect stress in the void regions for glass fiber filaments until failure occurs. In addition, exposure to high pH environments may cause aging or rupture associated with time.
51. Orientation of Graphite Crystallites with respect to fiber axis and stretching and heat-treatment.
52. Precursor is an organic material which is rich in carbon. Carbon fiber is made by pyrolizing (burning) the precursor fiber until it is mostly carbon (92%). Precursor fibers include PAN, pitch and rayon.
53. Prepreg is manufactured using many fiber bundles laid parallel to one another and impregnated with resin. The prepreg tape is on a paper backing and ranges in width from 3" to many feet. This is used to produce high quality parts and does not need application of wet resin to place the fiber and used for laminate construction.
54. Pultrusion process

55. This helps in providing uniform distribution of resin along the strengthening pattern.
56. Aramid fiber (Kevlar) and polyethylene (Spectra).
57. Compression members because of low compressive strength. Compressive and off-axis properties are critical for AFRP.
58. Excellent
59. Aramid and carbon.
60. Compressive behavior. The low compressive strength of fibers results in local crumpling and fibrillation of individual fibers, thus leading to low strength under conditions of compression and bending. These fibers can be hybridized with glass or carbon fiber for use in FRP shell structures which have to carry high compressive and bending loads. Such hybridized fiber structures lead to a high vibration damping factor which may offer advantages in dynamically loaded FRP structures.
61. SpectraR
62. None of the current resin matrix bond well to this fiber. Plasma treatment has been used to etch the surface of the fibers for a mechanical bond to the resin matrix, but this is expensive and is not readily available in commercial production.
63. Axi-symmetric cylindrical components.
64. Polyester resin and vinylesters are the major matrix materials used in the Pultrusion process.
65. Chemical reaction causing degradation of matrix properties and hence, the composite property degradation.
66. Resistance of FRP to chemical and other environmental factors is known as durability, strength and stiffness reduces.
67. Yes, direct exposure will cause greater degradation than that for FRP embedded in concrete.
68. Short-term need for long-term weathering data has necessitated the creation of such analytical technique as accelerated aging to predict the durability of composite structures.
69. Due to the low strength of FRP reinforcing bars and tendons in transverse direction, the forces introduced by the grips can result in localized failure of the FRP within the grip zone. Clearly, the use of longer grips to reduce the stresses in the grip zone is impractical in many situations.
70. By surface deformation and treatment and by filament winding.
71. Because of low strength and stiffness.
72. Yes, by introducing strength reduction factor.
73. Horse hair.
74. Linear strain distribution along the depth of cross-section.
75. No. One is used for ductility and the other takes into account the uncertainties involved with FRP material characteristics.
76. Strength
77. 0.6
78. Average strength –3 standard deviation of stress. This ensures about 99.87% probability that test results are not below this value.
79. Guaranteed strength $= 2468 - 3 \times 68.95 = 2261.15$ MPa.
    Guaranteed rupture strain $= 2261.15/147 \times 10 = 0.015$ MPa.
80. Ratio of energy absorption (area under the moment curvature curve) at ultimate strength of section to the energy absorption at the service load. This is related to the ductility of structures. Ductility index refers to the ratio of inelastic energy absorbed to the total energy energy of the system.
81. The process of exposing materials to an environment for an interval of time.
82. The number of cycles of deformation or load required bringing about failure of a material, test specimen, or structural member is known as endurance limit. FRP reinforcing bars subjected to constant load overtime can suddenly fail after a time period called endurance time. This phenomenon is known as creep-rupture.
83. The greatest stress that can be sustained for a given number of load cycles without failure. GFRP
84. The susceptibility of material to fail due to stress raiser such as notches is known as notch sensitivity.

85. Compressive strengths of 55%, 78% and 20% of tensile strength have been reported for GFRP, CFRP and AFRP, respectively.
86. Compressive modulus is approximately 80% for GFRP, 85% for CFRP and 100% for AFRP of the tensile modulus.
87. Adhesion resistance of interface (chemical bond), frictional resistance of the interface against slip; and mechanical interlock due to irregularity of the interface.
88. Yes, in case of elevated temperature.
89. Linear logarithmic relationship.
90. Frequency of test cyclic loading.
91. Stress versus number of cycles to failure. It is helpful in predicting fatigue strength corresponding to a particular load cycle.
92. r_b/d_b ratio (ratio of radius of bend to diameter of bar).
93. No. Since FRP is brittle material.
94. The resistance of structures to deformation and cracking refers to serviceability of structures. Crack widths and deflection.
95. This takes into account the tension stiffening of concrete after development of crack into concrete.
96. Time dependent factor which takes into account the creep and shrinkage of concrete in computing long-term deflection.

CHAPTER 2

Material Characteristics of FRP Bars

2.1. Physical and Mechanical Properties

To develop a fundamental understanding of the behavior of FRP composite bars and the properties that affect their use in concrete structures, the physical and mechanical characteristics of these materials are presented in this chapter. However, these material characteristics should be considered as generalizations only and may not apply to all products that are commercially available. The following points must be kept in mind:

- FRP bar is anisotropic with longitudinal axis being the strongest axis.
- FRP bars can be manufactured using a variety of techniques such as pultrusion, braiding, and weaving (Bank, 1933, and Bakis, 1933).
- Unlike steel, mechanical properties of FRP composites vary significantly from one product to another.
- Factors such as volume and type of fiber and resin, fiber orientation, dimensional effects, and quality control during manufacture, play a major role in establishing product characteristics.
- Mechanical properties of FRP composites, like all structural materials are affected by such factors as loading history and duration, temperature, and moisture.
- Mechanical characteristics of FRP composites corresponding to each loading condition should be determined in consultation with the manufacturer.

The following sections describe the physical and mechanical properties and behavior of FRP composites.

2.2. Physical Properties

Density

The density of FRP bars ranges from 1.25 to 2.1 g/cm^3 that is about one-sixth to one-fourth that of steel (Table 2.1). The advantage of reduced weight lies in the fact that these materials will result in reduced transportation costs and may ease their handling on the project side.

Coefficient of thermal expansion

Depending on the types of fibers, resin, and volume fraction of fibers, the coefficient of thermal expansion varies in longitudinal and transverse directions. The longitudinal coefficient of thermal

Table 2.1. Typical density of FRP bars, g/cm^3.

Steel	GFRP	CFRP	AFRP
7.90	1.25–2.10	1.50–1.60	1.25–1.40

expansion is dominated by the properties of fibers while the transverse coefficient is dominated by the resin. For typical FRP bars and steel, the longitudinal and transverse coefficients of thermal expansion are presented in Table 2.2. In the table, negative coefficient of thermal expansion indicates that the material contracts with increased temperature and expands with decreased temperature. For reference, the concrete has a coefficient of thermal expansion that varies from 7.2×10^{-6} to 10.8×10^{-6}/°C and is usually assumed to be isotropic (Mindess and Young, 1981).

High temperature effects

The FRP reinforcement is usually not recommended for structures in which fire resistance is essential to maintain structural integrity. Due to lack of oxygen, FRP reinforcement embedded in concrete cannot burn; however, the polymers will soften due to excessive heat. The temperature at which a polymer will soften is known as the glass-transition temperature, T_g. The elastic properties of a polymer are significantly reduced due to changes in its molecular structure. The value of T_g depends on the type of resin and lies normally in the range of 65–120°C. In composites, if fibers have better thermal characteristics than the resin, they can continue to support some load, however, the tensile properties of the overall composite are reduced due to reduction in force transfer between fibers through bond to the resin. It has been observed from the test results that the temperature of about 250°C will reduce the tensile strength of glass FRP (GFRP) and carbon FRP (CFRP) bars in excess of 20% (Kumhara *et al.*, 1993). Similarly, other properties such as shear and bending strength are reduced significantly at temperature above T_g (Wang and Evans, 1995).

The properties of polymer at the surface of the bar are essential in maintaining the bond between FRP and concrete. At a temperature close to its T_g, the mechanical properties of polymer are significantly reduced and the polymer is not able to transfer stresses from the concrete to the fibers. It has been observed from a study that a bar having T_g of about 60°C–124°C reports a reduction in pullout (bond) strength of 20%–40% at a temperature of approximately 100°C, and a reduction of 80%–90% at a temperature of about 200°C (Katz *et al.*, 1998 and 1999).

Table 2.2. Typical coefficient of thermal expansion (CTE) for reinforcing bars.*

Direction	CTE × 10^{-6}/°C			
	Steel	GFRP	CFRP	AFRP
Longitudinal, α_L	11.7	6.0–10.0	(−9.0)–(0.0)	(−6.0)–(−2.0)
Transverse, α_T	11.7	21.0–23.0	74.0–104.0	60.0–80.0

*Typical values for fiber–volume fraction ranging from 0.5 to 0.7.

Locally such behavior of FRP under high temperature can result in increased crack widths and deflections. Structural collapse can be avoided if high temperatures are not experienced at end regions of FRP bars (in pre-stressed concrete structures), allowing anchorage to be maintained. Structural collapse can occur if the entire anchorage is lost due to softening of the polymer or if the temperature rises above the temperature threshold of the fibers themselves. Temperature threshold of fibers can occur at temperatures near 980°C for glass fibers and 175°C for aramid fibers. However, the carbon fibers are capable of resisting temperature in the excess of 1600°C. It may be noted that the behavior and endurance of FRP reinforced concrete structures under exposure to fire and temperature is still not well understood. Readers may use ACI 216 R for an estimation of temperature at various depths of a concrete section. However, further research is needed in this area.

2.3. Mechanical Properties and Behavior

2.3.1. Tensile Behavior

In tensile loading, the FRP materials do not exhibit yielding (plastic behavior) before rupture. Thus the tensile behavior of FRP bars consisting of one type of fiber material is characterized by a linear elastic stress–strain relationship until they fail. Typical tensile properties of commonly used FRP bars are presented in Table 2.3. The following factors affect the tensile behavior of the FRP bars:

- The ratio of the volume of fiber to the overall volume of the FRP (fiber–volume fraction), i.e., stiffness and strength will vary with various fiber–volume fractions even in bars with the same diameter, appearance, and constituents.
- The rate of curing, manufacturing process, and quality control methods.
- Unlike steel bars, some FRP bars exhibit a substantial effect of cross-section area on the tensile strength. For example, GFRP bars from three different manufacturers show tensile strength reductions of up to 40% as diameter increases proportionally from 9.5 to 22.2 mm [Faza and Gangarao, 1993].

Table 2.3. Usual tensile properties of reinforcing bars.*

Material / Properties	Steel	GFRP	CFRP	AFRP
Nominal yield stress, MPa	276–517	N/A	N/A	N/A
Tensile strength, MPa	483–690	483–1600	600–3690	1720–2540
Elastic modulus, GPa	200.0	35.0–51.0	120.0–580.0	41.0–125.0
Yield strain, %	1.4–2.5	N/A	N/A	N/A
Rupture strain, %	6.0–12.0	1.2–3.1	0.5–1.7	1.9–4.4

*Typical values for fiber–volume fraction ranging from 0.5 to 0.7.

- Effect of cross-section change on twisted CFRP bars is not appreciable, whereas the sensitivity of aramid FRP (AFRP) bars to cross-section size varies from one commercial product to another. For example, in braided AFRP bars, there is less than 2% strength reduction as bars increase in diameter from 7.3 to 14.7 mm, whereas strength reduction in a unidirectionally pultruded AFRP bar with added AFRP bar wraps is approximately 7% for diameter increasing from 3 to 8 mm.

Guaranteed strength and strain

The tensile properties of a particular FRP bar should be obtained from the bar manufacturer. A normal (Gaussian) distribution is assumed to represent the strength of a population of a bar specimen. Based on this distribution, guaranteed tensile strength, f_{fu}^* is equal to mean tensile strength of a sample of test specimens minus three times the standard deviation ($f_{fu}^* = f_{u,\text{ave}} - 3\sigma$). Similarly, guaranteed rupture strain, ε_{fu}^* is defined by mean rupture strain minus three times the standard deviation ($\varepsilon_{fu}^* = \varepsilon_{u,\text{ave}} - 3\sigma$). However, the specified tensile modulus is reported as the average value of the modulus of all specimens. It may be noted that these guaranteed values of strength and strain provide a 99.87% probability that the indicated values are exceeded by similar FRP bars, provided at least 25 specimens are tested. If fewer specimens are tested or a different distribution is used–texts and manuals on statistical analysis should be consulted to determine the confidence level of the distribution parameters (MIL-17, 1999). In any case, the manufacturer should provide a description of the method used to obtain the tensile properties.

It should be noted that an FRP bar cannot be bent once it has been manufactured except in the case of an FRP bar with thermoplastic resin that could be reshaped with addition of heat and pressure. However, FRP bars can be fabricated with bends. The tensile strength of FRP bars with bends is about 40%–50% lower than that of corresponding straight bar because of fiber bending and stress concentrations.

2.3.2. Compressive Behavior

Tests on FRP bars with a length to diameter ratio from 1:1 to 2:1 have indicated that the compressive strength is lower than the tensile strength (Wu, 1990). The FRP bars subjected to longitudinal compression can have different failure modes such as transverse tensile failure, fiber microbuckling or shear failure. The failure mode depends on the type of fiber, fiber–volume fraction, and type of resin. It has been observed that the compressive strengths of GFRP, CFRP, and AFRP bars are about 55%, 78%, and 20% of their tensile strength, respectively (Mallick, 1988, Wu, 1990). In general, compressive strengths are higher for bars with higher tensile strengths, except in the case of FRP where the fibers exhibit nonlinear behavior in compression at a relatively low level of stress.

Like strength, compressive modulus of elasticity of FRP reinforcing bars (FRP rebars) is slightly lower than its tensile modulus. The slightly lower values of modulus of elasticity may be attributed to the premature failure in the test, resulting from end brooming and internal fiber microbuckling under compressive loading. According to reports, the compressive modulus of elasticity is approximately 80% for GFRP; 85% for CFRP; and 100% for AFRP of their corresponding tensile modulus of elasticity. Note that there are no standard test methods to characterize the compressive behavior of FRP bars. The manufacturer should provide a description of the test method adopted, to obtain the compressive properties.

2.3.3. Shear Behavior

Most FRPs are relatively weak in interlaminar shear where layers of unreinforced resin lie between layers of fibers. Since there is usually no reinforcement across layers, the interlaminar shear strength is governed by relatively weak polymer matrix. Thus, the orientation of fibers in an off-axis direction across the layers of fiber will increase the shear resistance, depending upon the degree of offset. In the case of FRP bars, shear resistance could be improved by braiding or winding fibers transverse to the main fibers. Off-axis fibers can also be placed in the pultrusion process by introducing a continuous strand mat in the roving/mat creel. There is no standard test method to obtain the reported shear values.

2.3.4. Bond Behavior

Bond characteristics of an FRP bar depends on the design, manufacturing process, mechanical properties of the bar, and environmental conditions. When a reinforcing bar is anchored in the concrete, the bond force can be transferred by:

- Adhesion resistance of the interface, also known as chemical bond;
- Frictional resistance of the interface against slip; and
- Mechanical interlock due to irregularity of the interface.

However, in FRP bars, it is postulated that the bond force is transferred through the resin to the reinforcement fibers with the possibility of bond-shear failure even in the resin. When a bonded deformed bar is subjected to increased tension, the adhesion between the bar and the surrounding concrete breaks down, and deformations on the surface of the bar cause inclined contact forces between the bar and surrounding concrete. The stress at the surface of the bar resulting from the force component in the direction of bar can be as considered the bond stress between the bar and concrete.

Unlike reinforcing steel, the bond of FRP rebars is not significantly influenced by the concrete compressive strength, provided adequate concrete cover exists to prevent longitudinal splitting. The bond properties of FRP bars can be evaluated by different tests such as pullout tests, splice tests, cantilever beam test, and T-beam tests. The bond stress of an FRP bar should be based on test data provided by manufacturer.

2.4. Time-Dependent Behavior

The time dependent behaviors such as creep characteristics and fatigue characteristics have been explained in the following sections.

2.4.1. Creep Rupture

The creep rupture (static fatigue) is a phenomenon characterized by the sudden failure of FRP bars subjected to a constant load over a time period called the endurance time. Creep rupture is not a concern for steel bars in reinforced concrete except in extremely high temperatures, such as those encountered in a fire. Endurance time decreases as the ratio of sustained tensile stress to the short-term strength of the FRP bar increases. Endurance time can decrease under aggressive environmental conditions such as high temperature, ultraviolet radiation exposure, high alkalinity, wet and dry cycles, or freezing-thawing cycles. In general, carbon fibers are least susceptible to creep rupture whereas aramid fibers are moderately susceptible, and the glass fibers are the most susceptible. From the experimental results, it has been indicated that a linear relationship exists between creep-rupture strength and logarithm of time for times up to nearly 100 hr. The ratios of the stress levels at creep rupture to the initial strength of the GFRP, AFRP, and CFRP bars after 500 000 hr (>50 yr) were linearly extrapolated to be 0.29, 0.47, and 0.93, respectively. Creep rupture data characteristics of 12.5 mm diameter commercial CFRP twisted strand in an indoor environment are available from the manufacturer (Tokyo Rope, 2000). The rupture strength at a projected 100 yr endurance time is reported to be 85% of the initial strength. For experimental characterization of creep rupture, the designer can refer to the test method currently proposed by the committee of Japan Society of Civil Engineers (1997), "Test Method on Tensile Creep-Rupture of Fiber Reinforced Materials, JSCE-E533-1995."

2.4.2. Fatigue

Some general observations on the fatigue behavior of FRP materials can be made, even though the bulk of the data is obtained from FRP specimens intended for aerospace applications rather than construction. Unless stated otherwise, the cases that follow are based on flat, unidirectional coupons with approximately 60% fiber–volume fraction and subjected to tension–tension sinusoidal cyclic loading at:

- A frequency low enough not to cause self-heating;
- Ambient laboratory environment;
- A stress ratio (ratio of minimum applied stress to maximum applied stress) of 0.1; and
- A direction parallel to the principal fiber alignment.

Test conditions that raise the temperature and moisture content of FRP materials generally degrade the ambient environment fatigue behavior. Amongst all types of FRP composites, CFRP is the least prone to fatigue failure. On a plot of stress versus the logarithm of the number of cycles at failure (S-N) curve, the average downward slope of the data is usually 5%–8% of the initial static strength per decade of logarithmic life. At 1 million cycles, the fatigue strength is generally between 50% and 70% of the initial static strength and is relatively unaffected by realistic moisture and temperature exposures of concrete structures unless the resin or fiber–resin interface is substantially degraded by the environment.

Stress corrosion

It may be noted that individual glass fibers, such as E-glass and S-glass are generally not prone to fatigue failure. However, individual glass fibers have demonstrated delayed rupture caused by stress corrosion induced by growth of surface flaws in the presence of even minute quantities of moisture in the ambient laboratory environment tests. When many glass fibers are embedded into a matrix to form an FRP composite, a cyclic tensile fatigue effect of approximately 10% loss in the initial static capacity per decade of logarithmic lifetime has been observed. This fatigue effect is thought to be due to fiber–fiber interactions and not dependent on the stress corrosion mechanism described for individual fibers. So far, there is no clear fatigue limit, which can usually be defined. Moreover, environmental factors play an important role in the fatigue behavior of glass fibers due to their susceptibility to moisture, alkaline, and acidic solutions.

As far as aramid fibers are concerned, they behave similarly to carbon and glass fibers in fatigue except in the case of compression. However, tension–tension fatigue behavior of an impregnated aramid fiber bar is excellent. Strength degradation is approximately about 5%–6% per decade of logarithmic lifetime. The 2 million cycles fatigue strengths of commercial AFRP bars for concrete applications have been reported in the range of 54%–73% of initial bar strengths. Based on these findings it has been suggested by researchers that the maximum stress be set to 54%–73% of the initial tensile strength. In addition to environmental factors such as moisture and increased temperature; other factors such as gripping and presence of concrete surrounding the bar during the fatigue test need to be considered.

In the case of CFRP bars, the fatigue strength for CFRP bars encased in concrete decreases when environmental temperature increases from 20°C to 40°C. The endurance limit of CFRP bars encased in the concrete has been observed to be inversely proportional to loading frequency. The higher cyclic loading frequencies (range of 0.–8 Hz) correspond to higher bar temperatures due to sliding friction. Endurance limit at 1 Hz could be more than 10 times higher than that at 5 Hz. For CFRP bars, the endurance limit also depends on the mean stress and the stress ratio (ratio of minimum and maximum values of stress cycles).

It should be noted that design limitations on fatigue stress ranges for FRP bars ultimately depend on the manufacturing process of the FRP bar, environmental conditions, and the type of fatigue load being applied. The designer should always consult with the bar manufacturer for fatigue response properties.

2.5. Durability

Durability refers to resistance of FRP bars to the exposure of aggressive environmental conditions such as moisture, ultraviolet rays, increased temperature, alkaline or acidic solutions, and saline solutions. Strength and stiffness may increase, decrease, or remain the same depending on the type of material and exposure conditions. Tensile and bond properties of the primary parameters are of interest for reinforced concrete constructions. Some of the typical durability results on various FRP bars are given as follows:

1. Aqueous solutions with high values of pH degrade the tensile strength and stiffness of GFRP bars. Higher temperatures and longer exposure times exasperate the problem.
2. The degree to which the resin protects the glass fibers from the diffusion of deleterious hydroxyl (OH^-) ions figures prominently in the alkali resistance of GFRP bars.
3. Vinyl ester resins have superior resistance to moisture ingress in comparison with other commodity resins.
4. About 0%–75% reductions in the tensile strength of GFRP bars have been reported, whereas corresponding reduction in stiffness has been observed to be about 0%–20%.
5. Tensile strength and stiffness of AFRP bars in elevated temperature alkaline solutions either with or without tensile stress applied have been reported to decrease between 10% and 50% and 0% and 20% of the initial values, respectively.
6. In the case of CFRP, both the strength and stiffness have been reported to decrease between 0% and 20%.

7. Extended exposure of FRP bars to ultraviolet rays and moisture before their placement in concrete could adversely affect their tensile strength due to degradation of the polymer constituents, including aramid fibers and all resins.
8. Some results from combined ultraviolet and moisture exposure tests with and without stress applied to the bars have shown tensile strength reductions of 0%–20% of initial values in CFRP, 0%–30% in AFRP and 0%–40% in GFRP.

2.6. Recommended Materials and Construction Practices

Mechanical properties of FRP bars vary significantly from one manufacturer to another. The physical and mechanical characteristics of such FRP bars depend on the factors such as fiber volume fraction and type of fiber, resin, fiber orientation, dimensional effects, quality control, and manufacturing process. FRP bars made of continuous fibers (aramid, carbon, and glass or any combination) should conform to quality standards as given in the following section.

2.6.1. Strength and Modulus Grades of FRP Bars

The FRP reinforcing bars are available in different grades of tensile strength and modulus of elasticity. The tensile strength grades are based on the tensile strength of bar, with the lowest grade being 414 MPa. Finite strength increments of 69 MPa are recognized according to the following designation:

- Grade F60: corresponds to a $f_{fu}^* \geq 414$ MPa;
- Grade F70: corresponds to a $f_{fu}^* \geq 483$ MPa; and
- Grade F300: corresponds to a $f_{fu}^* \geq 2069$ MPa.

The designer may select any FRP strength grade between F60 and F300 without having to choose a specific commercial FRP bar type. A modulus of elasticity grade is prescribed similar to strength grade based on the fiber type. The designer can select the minimum modulus of elasticity grade that corresponds to the chosen fiber type for the member or project. For example, an FRP bar specified with a modulus grade of E5.7 indicates that the modulus of bar should be at least 39.3 GPa. Manufacturers producing FRP bars with modulus of elasticity in excess of the minimum specified will have superior FRP bars that can result in savings on the amount of FRP reinforcement used for a particular application. Table 2.4 shows the modulus of elasticity grades for different types of FRP bars. For all these FRP bars, rupture strain should not be less than 0.005 mm/mm.

Table 2.4. Minimum modulus of elasticity by fiber type for reinforcing bars.

Reinforcing bars	Modulus grade (GPa)
GFRP	E5.7 (39.3)
AFRP	E10.0 (68.9)
CFRP	E16.0 (110.3)

2.6.2. Surface Geometry, Bar Sizes, and Bar Identification

FRP reinforcing bars are produced through a variety of manufacturing processes and each manufacturing method produces a different surface condition. The physical characteristics of the surface of the FRP bar are important properties for mechanical bond with the concrete. The surface deformation patterns for

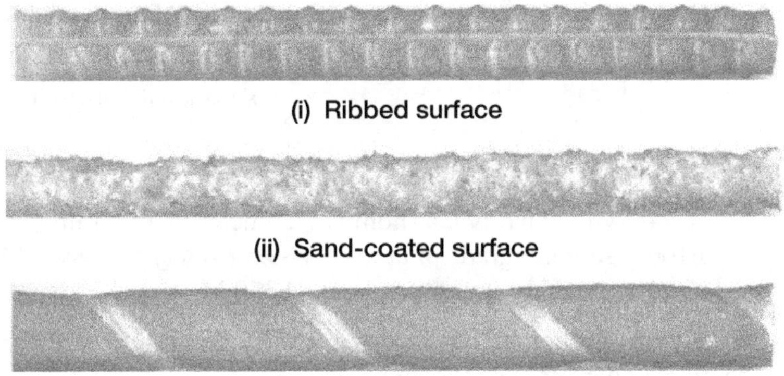

(i) Ribbed surface

(ii) Sand-coated surface

(iii) Wrapped and Sand-coated

Figure 2.1. Surface deformation pattern for typical FRP bars available in the market

commercially available FRP bars such as (i) ribbed, (ii) sand-coated, and (iii) wrapped and sand-coated are shown in Fig. 2.1.

The FRP bar sizes are designated by a number corresponding to the approximate nominal diameter in eighths of an inch, similar to standard ASTM steel reinforcing bars. As shown in Table 2.5, there are twelve standard sizes. The nominal diameter of a deformed FRP bar is equivalent to that of a plain round bar having the same area as the deformed bar. When the FRP bar is not of conventional solid round shape (that is rectangular or hollow), the outside diameter of the bar or the maximum outside dimension of the bar will be provided in addition to the equivalent nominal diameter. The nominal diameter of these unconventional bars would be equivalent to that of a solid plain round bar having the same area.

Table 2.5. ASTM standard reinforcing bars.

Bar size designation		Nominal diameter, mm	Area, mm^2
Standard	Metric conversion		
No. 2	No. 6	6.4	31.6
No. 3	No. 10	9.5	71
No. 4	No. 13	12.7	129
No. 5	No. 16	15.9	199
No. 6	No. 19	19.1	284
No. 7	No. 22	22.2	387
No. 8	No. 25	25.4	510
No. 9	No. 29	28.7	645
No. 10	No. 32	32.3	819
No. 11	No. 36	35.8	1006
No. 14	No. 43	43.0	1452
No. 18	No. 57	57.3	2581

Due to various grades, sizes, and types of FRP bars available, it is necessary to provide some means of easy identification. An example of identification symbols for solid bars is given below:

XXX-G#4-F100-E6.0

where

XXX = manufacturer's symbol name;
G#4 = GFRP bar No. 4 (nominal diameter of 12.5 mm);
F100 = strength grade of at least 689.5 MPa ($f^*_{fu} \geq 689.5$ MPa);
E6.0 = modulus grade of at least 41 370 MPa.

In the case of a hollow or unconventionally shaped bar, an extra identification should be added to the identification symbol as shown below:

XXX-G#4-F100-E6.0-0.63

where 0.63 = maximum outside dimension is 15.9 mm.

Note: In the case of carbon and aramid fibers, the identification letters are 'c' and 'a,' respectively. Markings should be used at the construction site to verify that the specified type, grades, and bar sizes are being used.

2.7. Construction Practices

The construction operations should be performed in a manner designed to minimize damage to the bars. Like in the case of epoxy-coated steel bars, FRP bars should be handled, stored, and placed more carefully than uncoated steel reinforcing bars.

2.7.1. Handling and Storage of Materials

Since FRP rebars are susceptible to surface damages, puncturing their surface can significantly reduce the strength of the FRP bars. The surface damage to the GFRP bars can cause loss of durability due to infiltration of alkalis. The following handling guidelines are recommended to minimize damage to both the bars and the bar handlers:

1. FRP rebars should be handled with work gloves to avoid personal injuries from either exposed fibers or sharp edges.
2. FRP bars should not be stored on the ground. Pallets should be placed under the bars to keep them clean and provide easy handling.
3. High temperatures, ultraviolet rays, and chemical substances should be avoided because they can damage FRP bars.
4. More often, FRP bars become contaminated with form releasing agents or other substances.
5. Substances that decrease bond should be removed by wiping the bars with solvents before placing FRP bars in concrete form.
6. It may be necessary to use a spreader bar so that the FRP bars can be hoisted without excessive bending.

7. When necessary, cutting should be performed with a high-speed grinding cutter or affine-blade saw. FRP bars should never be sheared. Dust masks, gloves, and glasses for eye protection are recommended when cutting.
8. Dust masks, gloves, and glasses for eye protection are recommended when cutting.

Note: Sufficient research is not available to make any recommendations on treatment of saw-cut bar ends.

2.7.2. Placement and Assembly of Materials

In general, placement of FRP bars is similar to that of steel bars and common practices should apply with some exceptions for the specifications prepared by the engineer as given below:

1. FRP reinforcement should be placed and supported using chairs (preferably plastic or noncorrosive). The requirements for support chairs should be included in the project specifications.
2. FRP reinforcement should be secured against displacement while the concrete is being placed.
3. Coated tie wire, plastic or nylon ties, and plastic snap ties can be used in tying the bars. The requirement for ties should be included in the project specifications.
4. Bending of cured thermoset FRP bars on site should not be permitted.
5. For other FRP systems, manufacturer's specifications should be followed
6. Whenever reinforcement continuity is required, lapped splices should be used. The length of lap splices varies with concrete strength, type of concrete, bar grades, size, surface geometry, spacing, and concrete cover. Details of lapped splices should be in accordance with the project specifications.

Mechanical connections are not yet available.

2.8. Quality Control and Inspection

The quality control of FRP bars should be done by testing on lot of FRP bars. Tests conducted by the manufacturer or a third-party independent testing agency can be used. All tests should be performed using the recommended tests and methods cited in the literature. Material characterization tests that include the following properties should be performed at least once before and after any change in the manufacturing process, procedure, or materials:

- Tensile strength, tensile modulus of elasticity, and ultimate strain;
- Fatigue strength;
- Bond strength;
- Coefficient of thermal expansion; and
- Durability in alkaline environment.

Thus to assess quality control of an individual lot of FRP bars, it is recommended to determine tensile strength, tensile modulus of elasticity, and ultimate strain. The manufacturer should furnish upon request, a certificate of conformance for any given lot of FRP bars, with a description of the test protocol.

Exercise Problems

1. For the FRP dowel bar layout shown in Fig. P2.1, determine the number of 20 mm diameter FRP dowels required to transfer a vertical force of 60 kN across the construction joint, if the dowel bars are capable of carrying an average shear stress of 20 MPa. How much stronger is the joint than it needs to be?

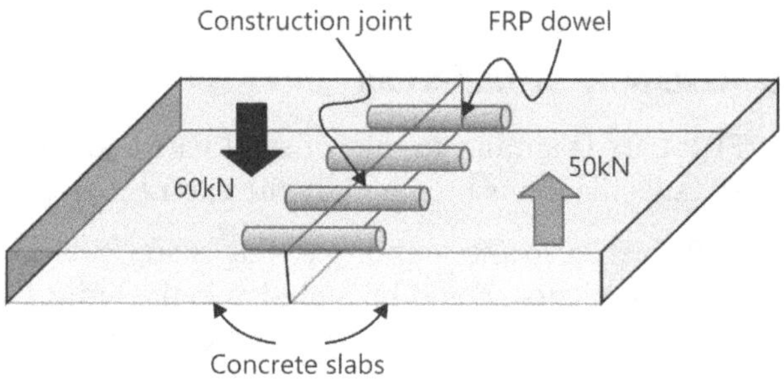

Figure P2.1.

2. Figure P2.2 shows typical stress–strain curves for steel and GFRP materials. Using Hooke's Law, determine the modulus of elasticity (Young's modulus) of both steel and GFRP. If both materials are subjected to a strain of 1.5%, what would be the stress in each material? If a 1000 mm long circular steel rod of 16 mm is subjected to an axial load of 30 kN, what would be its axial elongation? What would be the elongation if the same rod were made from GFRP?

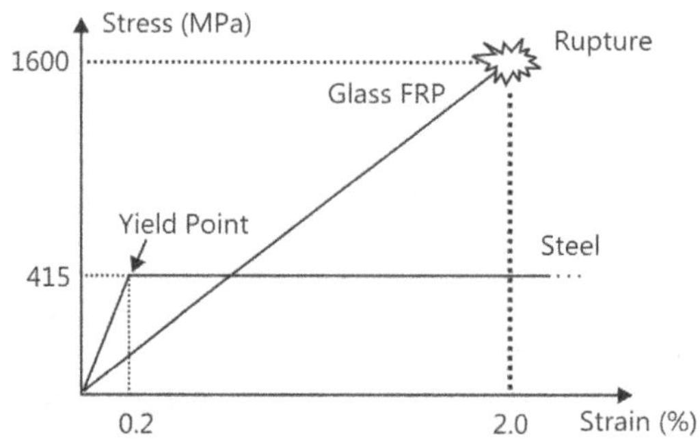

Figure P2.2.

3. In a specific application of rehabilitation, a 60 mm diameter cracked steel tube as shown in Fig. P2.3 has been rehabilitated using a CFRP wrap. Determine the average shear stress over the bonded area, when a tensile axial load of 200 kN is applied to both the ends of the tube.

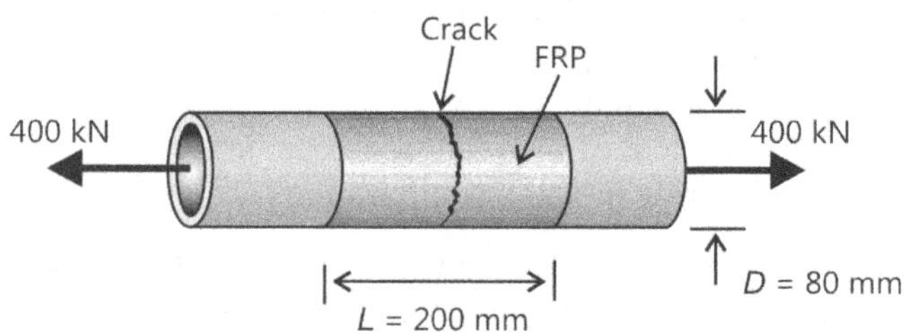

Figure P2.3.

4. As shown in Fig. P2.4, the concrete beam is reinforced with three 10 mm diameter CFRP reinforcing bars, and has the cross-sectional properties shown. The beam is subjected to a maximum in-service bending moment of 30 kN·m. The modulus of elasticity of the concrete is taken as Ec = 25 GPa, and that of CFRP is taken as E_{CFRP} = 150 GPa. Compute the maximum bending stress in the concrete. Is the FRP reinforcement capable of withstanding this moment, if its ultimate tensile strength is 2000 MPa?

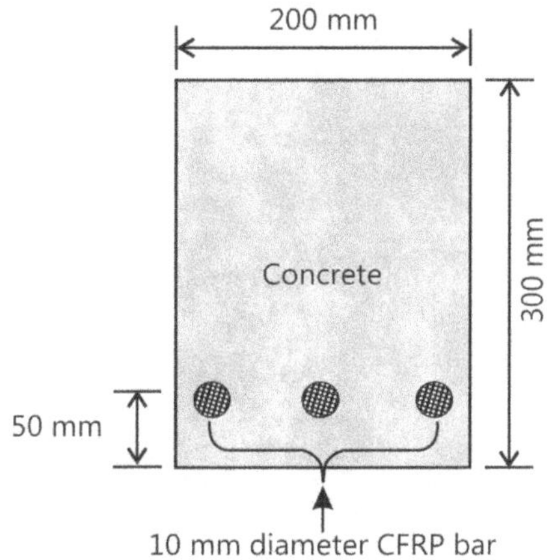

Figure P2.4.

5. An aluminum beam with the cross-sectional properties as shown in Fig. P2.5, is subjected to a moment of 80 kN·m. What would be the percentage decrease in the maximum stress in the aluminum, with the addition of a 150 mm × 1.4 mm CFRP plate to: (a) the bottom face of the beam only and (b) both the top face and the bottom face of the beam (assuming CFRP is fully effective in compression, which is not strictly true)?

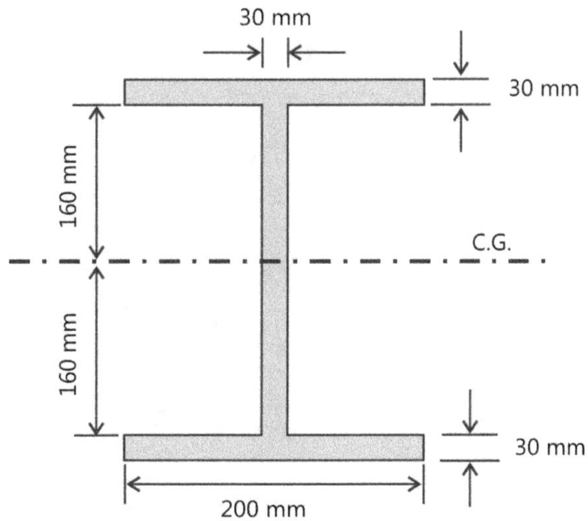

Figure P2.5.

CHAPTER 3

History and Uses of FRP Technology

In this chapter a brief description of the current development and use of FRP in leading countries of the world such as Japan, European countries, Canada and the USA have been presented to make the reader aware of some important applications of FRP composites for the current and future development of infrastructure of the world and future potential of FRP composites for civil engineering infrastructure.

3.1. FRP Composites in Japan

The applications of FRP in Japan are divided into two categories. The first consists of reinforcing bar and grid type FRP used in place of reinforcing steel and prestressing steel. The second comprises fiber composite sheets that are used to repair and strengthen concrete structures. In both cases, FRP offers some unique properties that the steel reinforcement does not, such as corrosion resistance, light weight, high strength and nonmagnetism. These properties help to provide high performance concrete structures and make it easier to build them.

3.1.1. Development of FRP Materials

The history of FRP reinforcement in Japan began in 1970 with trial and error experimentation in manufacturing technology sector. In 1980s, extremely fine FRPs were bundled together, impregnated with resin as binder in a state of tension, and molded or pressed (pultrusion method). Figure 3.1 shows various classification of various FRP reinforcements by shape.

Some of the more common reinforcing bar types include those with fibers in straight, strand, and braided configurations, and a type with fibers woven into the surface or with sand adhered on the surface to increase bonding. The grid type consists of multiple layers of fiber bundles, which are impregnated with resin, woven into each other at right angles in a state of tension, and pressed. In addition, there are three dimensional textiles whose fibers are interwoven perpendicular to each other in three directions, and those that are made by impregnating resin after the textiles are made. Figure 3.2 shows the typical FRP reinforcements used in Japan. The largest percentage of FRP used in Japan is carbon fiber, followed by aramid and glass fibers in that order. In Fig. 3.2, the black material is carbon fiber, and the yellow one is aramid. Epoxy is the most common type of binder.

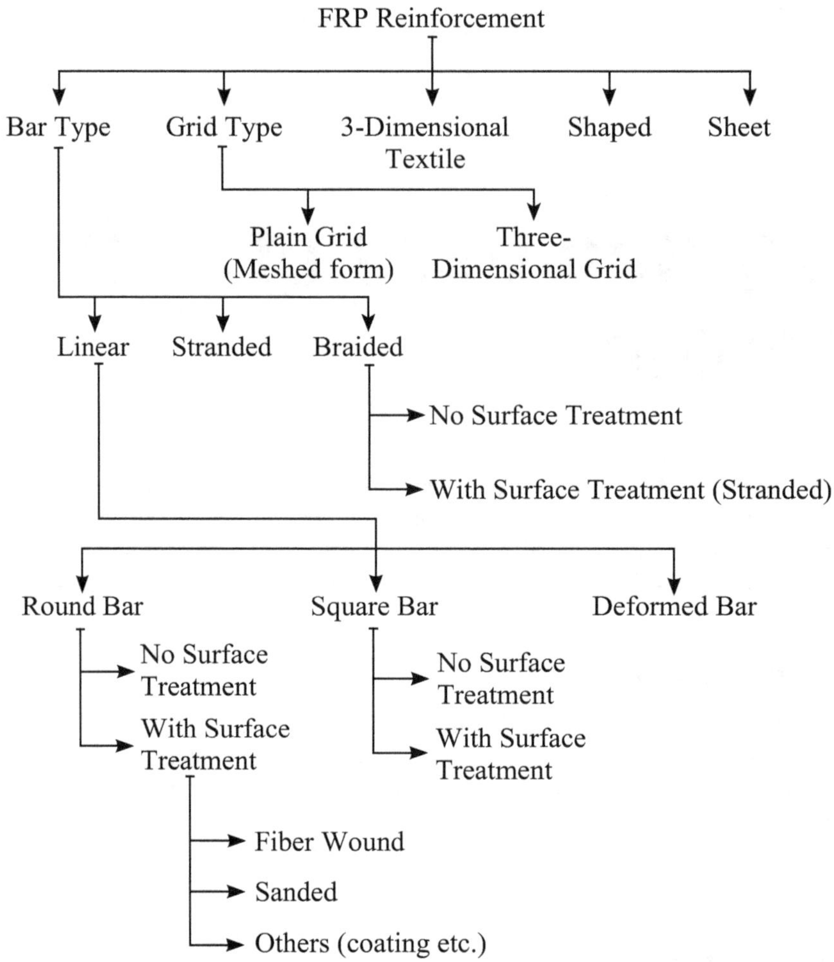

Figure 3.1. Classification of FRP reinforcement by shape.

Sheet type FRPs have already been used widely in the aeronautics and space industries, as well as in the sport and recreation fields. However, in this product, resin is impregnated between fine fibers beforehand after they are shaped in the factory, and then they are heated and made into sheet type FRPs. When such materials are used in building engineering and civil engineering applications, it is difficult to heat them at the construction site, so resinless sheets have been developed which are treated at the work site with a resin that hardens at the normal outdoor temperatures. Fiber sheets used in retrofitting work have fine fibers aligned in a single direction and are made into thin planes ranging in thickness from 0.1 to 0.4 mm.

To improve the flexural (bending) strength of columns and beams, these sheets are applied in axial direction of the elements. However, they are applied perpendicular to axial direction when the aim is to increase shear resistance. While there are 10 or more types of fiber sheets made in Japan, the majority use either carbon or aramid fibers. Most recently, FRP sheets have been developed using relatively inexpensive glass and polyacetal fibers.

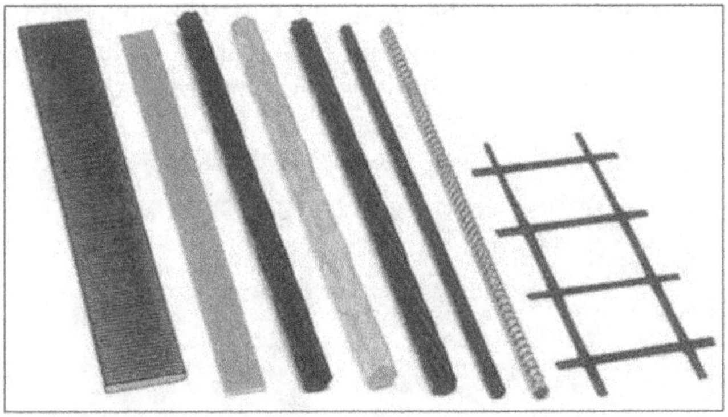

Figure 3.2. FRP reinforcements used in Japan.

Different properties of FRP made in Japan are dictated by the type of fiber and binder, fiber content, and cross-sectional configuration, among other things. However, there are some properties that have all types in common—lightweight, high tensile strength, elastic up to failure, high resistance to corrosion, nonmagnetism, and lower Young's modulus (reinforcing bar and grid types) than steel. On the other hand, since FRPs are brittle material that do not show yielding phenomenon, the redistribution of the bending moment would not be allowable; moreover, tensile strength declines in the curved parts of FRP and fire resistance decreases. These potential problems which are not found in RC materials must be given special attention in the design stage.

3.1.2. Development of Design Methods in Japan

In Japan, a clear distinction is made between building engineering and civil engineering and these two fields have had different approaches to developing design methods. In 1993, the Japan Society of Civil Engineers (JSCE) established the first design guidelines in the world for FRP reinforced and prestressed concrete structures. Sheet-type FRP was first studied in 1978 as a means to increase seismic resistance. The reinforcing bar and grid-type FRPs are used in the following cases to make the most of their respective characteristics:

1. In civil engineering, these materials are often used as tendons and reinforcement for girders and slabs of concrete bridges.
2. They are used as external cables, and in stay cables for cable-stayed and suspension bridges.
3. They help strengthen concrete marine structures—such as piers and quay walls—reinforce concrete canals, and are used as ground anchors and as reinforcement for shotcrete.
4. FRP is also used to strengthen concrete walls of shields in vertical shafts since they can be easily cut by tunnel boring equipment.
5. Because FRPs are nonmagnetic, they are used in guideways of high-speed transportation systems and as reinforcing materials in airport compass check aprons.

3.1.3. Typical FRP Reinforced Concrete Structures in Japan

Figure 3.3 shows an example of a stress-ribbon bridge which is provided with an extremely thin floor (deck) for aesthetic purpose. In this bridge, AFRP and CFRP tendons are used in the floor, ground anchors and as reinforcements for precast concrete slabs due to their short construction time, high durability and ease of maintenance. In another example, GFRP being used in shotcrete reinforcement of underground petroleum facilities (see Fig. 3.4). Since FRPs are durable and highly workable due to their lightweight and moldability to surface irregularities, they are used as substitute to steel mesh.

Figure 3.3. Stress-ribbon bridge.

Figure 3.4. Underground petroleum storage facilities.

Some typical applications of FRPs in building engineering include—use as tendons in entrance gates and beams, as reinforcement in curtain walls and in roofs for indoor pools and as reinforcement for columns, walls, floors of observation stations in the Antarctica. Moreover, the nonmagnetic properties of FRPs make them well suited for reinforcing concrete foundations of wooden ground-based magnetism observation stations and for strengthening the area around metallic sensors that detect the movement and number of vehicles in parking lots. In addition, a lack of permeability to electromagnetic waves makes FRPs useful for reinforcing curtain walls to prevent electromagnetic disturbance or damage. An example of trusses containing carbon FRP hollow pipes that were used in the roof of an indoor swimming pool is shown in Fig. 3.5. Since FRP pipes have lightweight, they could be assembled manually and lifted into place easily. This helps to reduce the time and cost of construction. The use of FRP also reduces the frequency of maintenance operations, which not only helps to lessen the life cycle costs, but should also help to reduce and life cycle carbon-di-oxide (CO_2) emissions through steel material manufacturing.

Figure 3.5. FRP truss structure for roof of pool.

3.1.4. FRP for Retrofitting and Repair

The main advantage of FRP for retrofitting is that retrofitting work using FRP composite sheets can be done manually without heavy equipment. In addition, the other advantages include:

- a shortened construction period;
- any increase in structural weight is so light that can be ignored in the design;
- there is no adverse effect on the stiffness of structural members in the case of shear strengthening;
- greater durability of wrapped members;
- less total cost including construction conditions and time than when using steel jacketing; and
- the project can be executed while building in use.

Thus, this method is attracting attention as a viable alternative to steel and concrete jacketing. The fiber sheets can manifest their maximum strength when they are adequately impregnated with resin. Thus the skill with which the project is executed will have tremendous impact on the strengthening effect.

In building engineering, sheet-type FRP is commonly used for strengthening the earthquake resistance of columns and beams, whereas, in civil engineering, it is frequently used to enhance the earthquake resistance of bridge piers and to increase the shear and bending strength of bridge girders and slabs to respond to increases in design vehicle load.

The other uses of FRP include flexural reinforcement of floors in building, strengthening of flexural and shear strength of smokestacks, retrofitting of walls, strengthening of openings, retrofitting of tunnels, reinforcement of marine structures, such as piers and strengthening of retaining walls and foundations. Figures 3.6 and 3.7 are typical examples of seismic rehabilitation (shear strengthening) using fiber composite sheets. Figure 3.6 shows a column of building, whereas Fig. 3.7 shows a photo of high piers of an expressway bridge.

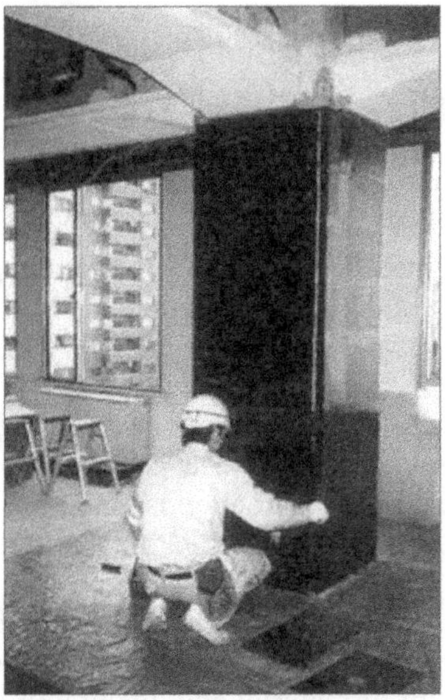

Figure 3.6. Fiber sheet jacketing of building column.

Figure 3.7. Fiber-sheet jacketing of high-pier.

Besides the above methods of retrofitting, several other methods have been developed including:

- Impregnating resin into strand-like fibers as they are wrapped around columns with a machine;
- Glued FRP plates, which were pre-shaped into U or L shapes, with fiber sheets at the job site and filling the gap between them and the concrete members with either mortar or resin;
- On-site assembling of precast mortar boards with finishes adhered with fiber sheets on the inside, and grouting mortar in to the gap between the concrete and sheets;
- Passing electric current through carbon fibers to speed the hardening process of resin; and
- Doing reinforcement work without removing the mortar finish or the concrete of the wall near the columns because the work is done without impending use of the building.

3.1.5. Future Uses of FRP

As people's lifestyles and activities have been dramatically diversifying in recent years, so has been the case with public demand for architectural structures and standards. To reflect such trends, structural design has been moving toward "performance-based" design. Since, FRP has some very desirable properties—such as high resistance to corrosion, lightweight, high strength and nonmagnetic—that steel materials do not, its use with concrete requiring those properties means that concrete structures can now show better performance compared with that of steel reinforcing process. In addition, FRP can help meet new demand from the public. However, it should be noted that these new materials are not complete replacements for reinforcing steel—sometimes they may co-exist with steel or be used in areas where steel is inadequate or where better performance is required. Future research requires data regarding effective use and performance evaluation of FRP so that more applications may be found for these materials.

3.1.6. FRP Construction Activities in Europe

Extensive research on the use of FRP in concrete structures started in Europe about 32 years ago and resulted in the world's first highway bridge—using FRP post-tensioning cables—in Germany in 1986. This section presents a survey of FRP activities in Europe. In The Netherlands in 1983, AKZO—a chemical producer and HBG, a contractor—jointly developed aramid fiber (AFRP) based prestressing elements, called Arapree in the shape of strips and bars. The practical applications of Arapree were mainly limited to pretensioned posts for a highway noise barrier and a fish-passage ladder alongside a hydroelectric power plant on the River Alphen. In the field of strengthening with externally bonded carbon fiber based (CFRP) plates, pioneering work was performed at the EMPA Institute near Zürich, Switzerland.

The BRITE/EURAM project (1991–1996), titled "Fiber Composite Elements and Techniques as Non-Metallic Reinforcement for Concrete" was executed by the Universities of Braunschweig and Ghent and industrial partners from Germany, and The Netherlands. This project dealt with the investigation of performance characteristics and structural aspects of FRP for prestressed and reinforced concrete members.

The EUROCRETE project, a pan-European collaborative research program with partners from the United Kingdom, The Netherlands, Switzerland, France, and Norway began in December 1993 and ended in 1997. This project dealt with the development of material and its durability, determination of structural behavior, and development of design guidance, and techno-economic and feasibility studies. The project included construction of demonstration structures.

3.2. Reinforced and Prestressed Concrete: Some Applications

As a result of EUROCRETE project, the use of GFRP bars was started. Demonstration structures were built including two footbridges in the U.K., and Norway. The Fidgett Footbridge at Chalgrove in Oxfordshire was the first concrete footbridge in Great Britain and probably in the Europe. It consists of 5 m long by 1.5 m wide slab, with a thickness of 300 mm.

Figure 3.8 shows the Oppegaard Footbridge situated on a golf course near Oslo. It has a span of approximately 10 m and is designed to carry the weight of a tractor and trailer as well as pedestrian loads. The bridge consists of two shallow arched concrete beams with a horizontal tie. Eurocrete glass FRP bars were used for the main reinforcement in the beams and Plytron, a thermoplastic material, was used for the shear reinforcement. Because of the relatively low elastic modulus of the Plytron, there was a need to reduce the requirement for shear reinforcement. Hence, the beams were prestressed longitudinally with external parafil tendons supplied by VSL/Linear Composites. One tendon was placed in each horizontal tie. The Parafil tendons are made from a parallel arrangement of aramid (Kevlar) fibers sheathed with an extruded polyethylene cover.

Figure 3.8. Oppegaard Footbridge near Oslo, Norway.

Figure 3.9 shows thermal insulation elements—such as Schöck Isokorb—and corrosion resistant—ComBar or GFRP Bars—which are primarily used to isolate the warm, interior concrete deck of a building from the exterior parts of the construction, e.g., a balcony. Schöck Isokorb consists of polystyrene foam and tensile, compressive and shear reinforcement made of reinforcing steel.

Cable-stayed Storchen Bridge in Winterthur, Switzerland (see Fig. 3.10) consists of two 35 m long CFRP stay cables and 22 steel stay cables. The total length of this road bridge is 124 m. Similarly, the footbridge in Herning, Denmark (see Fig. 3.11) is a cable stayed bridge 80 m long and constructed with the exclusive use of CFRP stay cables. This bridge primarily consists of a deck, supported by 16 stay cables anchored to a central pylon and facilitates pedestrian and emergency vehicle traffic crossing a railway switchyard. The bridge has a 3.5 m wide walkway and 5 m wide deck. The bridge deck is post-tensioned with six CFRP tendons, 12.5 mm

Figure 3.9. Thermal insulating element.

seven-wire strands from Tokyo Rope and a 40 m segment of the bridge deck is reinforced with CFRP Bars and stirrups. The other 40 m segment is reinforced with combination of conventional steel and stainless steel reinforcement. The bridge is a key element of an R&D project, initiated by the Danish Road Directorate in 1997.

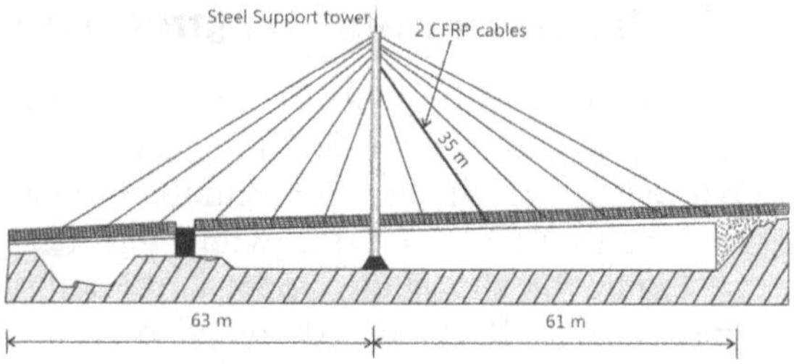

Figure 3.10. Cable-stayed Storchen Bridge in Winterthur, Switzerland.

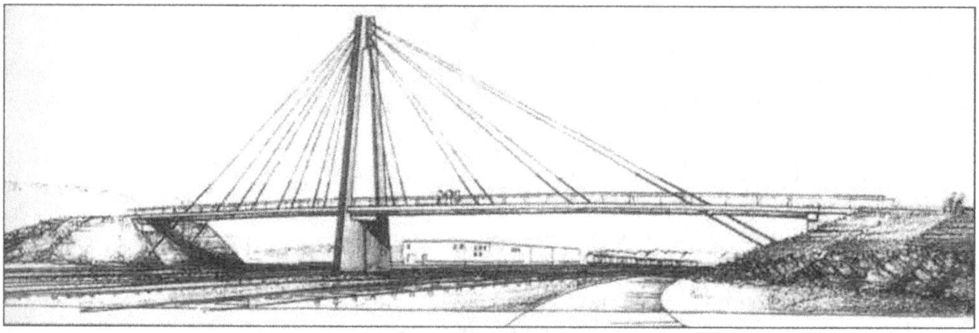

Figure 3.11. Artistic impression of footbridge in Herning, Denmark.

3.2.1. Rehabilitation and Strengthening

Most applications of FRP in Europe deal with externally bonded reinforcement for repair and strengthening of damaged structures—to upgrade structures in seismic regions, to protect and strengthen historical buildings, to modify the concept or use of a structure, to increase structural safety, or to meet serviceability criteria. Since FRP materials offer advanced properties and are easy-to-use, practical applications are increasing exponentially. However, the design guidelines for this novel strengthening technique are still scarce. Various projects have been completed in Switzerland, Austria, Italy, Belgium, Greece, Sweden, France, and Germany.

In association with EMPA's pioneering work in Switzerland, Sika has been strongly involved with FRP plate bonding for quite some time. Their research has resulted in more than 1000 applications of CFRP in Switzerland alone. Figure 3.12 shows prefab CFRP L-shaped angles for shear strengthening, which is a product of the Sika Company. Figure 3.13 shows strengthening of Tanberg Bridge in Austria using CFRP sheets.

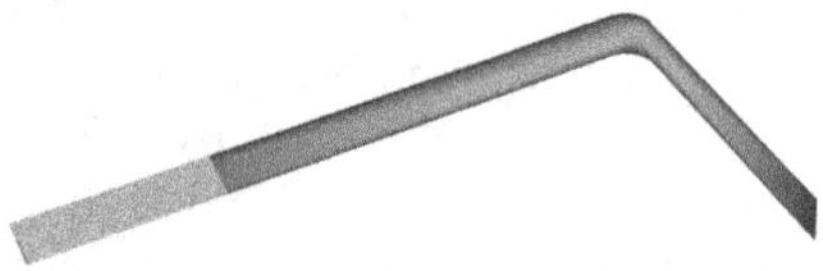

Figure 3.12. L-shaped CFRP angle element (CarboShear L) for shear strengthening.

Figure 3.13. Strengthening of the Tanberg Bridge, Austria.

3.2.2. Design Guidelines

At the European level, unified design guidelines for FRP reinforcement are under development with the establishment of a task group. Since the merger of CEB (Euro-International Concrete Committee) and FIP (International Federation for Prestressed Concrete), the task group is integrated in the new FIB (International Federation for Structural Concrete). Within Task Group 9.3 of FIB Commission 9, design guidelines are elaborated for concrete structures reinforced, prestressed, or strengthened with FRP, based on the design format of the CEB-FIP Model Code and Eurocode 2.

3.3. FRP Prestressing in the USA

Since FRP tendons have a high tensile strength, low creep, moderate modulus of elasticity and a high resistance to the corrosive elements, they become the prime candidates for prestressing tendons in concrete structures. The following is an overview of the development of FRP prestressed concrete structures in the United States, i.e., the successes that have been accomplished and the difficulties they are get to be overcome. Figure 3.14 shows several alternative FRP reinforcement configurations. The top sample is GFRP sand coated bar from Hughes Brother Inc. This type of bar is principally made for the reinforced concrete applications. The second sample is a stranded CFRP tendon (manufactured by the Tokyo Rope Corporation). It is similar to a seven wire steel strand with a loose epoxy bonding between the strands. The other strand systems include the Japanese NACC tendons and a seven wire strand developed at South Dakota School of Mines and Technology. The bottom bar is an aramid fiber based Technora rod which has an overwrap to assist in bond development. A Kevlar rope similar to the Parafil tendon (manufactured by Linear Composites in the UK) uses parallel Kevlar fibers and spike-in-cone anchor system to terminate the unbonded tendon. Parafil is unique because it does not use a matrix to confine the fibers.

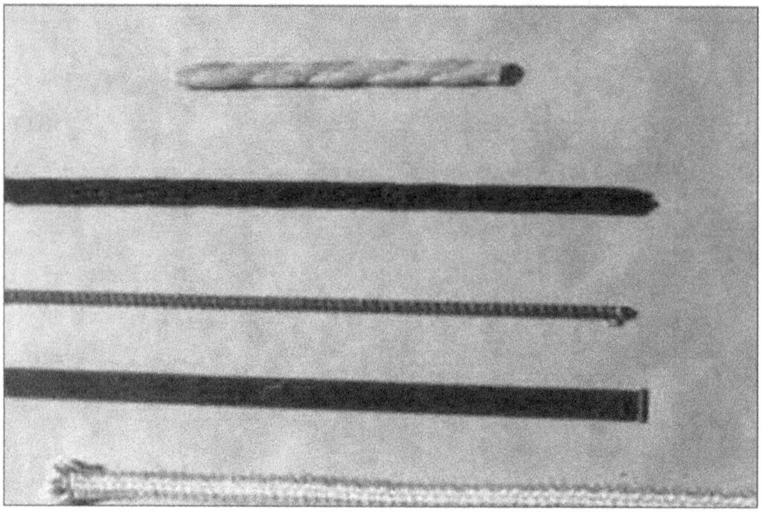

Figure 3.14. Alternative FRP Reinforcements used in the US.

3.3.1. Historical Development of FRP Tendons

In the US, the FRP reinforcements for concrete were first developed in the 1960s at the US Army Corps of Engineers Waterways Experiment Station in Vicsburg, Mississippi. This pioneering work focused on GFRP reinforcement. However, the lower modulus of glass did not perform well for Portland cement concrete structures. Thereafter, a hiatus of approximately 20 years occurred, as the composite industry improved its product and the need for FRP reinforcement emerged.

The driving factor for reconsideration of FRP reinforcement was the application of salt on bridges and roads to obtain all-weather driving conditions. The salt led to a serious deterioration of the steel reinforcement in concrete structures and bridge decks. FRP reinforcement reappeared as a possible candidate to provide corrosion resistant reinforcement for these applications.

It must be noted that FRP prestressing systems were introduced in the 1980s in Europe and Japan. Fiberglass post-tensioning tendons were developed for use in bridge projects in Germany in 1986. In the UK, unbonded parafil tendons were used to post-tension rail-road and highway bridge girders. Small scale beams were successfully tested at Cornel University using bonded prestressing tendons. South Dakota Schools of Mines and Technology developed glass based FRP tendons.

These early projects identified the need for complete tendon systems including anchors and coupling devices. The Japanese national project took the lead by developing the complete tendon systems and made these tendons and anchorage systems available to researchers and demonstration projects in the US. However, the high cost of the FRP tendons and their perceived difficulties of brittle behavior and sensitivity to handling have impeded the widespread implementation.

3.3.2. Research and Demonstration Projects

This section deals with a number of research and demonstration projects to illustrate the benefits of FRP systems. These projects include an FRP prestressed bridge deck in South Dakota and condominium repairs in Florida. Some other notable projects reported here include the pile tests developed at the

University of South Florida, a US Navy pier and water-front project, the Federal Highway Administration program for prestressed girders, and the Bridge Street Bridge project in Southfield, Michigan.

Prestressed Piles: The University of South Florida explored the possibility of using glass FRP tendons for prestressed concrete pile construction because of their lower cost, i.e., the material costs of glass fiber is 25% less than that of the corresponding carbon or aramid fiber systems. The Florida program involved manufacturing and driving several piles and then placing small-scale piles in a tidal pool environment for an assessment of the durability of the product. The project encountered a number of significant difficulties that led to an improved understanding of the performance of FRP materials. These problems are given as follows:

1. FRP confinement reinforcement was used at the ends of the piles.
2. The lower modulus of the FRP led to the splitting of the pile heads during driving (see Fig. 3.15).
3. Exposure of the piles to the tidal environment resulted in the failure of GFRP in as little as six months.

Moreover, the Florida project also brought out the following information regarding the selection of glass fibers and epoxy resin:

i. The Florida project, more than any other, alerted engineers to the potential problems of material selection and alkali corrosion of stressed glass fibers in a wet/dry concrete environment.
ii. The creep-rupture studies confirmed this alkali corrosion behavior for stressed glass fiber tendons.

Keeping in view the above problems associated with GFRP systems, some researchers and the fiberglass industry are continuing to search for a low cost GFRP system that is a nonalkali reactive combination of glass fibers, sizing, and resin. In the meantime, most researchers have shifted their attention to carbon fiber based tendons for concrete applications.

Figure 3.15. GFRP reinforced concrete piles.

Waterfront Facilities: The US Navy has a continuing program to examine FRP prestressing and reinforcement for waterfront facilities. During 1994–96, the Naval Facilities Engineering Service Center at Port Hueneme, California, designed and constructed a prototype pier with FRP prestressed piles and deck panels (see Fig. 3.16). Working in conjunction with the Composites Institute and a construction productivity research grant, they developed both pile and pier designs using FRP prestressing.

Capitalizing on the information gained in the Florida program, carbon fiber tendons were used for some of the prestressing in the structures and attention was given to the stiffness requirements for confinement of concrete during pile driving. To date, the structure is performing according to the design intent. The success of the Port Hueneme project demonstrated that proper design and material selection can overcome the difficulties identified in the Florida project.

Figure 3.16. Waterfront facilities of US Navy.

In addition to construction of the demonstration facility, the Navy is conducting basic research and monitoring long-term performance of FRP tendons. Basic research includes the activities such as bond and transfer length of FRP tendons and use of FRP shells for concrete piling. Future naval research includes the development of a new generation of pier structures. One criterion in this project is a design goal of a 75-year maintenance free service life. Composite materials are being investigated to meet this criterion.

FHWA Research Program: The Federal Highway Administration has sponsored a multi-year program to develop design guidelines and standards for FRP prestressed girders and bridge decks. Several results are available including specifications for performance testing of FRP tendons and fatigue performance of prestressed beams. Test specimens for this program were fabricated at Rocky Mountain Prestress in Denver, Colorado, US, a PCI certified plant and shipped to the University of Wyoming for testing (see Fig. 3.17).

Figure 3.17. Shipping of test specimen of FHWA research program to the University of Wyoming.

Southfield, Michigan, Bridge Project: Lawrence Technological University (LTU) has tested many double tee (DT) beams prestressed with internal and external CFRP tendons. By using external tendons, the total amount of prestress in the bridge may be increased beyond the limits established by cover and clearance of internal prestressing. This extra prestress with external prestressing will increase the capacity of the precast concrete segment. Full-scale tests of prototype bridges were conducted at LTU and the success of these tests has led to the design and acceptance of the external prestressing concepts for a demonstration bridge (Bridge Street Bridge, Southfield, Michigan) (see Fig. 31.8). This bridge has already been monitored for five years and has performed efficiently. This bridge consists of two structures, i.e., Structure A and Structure B (see Fig. 3.18). Structure A consists of five AASHTO I-girders, whereas Structure B consists of four double tee beams in each of the three spans of the bridge. Moreover, additional research projects have also been undertaken at LTU using CFRP tendon for multi-span continuous bridges and simply supported box-beams and bridges.

(a) Looking South (b) Side view of structure B.

Figure 3.18. Bridge Street Bridge, Southfield, Michigan, USA.

3.3.3. Future Prospects

As the body of knowledge on the use of FRP tendons expands and the profession becomes comfortable with the behavior of these materials, their use will expand, especially for saltwater and corrosive environments. Currently, the industry is hampered by the lack of commercial products and the premium cost for FRP materials. As a consequence, the prestressing industry is hesitant to incorporate FRP strands into their normal production lines. As indicated by the University of Wyoming work at the Rocky Mountain Prestress, the FRP tendons can be used in existing stress lines, however, numerous modifications and precautions are needed to incorporate FRP into existing stressing operations. The corrosion resistance properties, especially of carbon based tendons, make them particularly attractive for traditionally corrosive environments. Note that similar applications of FRP as a reinforcing bar and prestressing tendons have been done in the Canadian infrastructures working as demonstration projects.

Exercise Problems

1. Name the various demonstration projects of importance across the world using FRP composite materials in bridges.
2. What is the advantage of using FRP materials in prestressed concrete structures?
3. Analyse a case study project for demonstrating the cost effectiveness of FRP materials for external strengthening of concrete structures.
4. Which is the leading country as far as FRP manufacturing is concerned?
5. What different kinds of FRP sections are available as rebars?
6. What are the different products of FRP which can be used as external strengthening of concrete and masonry structures?
7. Do you think that FRP is potential future construction materials for bridges? If yes, then substantiate your answer with reasoning.
8. Where Bridge Street Bridge is located and what is its main features?
9. Give examples of bridges where FRP has been used as prestressing tendons.
10. Comment on the advantages of FRP under dynamic loading situations such as bridges.
11. Give examples of bridges where FRP has been used as main supporting cables.

CHAPTER 4

Design of RC Structures Reinforced with FRP Bars

In this chapter, the general design recommendations for flexural concrete elements reinforced with FRP bars are presented. Design equations are based on the principles of equilibrium, compatibility and the constitutive laws of materials. In addition, the brittle behavior of both the FRP and reinforcement and concrete allows consideration to be given to either FRP rupture or concrete crushing as mechanisms that control failure. In this chapter, design approaches and examples based on both the ACI 440-1R-06 and ISIS Canada guidelines will be discussed. In addition, design of FRP prestressed concrete bridge girders/beams has also been discussed in detail emphasizing the fundamentals of design approach.

4.1. Design Philosophy

ACI 440 Committee considered both the strength and working design approaches. Design recommendations presented in this chapter are based on the limit states of strength and serviceability and will be dealt with two design approaches as given in the following:

- Strength design approach;
- Check for fatigue endurance, creep rupture endurance, and serviceability requirement.

The FRP reinforced concrete member is designed based on its required strength and then checked for fatigue endurance, creep rupture endurance and serviceability criteria. It may be noted that in many instances, serviceability criteria or fatigue and creep rupture endurance limits may control the design of concrete members reinforced for flexure with FRP bars—especially aramid and glass FRP bars that exhibit low stiffness. To be consistent with ACI 440 committee recommendations for the design of FRP reinforced concrete structures and other documents related to FRP prestressing and external strengthening, retrofitting and/or rehabilitation using FRP, the load factors given in ACI 318 are recommended to be used to determine the required strength of a reinforced concrete member. Following are the basic design steps to be used to determine the nominal flexural and/or shear strength of FRP reinforced concrete structures.

4.1.1. Design Material Properties

Since, the initial material properties such as guaranteed tensile strength of FRP as provided by manufacturer do not include the effects of long-term exposure to the environment, the initial material properties used in design equations should be reduced based on the type and level of environmental

Table 4.1. Environmental reduction factor for various fibers and exposure conditions.

Exposure conditions	Fiber type	Environmental reduction factor, C_E
Concrete not exposed to earth and weather	Carbon	1.0
	Glass	0.8
	Aramid	0.9
Concrete exposed to earth and weather	Carbon	0.9
	Glass	0.7
	Aramid	0.8

exposure. The tensile properties of FRP are expressed by Eqs. (4.1) to (4.5) should be used in all design equations. The design tensile strength should be determined by

$$f_{fu} = C_E f_{fu}^* \tag{4.1}$$

where, f_{fu} = design tensile strength of FRP considering reduction for service environment
C_E = environment reduction factor (see Table 7.1 of ACI 440.1R-06 or Table 4.1 of the text)
f_{fu}^* = guaranteed tensile strength of an FRP bar defined as the mean tensile strength of a sample of test specimens minus three times the standard deviation ($f_{fu}^* = f_{fu,\text{ave}} - 3\sigma$)

Similarly, the design rupture strain should be determined as

$$\varepsilon_{fu} = C_E \varepsilon_{fu}^* \tag{4.2}$$

where, ε_{fu} = design rupture strain of FRP reinforcement; and
ε_{fu}^* = guaranteed rupture strain of FRP reinforcement defined as the mean tensile strain at failure of a sample of test specimens minus three times the standard deviation ($\varepsilon_{fu}^* = \varepsilon_{fu,\text{ave}} - 3\sigma$).

The design modulus of elasticity should be taken as the average value of the modulus of elasticity as reported by manufacturer or obtained from the test results by designers, that is E_f should be equal to $E_{f,\text{ave}}$.

It must be noted that the environmental reduction factors (Table 4.1) are conservative estimates depending on the durability of each fiber type and are based on the consensus of Committee 440. Temperature effects are included in C_E values. However, FRP bars should not be used in environment with a service temperature higher than the T_g of the resin used for their manufacturing.

The design tensile strength of FRP bars at a bend portion can be determined as,

$$f_{fb} = \left(0.05\frac{r_b}{d_b} + 0.3\right) f_{fu} \le f_{fu} \tag{4.3}$$

where, f_{fb} = design tensile strength of bend of FRP bar
r_b = radius of bend
d_b = diameter of reinforcing bar
f_{fu} = design tensile strength of FRP, considering deduction for service environment

Equation (4.3) is adapted from design recommendations by the Japan Society of Civil Engineers (1997). From limited research on FRP hooks, it is indicated that the tensile force developed by the bent portion of a GFRP bar is mainly influenced by the ratio of the bent radius to the bar diameter, r_b/d_b, the tail length and to lesser extent, the concrete strength. Most recently, the author (2006) has developed a uni-axial strength model for CFRP stirrups, which can be alternatively used for determining the tensile strength of bar at bend portion or for shear strength contribution of stirrups in beam in conjunction with effective stress in stirrups model as developed by the author (2006). These two alternate design equations is for predicting the strength of stirrups which bend along with effective stress in stirrups that used in actual beam situation, is given in the following Eqs. (4.4) and (4.5), respectively.

$$\left(\frac{f_u}{f_{fu}}\right)\frac{d_e\sqrt{\rho_f}}{\alpha_b} = 0.2757\left(\frac{l_d}{d_e}\right) + 3.9385 \geq 5.6 \tag{4.4}$$

where f_u refers to uni-axial stirrup strength (MPa); f_{fu} refers to the design strength of stirrup strand parallel to fibers (MPa); d_e is effective diameter of stirrup bar in mm; l_d is embedded length of stirrups in mm; ρ_f specific gravity of stirrup material; α_b is bond factor which is equal to 1.0 for stranded bar (e.g. CFCC 1 × 7 or CFTC 1 × 7) and 0.7 for solid stirrup bars, e.g., C-Bar. It should be noted that in no case the value of f_u should exceed f_{fu}.

$$\frac{f_{fv}}{f_{fu}} = 0.3\left(\frac{s}{d}\right) + 0.05 \leq 1.0 \quad \text{for } f_{fv} \leq f_u \tag{4.5}$$

where, f_{fv} is effective stress in stirrups at beam failure; f_{fu} is the design strength of stirrup bar and s is the center to center spacing of stirrups. Based on Eq. (4.5), it is recommended that stirrup spacing should not be more than $3.2d$ where d is effective depth of beam.

4.1.2. Flexural Design Philosophy

- Steel-reinforced concrete sections are commonly under-reinforced to ensure yielding of steel before crushing of concrete.
- The yielding of steel provides ductility and warning of failure to members.
- The concrete crushing failure mode is more desirable for flexural members reinforced with FRP bars.
- If FRP reinforcement ruptures, failure of member is sudden and catastrophic and hence, no ductility is exhibited.
- By experiencing concrete crushing, a flexural member does exhibit some plastic behavior before failure.

Assumptions: The flexural strength of cross-sections should be based on the following assumptions:

- Strain in concrete and the FRP reinforcement is proportional to the distance from the neutral axis, i.e., a plane section before bending remains plane after bending.
- The maximum usable compressive strain in the concrete is 0.003.
- The tensile strength of concrete is ignored.
- The tensile behavior of the FRP reinforcement is linearly elastic until failure.
- Perfect bond exists between concrete and FRP reinforcement.

Flexural strength: The strength design philosophy states that the design flexural capacity of a member must exceed the flexural demand, i.e.,

$$\phi M_n \geq M_u \tag{4.6}$$

where, M_n = nominal flexural capacity of section

$$M_u = 1.2M_D + 1.6M_L \tag{4.7}$$

Note: The nominal flexural capacity of an FRP reinforced concrete member can be determined based on strain compatibility, internal force equilibrium and the controlling failure mode.

Reinforcement ratio,

$$\rho_f = \frac{A_f}{bd} \tag{4.8}$$

Balanced reinforcement ratio:

$$\rho_{fb} = 0.85\beta_1 \frac{f_c'}{f_{fu}} \frac{E_f \varepsilon_{cu}}{E_f \varepsilon_{cu} + f_{fu}} \tag{4.9}$$

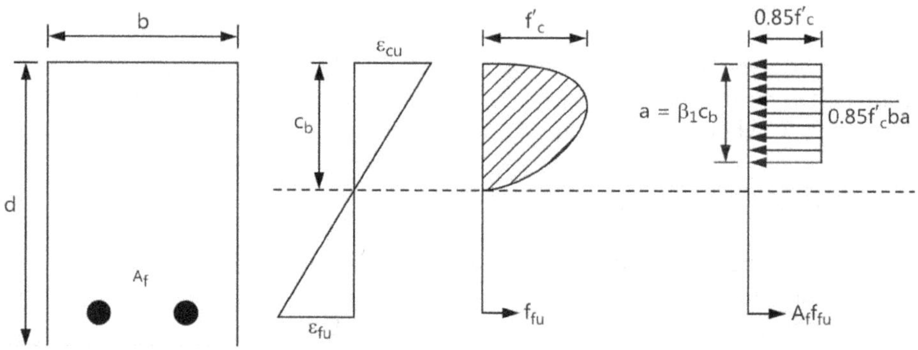

Figure 4.1. Balanced failure condition.

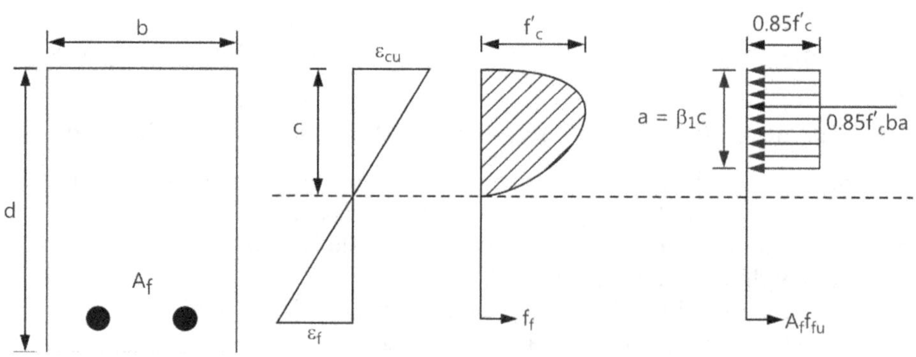

Figure 4.2. Failure governed by concrete crushing.

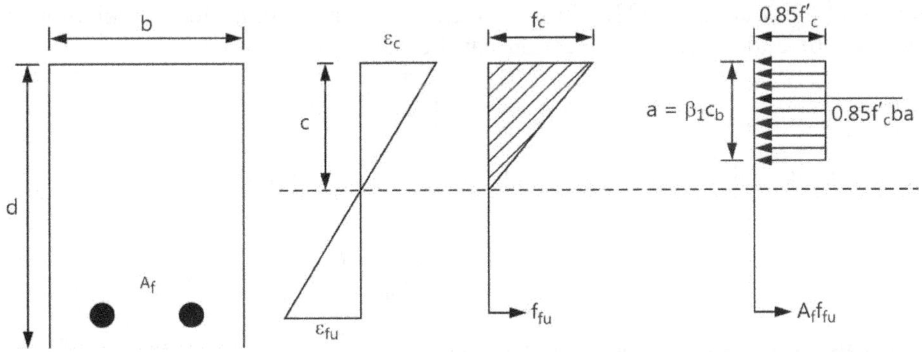

Figure 4.3. Failure governed by FRP rupture (concrete stress may be nonlinear).

4.1.3. Nominal Flexural Capacity

(A) Concrete crushing mode of failure ($\rho_f > \rho_{fb}$): Based on the equilibrium of forces and compatibility of strains, we have

$$\varepsilon_f = \frac{\varepsilon_{cu}(\beta_1 d - a)}{a} \tag{4.10}$$

$$a = \frac{A_f f_f}{0.85 f_c' b} \tag{4.11}$$

$$f_f = \frac{E_f \varepsilon_{cu}(\beta_1 d - a)}{a} \tag{4.12}$$

$$f_f = \left(\sqrt{\frac{(E_f \varepsilon_{cu})^2}{4} + \frac{0.85 \beta_1 f_c'}{\rho_f} E_f \varepsilon_{cu}} - 0.5 E_f \varepsilon_{cu} \right) \le f_{fu} \tag{4.13}$$

$$M_n = A_f f_f \left(d - \frac{a}{2} \right) \tag{4.14}$$

Alternatively,

$$M_n = \rho_f f_f \left(1 - 0.59 \frac{\rho_f f_f}{f_c'} \right) b d^2 \tag{4.15}$$

(B) FRP rupture mode of failure ($\rho_f < \rho_{fb}$): In the case of under-reinforced section, the ACI stress block is not applicable, because maximum concrete strain (0.003) may not be attained at failure. Hence, an equivalent stress block would be used that approximates the stress distribution in the concrete at the particular strain level reached. The analysis incorporates two unknown: the concrete compressive strain at failure, ε_c, and the depth to the neutral axis, c. In addition, the rectangular stress block factors, α_1 and β_1 are also unknown. The factor, α_1, is the ratio of the average concrete stress to the concrete strength. β_1 is the ratio of the depth of the equivalent rectangular stress block to the depth of the neutral axis. For a given section, the product $\beta_1 c$ varies depending on the material properties and FRP reinforcement ratio ρ_f. The maximum value of $\beta_1 c$ is equal to $\beta_1 c_b$ and is achieved when concrete strain reaches 0.003. Thus, a conservative and simplified calculation of the nominal flexural capacity of under-reinforced section can be based on Eq. (4.16).

$$M_n = A_f f_{fu}\left(d - \frac{\beta_1 c_b}{2}\right) \tag{4.16}$$

where, c_b is the depth to the neutral axis at balanced failure condition.

$$c_b = \left(\frac{\varepsilon_{cu}}{\varepsilon_{cu} + \varepsilon_{fu}}\right)d \tag{4.17}$$

4.1.4. Strength Reduction Factor for Flexure (ϕ)

Since, FRP reinforced members do not exhibit ductile behavior, a conservative strength reduction factor is adopted to provide a higher reserve of strength in the member.

$\phi = 1/1.3$ JSCE (1997)
$= 0.75$ (Benmokrane *et al.*, 1996) determined based on probabilistic approach.

As per ACI 440.1R-06, the value of ϕ can be computed using Eq. (4.18), which is shown graphically in Fig. 4.4. It may be noted that concrete crushing failure mode can be predicted based on calculations but the member as constructed may not fail accordingly, because if the concrete strength is higher than specified, the member can fail due to FRP rupture. Hence, to establish a transition between the two values of ϕ, a section controlled by concrete crushing is defined as a section in which $\rho_f \geq 1.4\rho_{fb}$ and a section controlled by FRP rupture is defined as one in which $\rho_f < \rho_{fb}$. Equation (4.18) gives a factor of 0.65 for sections controlled by concrete crushing, 0.55 for sections controlled by FRP rupture and provided a linear transition between the two.

$$\phi = \begin{cases} 0.55 & \text{for } \rho_f \leq \rho_{fb} \\ 0.3 + 0.25\dfrac{\rho_f}{\rho_{fb}} & \text{for } \rho_{fb} < \rho_f < 1.4\rho_{fb} \\ 0.65 & \text{for } \rho_f \geq 1.4\rho_{fb} \end{cases} \tag{4.18}$$

Design moment $= \phi M_n \geq M_u$ (required moment)

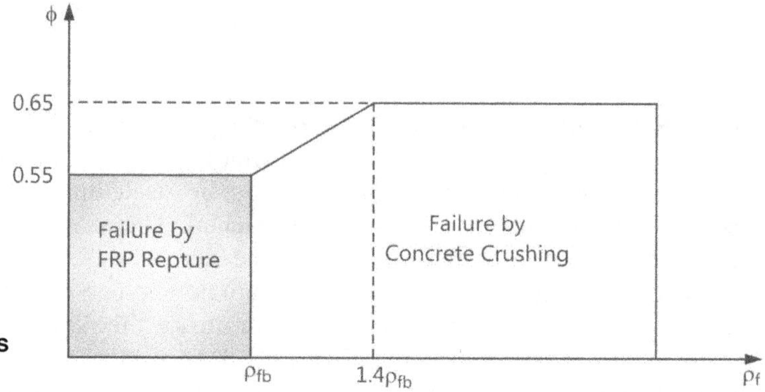

Figure 4.4. Strength reduction factor as function of the reinforcement ratio.

4.1.5. Check for Minimum FRP Reinforcement

If $\rho_f < \rho_{fb}$, a minimum amount of reinforcement ($A_{f,\min}$) should be provided as per Eq. (4.19) to prevent failure upon concrete cracking (i.e., $\phi M_n \geq M_{cr}$ where M_{cr} is the cracking moment). The minimum reinforcement area as given in Eq. (4.19) for FRP-reinforced members has been obtained by multiplying the existing ACI318-05 equation for steel reinforcement by 1.64 (i.e., 0.90/0.55 = 1.64). If the failure of the member is not controlled by FRP rupture (i.e., $\rho_f > \rho_{fb}$), the minimum amount of reinforcement to prevent failure upon cracking is automatically achieved. Hence, Eq. (4.19) is needed as a check only if $\rho_f < \rho_{fb}$.

$$A_{f,\min} = \frac{4.9\sqrt{f_c'}}{f_{fu}} b_w d \geq \frac{330}{f_{fu}} b_w d \tag{4.19}$$

4.1.6. Serviceability

In general, designing FRP reinforced cross-sections for concrete crushing failure satisfies serviceability criteria for deflection and crack width. Serviceability can be defined as satisfactory performance under service load condition. Serviceability can be described in terms of two parameters:

- **Cracking:** Excessive crack width is undesirable for aesthetic and other reasons (e.g., to prevent water leakage) that can damage or deteriorate the structural concrete.
- **Deflections:** Deflections should be within acceptable limits imposed by the use of the structures (e.g., supporting attached nonstructural elements without damage).

The serviceability requirements (ACI 318) are to be modified for FRP reinforced members due to differences in properties of steel and FRP—such as lower stiffness, bond strength and corrosion resistance. The substitution of FRP on an equal area basis would typically result in larger deflections and wider crack widths.

Cracking: FRP rods are corrosion resistant,therefore, the maximum crack-width limitation can be relaxed when corrosion of reinforcement is the primary reason for crack-width limitations. If steel is to be used in conjunction with FRP reinforcement, however, ACI 318 provisions should be used.

- The Japan Society of Civil Engineers (1997) takes into account the aesthetic point of view only to set the maximum allowable crack width of 0.5 mm.
- The Canadian Highway Bridge Design Code—Canadian Standards Association 1996 allows crack widths of 0.5 mm for exterior exposure and 0.7 mm for interior exposure when FRP reinforcement is used.
- ACI 318 provisions for allowable crack-width limits in steel-reinforced structures correspond to 0.3 mm for exterior exposure and 0.4 mm for interior exposure.

Recommendations

- It is recommended that Canadian Standard Association (1996) limits be used for most cases.
- These limitations may not be sufficiently restrictive for structures exposed to aggressive environments or designed to be watertight. Therefore, an additional caution is required for such cases.
- For structures with short life-cycle requirements or those for which aesthetics is not a concern, crack-width requirements can be disregarded unless steel reinforcement is also present.

Crack-width estimation: A reasonable estimate of crack-width (w) for FRP reinforced concrete structures can be made by modifying the well-known Gergely–Lutz equation (Eq. (4.20)) (1973)

$$w = 0.076\beta(E_s\varepsilon_s)\sqrt[3]{d_c A} \tag{4.20}$$

where E_s is in ksi and w is in mils (10^{-3} in.).

Since, the crack width is proportional to the strain in the tensile reinforcement rather than the stress (Wang and Salmon 1992), the Gergely–Lutz equation can be modified to predict the crack width of FRP-reinforced flexural members by replacing the steel strain, ε_s, with the FRP strain $\varepsilon_f = f_f/E_f$ and by substituting 29 000 ksi for the modulus of elasticity for steel as follows:

$$\text{w} = 0.076\beta\frac{E_s}{E_f} f_f \sqrt[3]{d_c A} \tag{4.21}$$

Equation (4.21) estimates the crack width accurately when FRP deformed bars having bond strength similar to that of steel are used. However, this equation can overestimate the crack-width when applied to a bar with a higher bond strength than that of steel and underestimate crack width when applied to a bar having lower bond strength than that of steel. Therefore, to make the expression more generic, it is necessary to introduce a corrective co-efficient for the bond quality. Thus, for FRP reinforced members, crack width can be calculated from Eq. (4.22)

$$w = \frac{2200}{E_f}\beta k_b f_f \sqrt[3]{d_c A} \tag{4.22}$$

For SI units,

$$w = \frac{2.2}{E_f}\beta k_b f_f \sqrt[3]{d_c A} \tag{4.23}$$

where f_f and E_f in MPa, d_c (effective concrete cover) in mm and A is an effective tension area of concrete in mm^2. Based on ACI committee 440, a value of 1.2 can be adopted when k_b is not known.

$$\beta = \frac{h - kd}{d(1-k)} \tag{4.24}$$

$$d_c = h - d \tag{4.25}$$

$$A = \frac{2 d_c b}{\text{No. of bars}} \tag{4.26}$$

As per ACI 440.1R-06 an alternative equation (Eq. (4.27)) for crack width calculation has been recommended as given in the following equation:

$$w = 2\frac{f_f}{E_f}\beta k_b \sqrt{d_c^2 + \left(\frac{s}{2}\right)^2} \tag{4.27}$$

In Eq. (4.27), w is maximum crack width, mm f_f reinforcement stress, MPa; β is the ratio of distance between neutral axis and tension face to the distance between neutral axis and centroid of reinforcement (see Eq. (4.24)); d_c is thickness of cover from the tension face to center of closest bar, mm and s is longitudinal bar spacing, mm.

The term, k_b is coefficient that accounts for the degree of bond between FRP bar and surrounding concrete. The term k_b is taken equal to 1 for FRP bars having bond behavior similar to uncoated steel bars; k_b is larger than 1 for FRP bars having bond behavior inferior to steel; while k_b is taken less than 1 for FRP bars having bond behavior superior to steel. The average value of k_b has been found to range from 0.60 to 1.72 (with mean value = 1.10) depending upon type of FRP, manufacturers, fiber types, resin formulations and the type of surface treatments. As per ACI 440-1R-06, in the absence of experimental data, the value of k_b can be assumed as 1.40.

Deflections:

1. In general, the ACI 318 provisions for deflection control are concerned with deflections that occur at service levels under immediate and sustained static loads and do not apply to dynamic loads—such as earthquakes, transient winds, or vibration of machinery.
2. Two methods are presently given in ACI 318 for control of deflections of one-way flexural members:
    - The indirect method of mandating the minimum thickness of the member (Table 9.5 (a) in ACI 318, Table 8.2 of ACI 440.1R-06) and Table 4.2 of this text.
    - The direct method of limiting computed deflections (Table 9.5 (b) in ACI 318).

Minimum depth: The minimum depth recommended as per ACI 440-1R-06 standards for FRP reinforced beams and slabs are provided in Table 4.2. The minimum thickness as given in Table 4.2 serves as a guideline for fixing the initial dimensions of the cross-section and does not guarantee that all deflection considerations will be satisfied. In fact, final deflection checks should be made on calculating the overall short and long-term deflections of the beam/slab and comparing with that of allowable deflection to satisfy the limit state of serviceability of deflection.

It may be noted that the values of Table 4.2 are based on a generic maximum span-depth ratio limitation corresponding to the limiting curvature associated with a target deflection-span ratio as expressed by Eq. (4.28). This equation assumes no tensile contribution from concrete between cracks (referred as tension stiffening). In the development of Table 4.2, the effects of tension stiffening were considered by multiplying the values of l/h obtained from Eq. (4.28) with ratio of effective and fully cracked moment of inertia expressed by Eqs. (4.30b) and (4.29a), respectively. Moreover, values

Table 4.2. Recommended minimum depth of FRP reinforced beams and slabs.

Type of structure	Minimum Depth, h			
	Simply supported	**One end continuous**	**Both ends continuous**	**Cantilever**
Beams	$l/10$*	$l/12$	$l/16$	$l/4$
Slabs	$l/13$	$l/17$	$l/22$	$l/5.5$

*In this table, l refers to the effective span of the beam and/or slab.

presented in Table 4.2 are based on an assumed service deflection limit of $l/240$ under total service load and assumed reinforcement ratio of $2.0\rho_{fb}$ and $3.0\rho_{fb}$ for slabs and beams, respectively.

$$\frac{l}{h} = \frac{48\eta}{5K_1}\left(\frac{1-k}{\varepsilon_f}\right)\left(\frac{\Delta}{l}\right)_{max} \tag{4.28}$$

where, $\eta = \dfrac{d}{h}$;

k = neutral axis depth coefficient = c/d;

$\left(\dfrac{\Delta}{l}\right)_{max}$ = Limiting service load deflection-span ratio;

K_1 = A parameter that accounts for boundary conditions and may be taken as equal to 1.0, 0.8, 0.6 and 2.4 for uniformly loaded simply supported, one end continuous, both end continuous and cantilevered spans, respectively.

ε_f = Strain in FRP reinforcement under service loads, evaluated at midspan except for cantilevered spans

Effective moment of inertia:

- When a section is uncracked, its moment of inertia is equal to the gross-moment of inertia, I_g;
- When applied moment, M_a > Cracking Moment, M_{cr}, cracking occurs, which reduces the member stiffness and the moment of inertia is based on cracked section, I_{cr}.

For a rectangular section:

$$I_{cr} = \frac{bd^3}{3}k^3 + n_f A_f d^2 (1-k)^2 \tag{4.29a}$$

$$k = \sqrt{2\rho_f n_f + (\rho_f n_f)^2} - \rho_f n_f \tag{4.29b}$$

The overall flexural stiffness, E_cI of a flexural member that has experienced cracking at service, varies between E_cI_g and E_cI_{cr} depending upon the magnitude of applied moment, M_a. Branson (1977) derived an expression (Eq. (4.30a)) to express the transition from I_g to I_{cr}. Equation (4.30a) was adopted by ACI 318 to compute effective moment of Inertia, I_e:

$$I_e = \left(\frac{M_{cr}}{M_a}\right)^3 I_g + \left[1-\left(\frac{M_{cr}}{M_a}\right)^3\right] I_{cr} \le I_g \tag{4.30a}$$

Equation (4.30a) reflects two different phenomena—the variation of EI stiffness along the member and the effect of concrete tension stiffening. This equation is based on the behavior of steel-reinforced beams at service load levels. Because FRP bars exhibit linear behavior up to failure, the equation offers a close approximation for FRP reinforced beams. Since the bond characteristics of FRP bars also affect the deflection of a member, Eq. (4.30a) can overestimate the effective moment of inertia of FRP reinforced beams. A modified Eq. (4.30b) is proposed, which takes into account the lower modulus of elasticity of FRP bars and the different bond behavior of FRP.

$$I_e = \left(\frac{M_{cr}}{M_a}\right)^3 \beta_d I_g + \left[1 - \left(\frac{M_{cr}}{M_a}\right)^3\right] I_{cr} \le I_g \tag{4.30b}$$

Equation (4.30b) is valid only if the maximum unfactored moment (M_a) in the member is equal to or greater than the cracking moment (M_{cr}). The recommended minimum thickness (Table 4.2) of members assumes this condition. The factor β_d is related to reduced tension stiffening observed with the FRP reinforced members. The degree of tension stiffening is affected by the amount and stiffness of the flexural reinforcement and by relative reinforcement ratio. As per ACI440-1R-06 standards, the value of β_d has been recommended as given in Eq. (4.31).

$$\beta_d = \frac{1}{5}\left(\frac{\rho_f}{\rho_{fb}}\right) \le 1.0 \tag{4.31}$$

Calculation of deflection (direct method):

1. The short-term deflections—instantaneous deflections under service loads of an FRP one-way flexural member can be calculated using the effective moment of inertia of FRP reinforced beam and usual structural analysis techniques.
2. Long-term deflection can be two to three times the short-term deflection and both the short-term and long-term deflections under service loads should be considered in the design.
3. The long-term increase in deflection is a function of member geometry—reinforcement area and member size, load characteristics—age of concrete at the time of loading and magnitude and duration of sustained loading and material characteristics—creep and shrinkage of concrete, formation of new cracks and widening of existing cracks.
4. Limited data on long-term deflections of FRP reinforced members (Kage *et al.*, 1995; Brown 1997) indicate that creep behavior of FRP reinforced member is similar to that of steel-reinforced member.
5. The time versus deflection curves of FRP reinforced and steel-reinforced members have the same shape, which indicate that the same approach for estimating the long-term deflection can be used.
6. Experiments have shown that initial short-term deflection of FRP-reinforced members is three to four times greater than those of steel-reinforced members for the same design strength.
7. After one year, FRP reinforced members deflected 1.2–1.8 times that of steel-reinforced members, depending on the type of FRP bar (Kage *et al.*, 1995).

According to ACI 318, Section 9.5.2.5, the long-term deflection due to creep and shrinkage, $\Delta(cp + sh)$, can be computed using Eq. (4.32).

$$\Delta(cp + sh) = \lambda(\Delta_i)\text{sus} \tag{4.32}$$

$$\lambda = \frac{\xi}{1 + 50\rho'} \tag{4.33}$$

where ρ' is compression reinforcement ratio and is taken as zero because FRP compression reinforcement is not considered. Thus, $\lambda = \xi$.

Brown (1997) indicated that long-term deflection varies with the compressive stress in the concrete. This issue is not addressed by the equations in ACI 318, which only multiplies the initial deflection by the time dependent factor, ξ. Brown concluded that the creep co-efficient should be adjusted twice—first, to account for the compressive stress in the concrete and second to account for the larger initial deflection.

$$\xi = \frac{\xi_{\text{FRP}}}{\xi_{\text{Steel}}} \tag{4.34}$$

ξ varies from 0.46 for AFRP and GFRP to 0.53 for CFRP. In another study, modification factor for ξ based on a failure controlled by concrete crushing varied from 0.75 after 1 year to 0.58 after 5 years. Based on these results, a modification factor of 0.6 is recommended. Therefore, Eq. (4.29) can be used to determine the long-term deflection of FRP-reinforced members.

$$\Delta(cp + sh) = 0.6\xi(\Delta_i)\text{sus} \tag{4.35}$$

Creep rupture stress limits: Since, stresses due to creep and fatigue under service loading condition will be within elastic range, the stresses can be computed through an elastic analysis as depicted in Fig. 4.5. To avoid failure of an FRP reinforced member due to creep-rupture of the FRP, stress level in the FRP reinforcement ($f_{f,s}$) under sustained loading—dead load plus sustained portion of live load—should be less than creep-rupture stress limit for the corresponding FRP bar type (Table 4.3).

Let M_s = unfactored moment due to all sustained loads

$$f_{f,s} = M_s \frac{n_f d(1-k)}{I_{cr}} \tag{4.36}$$

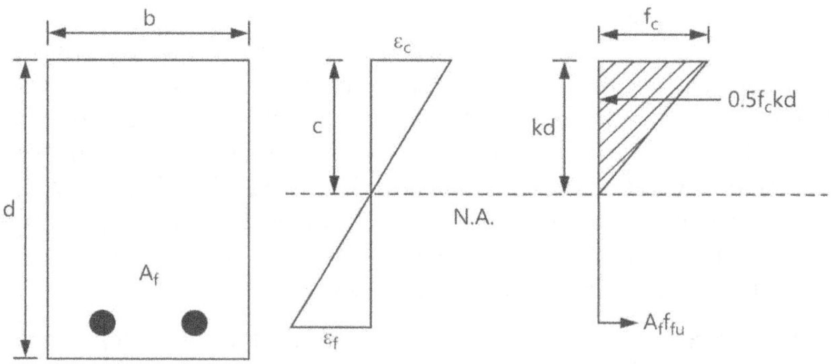

Figure 4.5. Strain and stress distribution at service load condition.

Table 4.3. Creep rupture stress limits in FRP reinforcement.

Fiber type	GFRP	AFRP	CFRP
Creep rupture Stress limit $F_{f,s}$	$0.20f_{fu}$	$0.30f_{fu}$	$0.55f_{fu}$

Fatigue stress limits: If the structure is subjected to fatigue regimes, the FRP stress should be limited to the values stated in Table 4.3. The FRP stress can be calculated using Eq. (4.36) with M_s equal to the moment due to all sustained loads plus the maximum moment induced in the fatigue loading cycle.

4.2. Shear

Several issues need to be addressed while using FRP as shear reinforcement, viz.:

- FRP has relatively low modulus of elasticity
- FRP has high tensile strength
- Tensile strength of bent portion of an FRP is significantly lower than the straight portion
- FRP has low dowel resistance.

4.2.1. Shear Design Philosophy

The design of FRP shear reinforcement is based on the strength design method. The strength reduction factor given by ACI 318 for reducing nominal shear capacity of steel-reinforced concrete members should be used for FRP reinforcement as well.

Shear strength of FRP-reinforced members: The concrete shear capacity V_c of flexural members using FRP as main reinforcement can be evaluated using Eq. (4.37).

$$V_c = \frac{2}{5}\sqrt{f_c'}b_w c \quad \text{(SI Units)} \tag{4.37a}$$

$$V_c = 5\sqrt{f_c'}b_w c \quad \text{(British Units)} \tag{4.37b}$$

where, $c = kd$ and k is neutral axis depth coefficient as given in Eq. (4.29b).

The shear resistance (V_f) provided by FRP stirrups perpendicular to axis of the member can be expressed by Eq. (4.38).

$$V_f = \frac{A_{fv} f_{fv} d}{s} \tag{4.38}$$

where,

$$f_{fv} = 0.004E_f \leq f_{fb} \tag{4.39}$$

The spacing of vertically arranged stirrups can be computed from Eq. (4.40).

$$\frac{A_{fv}}{s} = \frac{(V_u - \phi V_c)}{\phi f_{fv} d} \tag{4.40}$$

1. When inclined FRP stirrups are used as shear reinforcement, Eq. (4.41) is used to calculate the contribution of the FRP stirrups.

$$V_f = \frac{A_{fv} f_{fv} d}{s}(\sin\alpha + \cos\alpha) \tag{4.41}$$

2. When continuous rectangular spirals are used as shear reinforcement—in this case, s is pitch and α is the angle of inclination of the spiral. Equation (4.42) gives the contribution of the FRP spirals.

$$V_f = \frac{A_{fv} f_{fv} d}{s} (\sin \alpha) \tag{4.42}$$

4.2.2. Shear Failure Modes

1. Shear-tension failure mode—controlled by the rupture of FRP shear reinforcement. This failure mode is brittle.
2. Shear-compression failure mode—controlled by the crushing of the concrete web. This failure mode results in larger deflection.

The following aspects should be noted regarding the shear failure modes:

1. Shear failure mode depends on the shear-reinforcement index $\rho_{fv}\phi_v E_f$, where $\rho_{fv} = A_{fv}/b_w s$ (ratio of FRP shear reinforcement). As the value of $\rho_{fv} E_f$ increases, the shear capacity in shear-tension increases and mode of failure changes from shear-tension to shear-compression.
2. In no case should effective strain in FRP shear reinforcement exceed 0.004 nor should the design strength exceed the strength of the bent portion of the stirrup f_{fb}.

4.2.3. Minimum Shear Reinforcement

A minimum amount of shear reinforcement is required when V_u exceeds $\phi V_c/2$. This requirement is to prevent or restrain shear failure where the sudden formation of cracks can lead to excessive distress.

$$A_{fv,\min} = \frac{50 b_w s}{f_{fv}} \tag{4.43}$$

where b_w and s are inches and f_{fv} is in psi. In SI units, Eq. (4.44) can be used to determine the minimum shear reinforcement area.

$$A_{fv,\min} = 0.35 \frac{b_w s}{f_{fv}} \tag{4.44}$$

where b_w and s are in mm and f_{fv} in MPa.

4.2.4. Shear Failure Due to Crushing of the Web

Studies by Nagaska, Fukuyama and Tanigaki (1993) indicate that for FRP reinforced sections—the transition from rupture to crushing failure mode occurs at an average value of $0.3f_c'b_w d$ for V_{cf} but can be as low as $0.18f_c'b_w d$. When V_{cf} is less than $0.18f_c'b_w d$, shear-tension can be expected, whereas, when V_{cf} exceeds $0.3f_c'b_w d$, crushing failure is expected. The correlation between rupture and crushing failure is not fully understood and is more conservative to use the ACI 318 limit of $8\sqrt{f_c'}b_w d$ rather than $0.3f_c'b_w d$. It is therefore, recommended to follow the ACI 318 limit.

4.2.5. Detailing of Shear Stirrups

1. The maximum spacing of vertical steel stirrups given in ACI 318: smaller of $d/2$ or 24 in. This limit ensures that each shear crack is intercepted by at least one stirrup.
2. A minimum r_b/d_b ratio of 3 is recommended.

3. FRP stirrups should be closed at 90° hooks.
4. The tensile force in a vertical stirrup leg is transferred to the concrete through the tail beyond the hook as shown in Fig. 4.6.
5. For a tail length (l_{thf}) beyond $12d_b$, there is no significant slippage and no influence on the tensile strength of the stirrup leg.

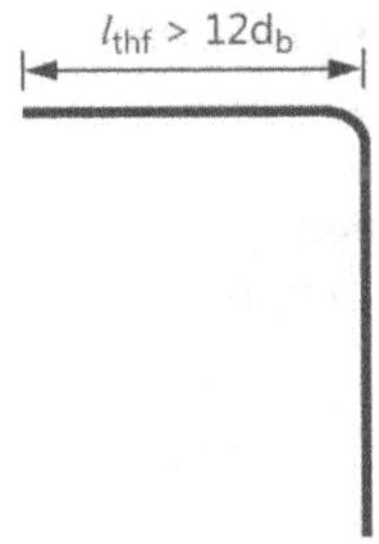

Figure 4.6. Required tail length for FRP stirrups.

4.2.6. Punching Shear Strength of FRP Reinforced, Two-Way Concrete Slab

The concentric punching shear capacity of FRP reinforced two-way concrete slabs that are either supported by interior columns or subjected to concentric loads that are either square or circular in shape can be determined by using Eq. (4.45).

$$V_c = 10\sqrt{f_c'}\, b_o c \text{ (British Units)}$$

$$V_c = \frac{4}{5}\sqrt{f_c'}\, b_o c \text{ (SI Units)} \tag{4.45}$$

where b_o = perimeter of critical section for slabs and footings, mm; c = cracked transformed section neutral axis depth, mm.

For slabs, the neutral axis depth, c, may be computed as follows:

$$c = kd \tag{4.46a}$$

$$k = \sqrt{2\rho_f n_f + (\rho_f n_f)^2} - \rho_f n_f \tag{4.46b}$$

where, ρ_f = Reinforcement ratio; n_f = Modular ratio $= \dfrac{E_f}{E_c}$

It may be noted that Eq. (4.46) provides a reasonable factor of safety for FRP reinforced two-way slabs across the range of reinforcement ratios and concrete strengths tested so far. However, further research is needed—to examine the punching capacity of FRP reinforced two-way slabs supported by edge and corner columns as well as the effects of column rectangularity and unbalanced moment transfer on the punching capacity of FRP reinforced two-way slabs supported on interior columns.

4.3. ISIS Canada Design Approach for Flexure

Like ACI-440.1R-06 design guidelines, ISIS Canada design approach incorporates limit states method of design. In the case of buildings, the loads and load combinations for FRP-reinforced concrete members should be determined—in the same manner as for reinforced concrete members using steel as reinforcements (CSA A23.3-94). In the case of bridges, the Canadian Highway Bridge Design Code CSA-S6-00 is recommended. Moreover, checks for all limit states of serviceability should be made.

Unlike ACI 440.1R-06 standards, the ISIS Canada design methodology adopts materials resistance factor for getting the design strength of materials. The material resistance factor for concrete is taken as 0.6 for buildings, 0.65 for precast concrete and 0.75 for bridges. The material resistance factor for FRPs depends on the type of FRP, variability of material characteristics, the effect of sustained load and durability conditions. Typical values of material resistance factors for CFRP, AFRP and GFRP are 0.8, 0.6 and 0.4, respectively.

Assumptions: The flexural design as per ISIS Canada design guidelines is based on the following assumptions:

1. FRPs are perfectly linear-elastic materials.
2. The failure of an FRP-reinforced section in flexure can be due to FRP rupture or concrete crushing.
3. The failure strain of concrete in compression is 0.0035.
4. The strain in concrete at any level is proportional to the distance from the neutral axis, i.e., plane sections remain plane before and after bending.
5. The concrete compressive stress-strain curve is parabolic and concrete has no strength in tension.
6. The perfect bond exists between FRP reinforcement and concrete.
7. The FRPs' strength in compression is neglected.

Failure modes: The potential failure modes for FRP-reinforced concrete sections are: (1) Balanced failure, (2) Compression failure and (3) Tension failure. The compression failure is considered to be most desirable, as this is less violent than tension failure. The tension failure is brittle in nature, as tensile rupture of FRP reinforcement will occur with less warning. The tension failure occurs, when reinforcement ratio is less than balanced reinforcement ratio for the section. The balanced failure refers to simultaneous FRP tensile rupture and concrete crushing. It is to be noted that balanced failure condition is drastically different for FRP-reinforced concrete than it is for members reinforced with steel. The FRP-reinforced members with balanced sections fail suddenly with cracking and appreciable deflection. The ratio of the balanced neutral axis depth, c_b to the effective depth of section, d, at the balanced reinforcement ratio, $\rho_{\text{frp}b}$, can be expressed by Eq. (4.47).

$$\frac{c_b}{d} = \frac{\varepsilon_{cu}}{\varepsilon_{cu} + \varepsilon_{frpu}} \tag{4.47}$$

The balanced reinforcement ratio is determined by equilibrium condition (Eq. (4.48)) wherein C and T represent, the total resultant compression and tension forces, respectively.

$$C = T \tag{4.48}$$

where,

$$C = \alpha_1 \phi_c f'_c \beta_1 c_b b \tag{4.49}$$

$$T = \phi_{\text{frp}} \varepsilon_{\text{frp}u} E_{\text{frp}} A_{\text{frp}} \tag{4.50}$$

In Eq. (4.49), α_1 and β_1 are stress block parameters as expressed by Eqs. (4.51) and (4.52), respectively. These stress block parameters are required to replace the actual compressive stress distribution (Fig. 4.7) of concrete, by an equivalent rectangular stress block (see Fig. 4.9). As per ISIS Canada design guidelines, stress block parameters α_1 and β_1 are used as those suggested in CSA A23.3-94/CHBDC for steel-reinforced concrete. Also, the terms, ϕ_c, ϕ_{frp}, $\varepsilon_{\text{frp}u}$, E_{frp} and A_{frp} are material resistance factor for concrete, material resistance factor for FRP, ultimate rupture strain of FRP, modulus of elasticity of FRP and cross-section area of FRP, respectively.

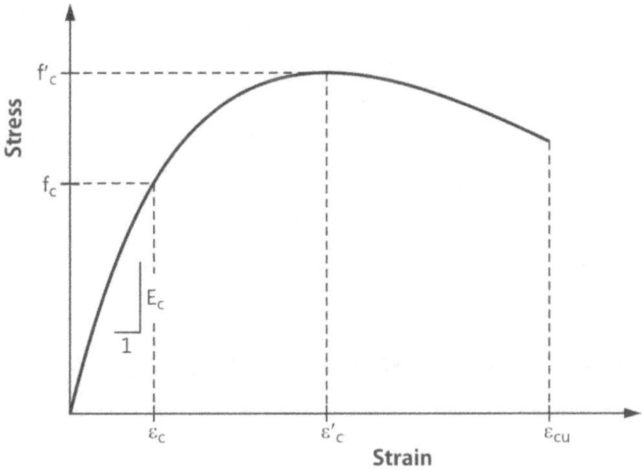

Figure 4.7. Assumed stress–strain behavior of concrete in compression.

$$\alpha_1 = 0.85 - 0.0015 f_c' \geq 0.67 \tag{4.51}$$

$$\beta_1 = 0.97 - 0.0025 f_c' \geq 0.67 \tag{4.52}$$

Using equilibrium condition (Eq. (4.48)) and rearranging the terms, the balanced failure reinforcement ratio, $\rho_{\mathrm{frp}b}$ is obtained as Eq. (4.53).

$$\rho_{frpb} = \frac{A_{frpb}}{bd} = \alpha_1 \beta_1 \frac{\phi_c}{\phi_{frp}} \frac{f_c'}{f_{frpu}} \left(\frac{\varepsilon_{cu}}{\varepsilon_{cu} + \varepsilon_{frpu}} \right) \tag{4.53}$$

4.3.1. Flexural Strength

Depending on whether the section is under-reinforced or over-reinforced, the different flexural strength equations will be used to evaluate the flexural strength of FRP reinforced beam as per ISIS Canada design guidelines as described in the following section. Two failure modes that are of practical interest are compression failure and tension failure.

Compression failure: The compression failure occurs in an over-reinforced section having actual reinforcement more than that corresponding to balanced failure condition. Analytically this failure mode could be predicted, when actual reinforcement ratio (ρ_{frp}) is greater than the balanced reinforcement ratio ($\rho_{\mathrm{frp}b}$). In such sections, crushing of concrete occurs in compression zone before rupture of FRP rebars. As shown in Fig. 4.8b, the crushing strain, ε_{cu} of concrete is taken as 0.0035 and nonlinear stress distribution in the concrete can be replaced by equivalent stress block parameters, α_1 and β_1 (see Eqs. 4.51 and 4.52) The values of compression and tension stress resultants can be estimated by Eqs. (4.54) and (4.55), respectively.

$$C = \alpha_1 \phi_c f_c' \beta_1 cb \tag{4.54}$$

$$T = \phi_{\mathrm{frp}} A_{\mathrm{frp}} f_{\mathrm{frp}} \tag{4.55}$$

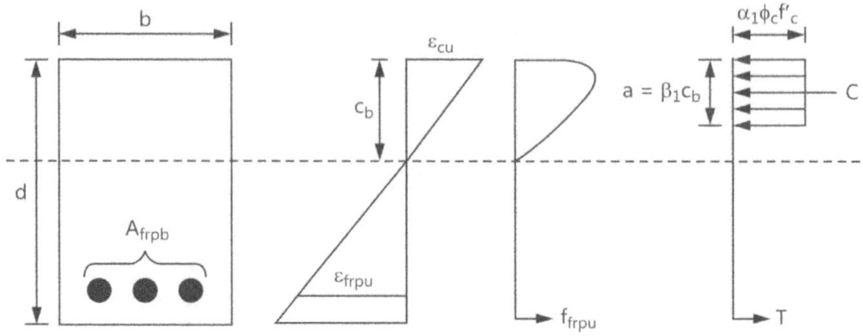

Figure 4.8a. Balanced failure mode ($\varepsilon_c = \varepsilon_{cu}$ and $\varepsilon_{frp} = \varepsilon_{frpu}$).

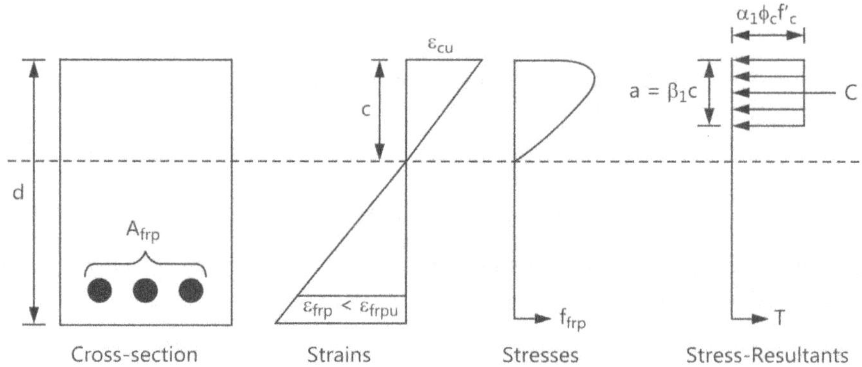

Figure 4.8b. Compression failure mode ($\varepsilon_c = \varepsilon_{cu}$ and $\varepsilon_{frp} < \varepsilon_{frpu}$).

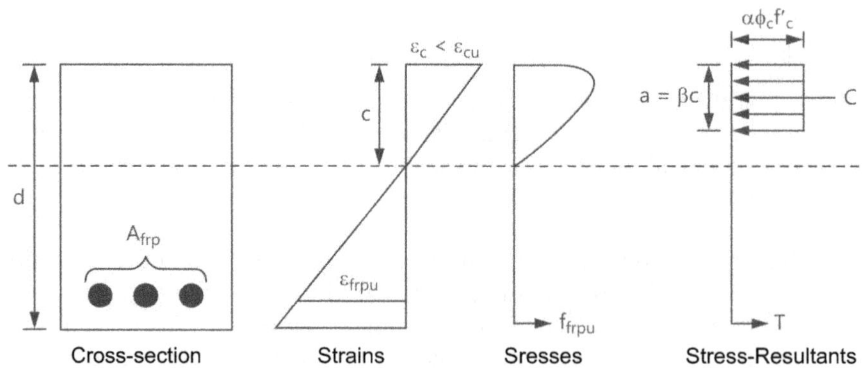

Figure 4.8c. Tension failure mode ($\varepsilon_c < \varepsilon_{cu}$ and $\varepsilon_{frp} = \varepsilon_{frpu}$).

In Eq. (4.55), the f_{frp} is unknown because in an over-reinforced section FRP does not reach its ultimate failure strain (ε_{frpu}). Equating tensile and compressive stress resultants, we have

$$\text{Depth of stress block, } a = \beta_1 c = \frac{\phi_{frp} A_{frp} f_{frp}}{\alpha_1 \phi_c f'_c b} \tag{4.56}$$

Using strain compatibility (see Fig. 4.8b), the actual strain in FRP at failure of the over-reinforced section could be derived as follows:

$$\frac{\varepsilon_{frp}}{\varepsilon_{cu}} = \frac{d-c}{c} \quad \Rightarrow \quad \varepsilon_{frp} = \varepsilon_{cu}\frac{\beta_1 d - \beta_1 c}{\beta_1 c} \tag{4.57}$$

The actual stress in FRP (f_{frp}) in over-reinforced section can be obtained using Eq. (4.58) or (4.59). In the above expression, the parameters, d and c, refer to effective depth of section and depth to the neutral axis, respectively

$$f_{\text{frp}} = E_{\text{frp}}\varepsilon_{\text{frp}} = E_{\text{frp}}\varepsilon_{cu}\frac{\beta_1 d - a}{a} \tag{4.58}$$

Substititing the value of stress block depth, a (from Eq. (4.56)) into Eq. (4.58) and rearranging the terms, we can express the stress in the FRP at ultimate compression failure by Eq. (4.59).

$$f_{frp} = \frac{1}{2}E_{frp}\varepsilon_{cu}\left[\left(1+\frac{4\alpha_1\beta_1\phi_c f_c'}{\rho_{frp}\phi_{frp}E_{frp}\varepsilon_{cu}}\right)^{1/2} - 1\right] \tag{4.59}$$

After obtaining the unknown stress (f_{frp}) in the FRP rebars and depth of stress block, a, the flexural strength of over-reinforced section can be evaluated as per Eq. (4.60).

$$M_r = \phi_{frp}A_{frp}f_{frp}\left(d - \frac{a}{2}\right) \tag{4.60}$$

Tension failure: The tension failure occurs by FRP rupture before crushing of concrete in an under-reinforced section having actual reinforcement ratio (ρ_{frp}) which is less than the balanced reinforcement ratio, (ρ_{frp}). In this case, the strain in concrete at failure is less than ε_{cu} = 0.0035, and strain in FRP is given by Eq. (4.61).

$$\varepsilon_{frp} = \varepsilon_{frpu} = \frac{f_{frpu}}{E_{frp}} \tag{4.61}$$

It may be noted that stress block parameters α_1 and β_1 expressed, respectively, by Eqs. (4.51) and (4.52) cannot be used as the concrete in the compression zone is not at ultimate conditions. Eqs. (4.51) and (4.52) are valid only when $\varepsilon_c = \varepsilon_{cu}$ Hence, in the case of tension failure associated with under-reinforced section, modified stress block parameters, α and β can be determined either from tabulated values available in ISIS Design Manual No. 3 or from Figs. 4.9 and 4.10, respectively. Both these figures give the values of α and β as functions of strain in concrete for different grades of concrete. Thus for a given grade of concrete, an initial trial depth of the neutral axis could be assumed and values of α and β could be determined. Hence, the tensile and compressive stress resultants can be determined as given in Eqs. (4.62) and (4.63).

$$T = \phi_{frp}A_{frp}f_{frpu} = \phi_{frp}A_{frp}\varepsilon_{frpu}E_{frp} \tag{4.62}$$

$$C = \alpha\phi_c f_c'\beta cb \tag{4.63}$$

Also, for equilibrium condition to be satisfied,

$$C = T \tag{4.64}$$

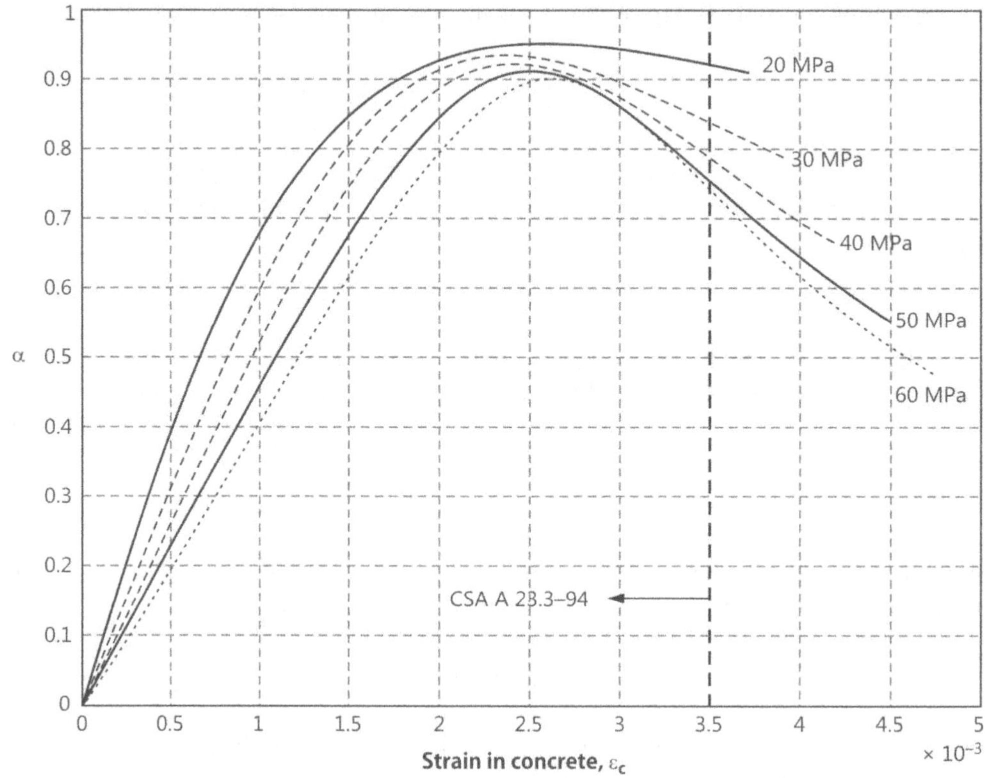

Figure 4.9. Equivalent stress-block parameter α for concrete.

If equilibrium condition (Eq. (4.64)) is not satisfied, then a new value of the depth of the neutral axis is assumed, α and β are re-evaluated and Eq. (4.64) is rechecked. This process is repeated till the equilibrium condition is satisfied. The updated neutral axis depth in each iteration can be determined using Eq. (4.65).

$$c = \frac{\phi_{frp} A_{frp} \varepsilon_{frpu} E_{frp}}{\alpha \phi_c f_c' \beta b} \tag{4.65}$$

where, α and β are determined at the following concrete strain determined from the strain compatibility equation (Eq. (4.66)).

$$\varepsilon_c = \varepsilon_{frpu} \frac{c}{d-c} \tag{4.66}$$

After finding the value of, c, and the associated values of α and β, the tensile and compressive stress resultants can be obtained. Finally, the moment of resistance (Eq. (4.67)) can be determined by taking moments about the compressive stress resultants.

$$M_r = \phi_{frp} A_{frp} f_{frpu} \left(d - \frac{\beta c}{2} \right) \tag{4.67}$$

Minimum flexural resistance: To ensure that the beam does not fail immediately after the cracking, the section should be provided with at least minimum tensile reinforcement. To satisfy this condition, three criteria have been suggested by ISIS Canada Design Manual No. 3. These three conditions or criteria are outlined as follows:

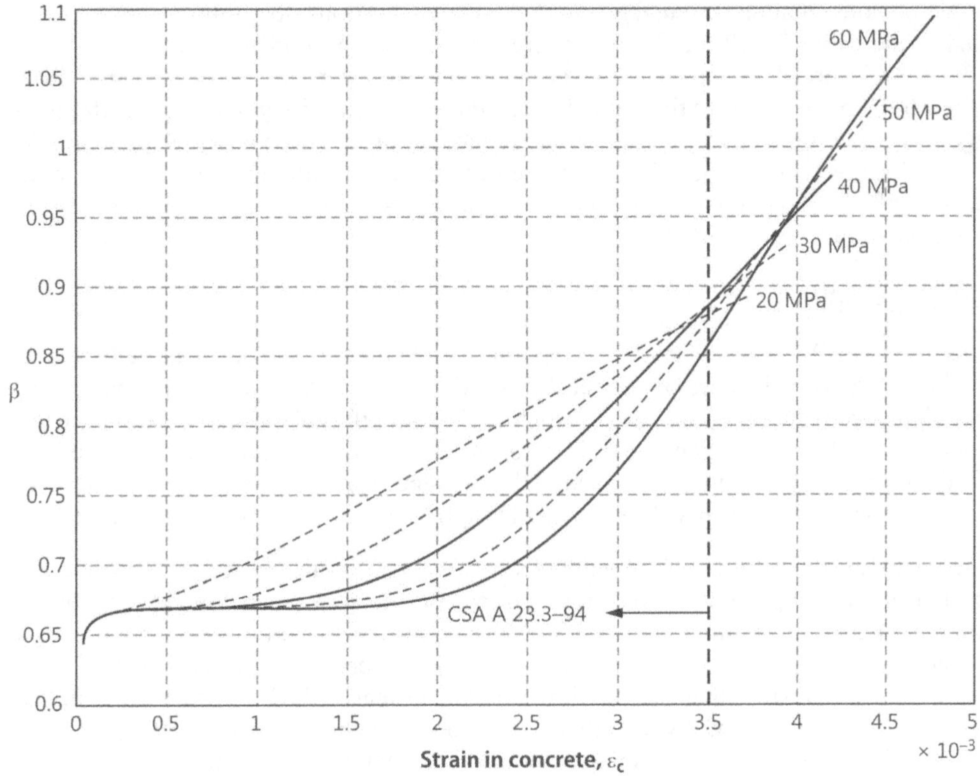

Figure. 4.10. Equivalent stress-block parameter *β* for concrete.

1. Failure of a member immediately after cracking, which occurs suddenly and without warning should be avoided. Thus, the moment of resistance, M_r, of an FRP reinforced concrete member should be at least 50% greater than the cracking moment, M_{cr}, as expressed by Eq. (4.68).

$$M_r \geq 1.5M_{cr} \tag{4.68}$$

   The cracking moment, M_{cr}, is given by Eq. (4.69).

$$M_{cr} = \frac{f_r I_t}{y_t}; \quad \text{where, } f_r = 0.6\sqrt{f_c'} \tag{4.69}$$

   In Eq. (4.69), f_r refers to the modulus of rupture in MPa; f'_c refers to cylinder strength of concrete in MPa; I_t refers to transformed moment of Inertia of cross-section, whereas y_t refers to the distance of extreme tensile fiber distance from the neutral axis.

2. If Eq. (4.68) is not satisfied, then the moment of resistance, M_r, must be at least 50% greater than the moment due to factored loads, M_f, as expressed by Eq. (4.70).

$$M_r \geq 1.5M_f \tag{4.70}$$

3. Finally, for a rectangular section, Eq. (4.71) can be used to satisfy the minimum reinforcement requirements.

$$A_{frp,\min} = \frac{5\sqrt{f_c'}}{12 f_{frpu}} bd \tag{4.71}$$

It must be noted that unlike steel reinforced concrete sections, the lumping of reinforcement is not permitted in the case of FRP reinforced concrete beams having reinforcement in two or more layers. Thus, the FRP reinforced sections with two or more than two layers of FRP rebars, strain in the outermost layer of reinforcements/rebars is the critical one. Hence, such multi-layered FRP reinforced sections can be easily designed on the basis of strain in the outermost layer of rebars by using strain compatibility. In the case of beams with compression reinforcement, the contribution of compression reinforcement to flexural strength of the section is usually neglected, as FRPs are weak in compression especially Aramid Fiber Reinforced Polymer (AFRP).

4.3.2. Serviceability

In the design of FRP reinforced concrete flexural members, the cracking and deflection are the main factors with regard to limit states of serviceability. Although the strength of FRP reinforcement is much more than that of conventional steel reinforcements, their modulus of elasticity is less. The lower modulus of elasticity of FRP reinforcements may lead to the formation of larger cracks and deflections in FRP reinforced concrete members in comparison to that for steel reinforced concrete members. The two main limit state of serviceability (as per ISIS Canada Design Manuals) are described in the following:

Cracking: Unlike steel reinforced concrete section where it is necessary to control crack widths for aesthetic reasons and to prevent corrosion, in the case of FRP reinforced concrete sections crack control is required primarily for aesthetic reasons and to prevent creep-rupture of FRP rebars. The corrosion of FRP rebars is not the issue in crack control. As a conservative approach, ISIS design guidelines suggest to control cracking, that the maximum strain in tensile FRP reinforcement at service should not exceed 0.2% as expressed by Eq. (4.72). To calculate the strain in FRP at service load levels, the concept of transformed section should be used for uncracked section. However, if the section is cracked, the effective moment of inertia can be used as discussed in sub-section later in this section.

$$\varepsilon_{frps} \leq 0.002 \tag{4.72}$$

Deflection: There are two ways to control the deflection of FRP reinforced concrete sections.

1. Direct computation of deflection using equations of structural mechanics wherein moment of inertia of section is replaced by effective moment of inertia (see Eq. (4.73)).
2. Indirect method of controlling deflection is by limiting span-to-depth ratio of member (see Eq. 4.76).

Effective moment of inertia: Under service loads, deflection requirements of uncracked section can be checked using the concept of transformed section whereas, for cracked section, the effective moment of inertia (Eq. (4.73)) should be used. Eq. (4.73) to calculate effective moment of inertia of a cracked section has been derived empirically based on test data available on FRP-Reinforced Concrete Members.

$$I_e = \frac{I_t I_{cr}}{I_{cr} + \left(1 - 0.5\left(\frac{M_{cr}}{M_a}\right)^2\right)(I_t - I_{cr})} \tag{4.73}$$

where, I_{cr} is the moment of inertia of the cracked section transformed to concrete with concrete in tension ignored, calculated using the following Eq. (4.74) below [mm^4]

I_t is the moment of inertia of a non-cracked section transformed to concrete [mm^4]

M_{cr} is the cracking Moment [N·mm]

M_a is the maximum moment in a member at the load stage at which deflection is being calculated [N·mm]

$$I_{cr} = \frac{bc^3}{3} + n_{frp} A_{frp} (d-c)^2 \tag{4.74}$$

where, b is the width of the compression zone [mm]
d is the effective depth of the section [mm]
n_{frp} is the m odular ratio E_{frp}/E_c

The neutral axis depth, c, can be calculated using the following equation (Eq. (4.75))

$$c = d\left[-n_{frp}\rho_{frp} + \sqrt{\left(n_{frp}\rho_{frp}\right)^2 + 2n_{frp}\rho_{frp}}\right] \tag{4.75}$$

where, ρ_{frp} is the FRP reinforcement ratio.

Minimum depth: To indirectly control the deflection of an FRP reinforced flexural member, the actual span-to-depth ratio of the member should not exceed the limiting value of span-to-depth ratio of FRP reinforced concrete members related with that for corresponding steel reinforced concrete members by Eq. (4.76).

$$\left(\frac{\ell_n}{h}\right)_{frp} = \left(\frac{\ell_n}{h}\right)_s \left(\frac{\varepsilon_s}{\varepsilon_{frp}}\right)^{\alpha_d} \tag{4.76}$$

where, ℓ_n is the member length [mm]
h is the member thickness [mm]
ε_s is the maximum strain allowed in the steel reinforcement in service
ε_{frps} is the maximum strain allowed in the FRP reinforcement in service
α_d is a dimensionless coefficient taken as 0.50 for a rectangular section

Note: The ratio $(\ell_n/h)_s$ is the equivalent ratio for steel-reinforced concrete and is obtained from Table 9.1 of CSA A23.3-94.

Deformability: Since, FRP reinforcement has a linear strain-stress relationship, hence, there is no plastic deformation—as observed in steel reinforcement in FRP reinforcement. However, because of the low modulus of elasticity of FRP reinforcements especially AFRP and GFRP, a FRP reinforced member exhibits large curvature at failure. As per the recommendation of ISIS Canada Design Manual No. 3, FRP reinforcement can be less than the balanced FRP reinforcement ratio, provided curvature at service loads is reasonably a low proportion of the curvature at ultimate. This concept is defined as deformability (DF) of FRP reinforced members and can be expressed by Eq. (4.77) for rectangular and T-beams in flexure.

$$DF = \left(\frac{\psi_u M_u}{\psi_s M_s}\right) \geq 4 \tag{4.77}$$

In Eq. (4.77), ψ_u and M_u are curvature and moment at ultimate conditions, respectively, while ψ_s and M_s are curvature and moment at service conditions which correspond to $\varepsilon_{frps} = 0.002$.

Table 4.4. Concrete cover to FRP flexural reinforcement.

Exposure condition	Beams	Slabs
Interior	$2.5d_b^*$ or 40 mm	$2.5d_b$ or 20 mm
Exterior	$2.5d_b$ or 50 mm	$2.4d_b$ or 30 mm

* d_b is the bar diameter.

Concrete cover requirements: The concrete cover is required to prevent cracking due to thermal expansion, swelling from moisture and to protect the FRP reinforcement from fire. As per ISIS Canada Design guidelines, the concrete cover requirements for beams and slabs for interior and exterior exposure conditions are presented in Table 4.4.

Spacing of bars: For easy placement of concrete through reinforcement cages and to avoid the cracking caused due to temperature, the minimum bar spacing for longitudinal main reinforcing bars in FRP reinforced concrete members should be taken as the maximum of the followings:

- 1.4 times the diameter of reinforcing bars;
- 1.4 times the maximum aggregate size;
- 30 mm;
- The concrete cover as given in Table 4.4.

The maximum spacing of the flexural reinforcement should be taken as the smaller of the followings:

- 5 times the slab thickness
- 500 mm

Additional features related to constructability: When designing with FRP reinforcement, the following additional points must be taken into considerations:

1. All FRP materials should be protected against ultra-violet radiation.
2. Storage and handling requirements for FRPs may vary significantly depending on the specific product being used.
3. FRPs should not come into contact with reinforcing steel in a structure.
4. FRP reinforcement is light and must be tied with plastic ties to formwork to prevent it from floating during concrete placing and vibrating operations.
5. Care must be taken when vibrating concrete, to ensure that the FRP reinforcement is not damaged, i.e., plastic protected vibrators could be used.

4.4. Design Approach for CFRP Prestressed Concrete Bridge Beams

In this section, a unified design approach for FRP concrete bridge beams prestressed using bonded pretensioning and unbonded post-tensioning tendons, arranged in multiple vertically distributed layers along with non-prestressing CFRP rods has been presented. Design equations to determine the flexural capacity, and to compute the stresses and strains in concrete and tendons are provided. In addition, based on parabolic stress-strain relation for concrete and bi-linear stress-strain relation for tendons, an analysis methodology has been presented to predict the non-linear response of FRP prestressed bridge beams and girders.

4.4.1. Theoretical Development of Design Equations

The design equations for FRP reinforced concrete (RC) beams prestressed with multiple layers of internally bonded and externally unbonded CFRP tendons are developed in the following sections. These equations are valid for evaluation of the flexural capacity of beams with typical cross-sections such as—double-T (DT), box and AASHTO-I beam sections. The design is based on balanced ratio, which characterizes the beam sections into under-reinforced and over-reinforced sections. The resultant compression force in concrete is obtained using equivalent stress block factors, which give the same resultant compression force and its location as obtained using parabolic stress-strain relationship.

Definition of material characteristics: The guaranteed ultimate tensile strength, guaranteed rupture strain and modulus of elasticity of tendons, must be specified by the designer and compared against the data provided by the manufacturers. The guaranteed strength refers to the mean break strength reduced by three standard deviations. Similarly, the guaranteed rupture strain refers to the mean rupture strain reduced by three standard deviations. The specified guaranteed tensile strength of FRP tendons are based on at least five test specimens. Following are the basic steps to be followed for the flexural design of prestressed beam with bonded and unbonded CFRP tendons strands.

1. Compute the required moment capacity: Let the dead load moment = M_D and Live load moment = M_L.

$$\text{Required moment capacity of the beam, } M_{\text{required}} = 1.2M_D + 1.6M_L \tag{4.78}$$

2. Select and proportion the cross-section: Select and proportion the cross-section of the beam and specify the number and arrangement of the bonded prestressing tendons, unbonded post-tensioning tendons and non-prestressing tendons in tension and compression zones. The maximum prestress force in the CFRP tendons should be limited to 50%–65% of the specified tensile strength of tendons.

3. Compute balanced ratio (ρ_b): The balanced ratio as expressed in the following lines is based on strain compatibility in the cross-section and signifies the reinforcement ratio at which simultaneous failure of concrete in compression and rupture of the bottom-bonded prestressing tendons occurs. This balanced ratio is based on four basic assumptions: (a) the ultimate compression strain (ε_{cu}) is 0.003, (b) the nonlinear behavior of concrete is modeled using an equivalent rectangular stress block, (c) tendon failure occurs at the ultimate tensile strain of tendon, ε_{fu} and (d) equivalent prestressing tendon is located at the centroid of multiple layers of prestressing tendons of the same material properties.

$$\rho_b = 0.85\beta_1 \frac{f_c'}{f_{fu}} \frac{\varepsilon_{cu}}{\varepsilon_{cu} + \varepsilon_{fu} - \varepsilon_{pbmi}} \tag{4.79}$$

where, β_1 = factor defined as the ratio of the depth of equivalent rectangular stress block to the distance from the extreme compression fiber to the neutral axis

f_c' = specified compressive cylinder strength of concrete

f_{fu} = specified tensile strength of bonded prestressing tendons

ε_{pbmi} = initial prestressing strain in bonded prestressing tendons of *m*th row (bottom row)

It is to be noted that the above balanced ratio is based on material properties of bonded prestressing tendons, with the assumption that bonded prestressing tendons are susceptible to early failure. Here, the actual reinforcement ratio of the section is defined as, the ratio of total weighted cross-sectional area of tendons to the effective concrete cross-sectional area. Weighting factor (α_i) is defined as ratio of the stress in a particular equivalent tendon, i.e., a tendon located at the centroid of tendons of the same material and having the cross-sectional area equal to the total cross-sectional area of corresponding tendons at the balanced condition to the specified ultimate strength of bonded pretensioning tendons. Here, balanced condition refers to condition at which crushing of concrete and rupture of bottom pretensioning tendons occurs simultaneously. The following expression should be used for calculating the reinforcement ratio. This expression is obtained using equilibrium of forces and compatibility of strains in the cross-section.

$$\rho = \frac{\sum_{i=1}^{p} A_{fi}\alpha_i}{bd_m} \tag{4.80}$$

where, $\alpha_i = \dfrac{f_{bi}}{f_{fu}}$

A_{fi} = cross-sectional area of reinforcement of a particular material (A_{fi} is positive for tensile reinforcement and negative for compression reinforcement)

b = flange width of the beam

f_{bi} = total stress in an equivalent tendon of a specific material at the balanced condition

p = total number of reinforcing materials

d_m = distance of centroid of bottom prestressing tendons from the extreme compression fiber

4. Compute cracking moment (M_{cr}) of section: The cracking moment can be found using the concept that stress in the extreme tensile fiber of the prestressed section under the superimposed moment (equal to cracking moment) should be equal to the modulus of rupture (f_r) of concrete:

$$M_{cr} = (f_r + \Sigma\sigma_{bp})S_b \tag{4.81}$$

where, $f_r = 0.498\sqrt{f_c'}$ (MPa)

$\Sigma\sigma_{bp}$ = resultant stress at extreme tension fiber of the beam due to effective pre-tensioning and post-tensioning forces.

S_b = section modulus corresponding to extreme tension fiber

5. Calculate flexural capacity: The flexural capacity of the DT-beam, box-beam and AASHTO-I beam provided with prestressing bonded and unbonded tendons arranged in vertically distributed multiple layers and classified as 'significantly under reinforced', 'under-reinforced' and 'over-reinforced' beams can be determined using the following approach. These beams are used in the construction of prestressed concrete bridges. Figure 4.11 shows the typical DT-beam, box-beam and AASHTO-I beam sections with tendons arranged in vertically distributed multiple layers.

A. *Significantly under-reinforced beams:* The reinforcement ratio (ρ) of significantly under-reinforced beams lies below $0.5\rho_b$. The failure of beams will occur due to the rupture of bottom prestressing tendons. The compressive stress distribution at ultimate for a typical significantly under-reinforced DT-beam will be triangular because compressive stresses in such section will be within linear elastic range. The strain distribution will be similar to that shown in Fig. 4.12.

Assume that there are *m* rows of pre-tensioning tendons and *k* rows of non-prestressing tendons arranged vertically in the webs of the DT-beam, box-beam and AASHTO-I beam sections. The first row lies at the top, while the *m*th and *k*th rows lie near the bottom of beams. The depth to neutral axis is defined as $n = k_u d_m$, where co-efficient k_u for the significantly under-reinforced section is defined by Eq. (4.81). This equation is derived using compatibility and equilibrium equations.

$$k_u = \frac{-B + \sqrt{B^2 - 4AC}}{2A} \tag{4.81}$$

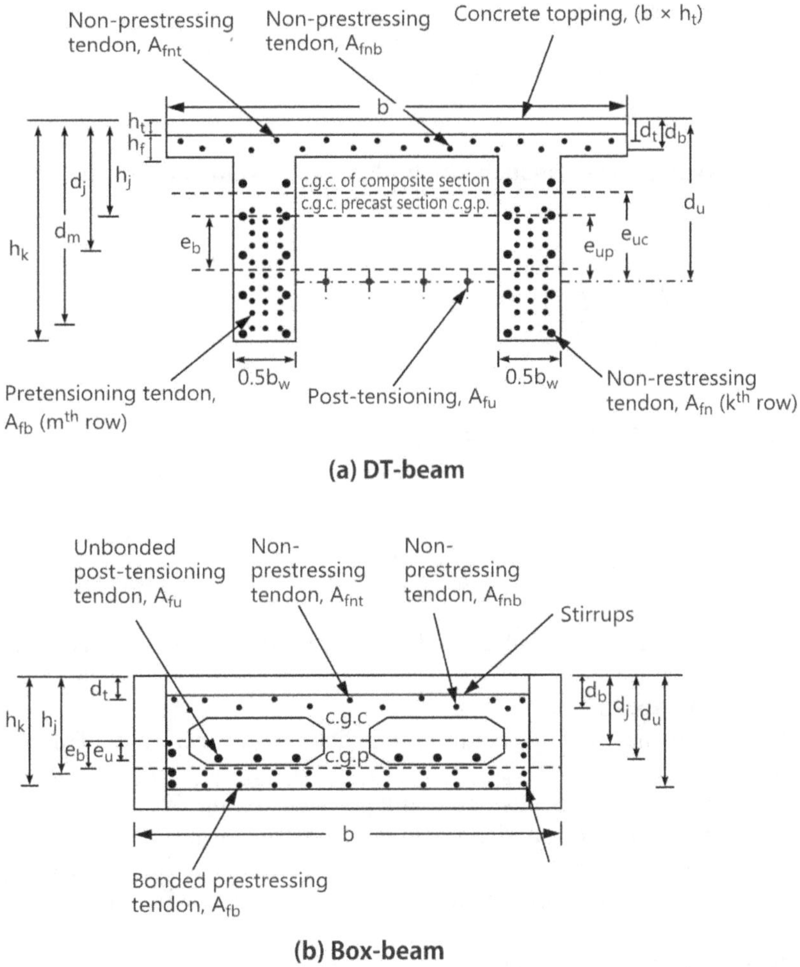

Figure 4.11. (Continued)

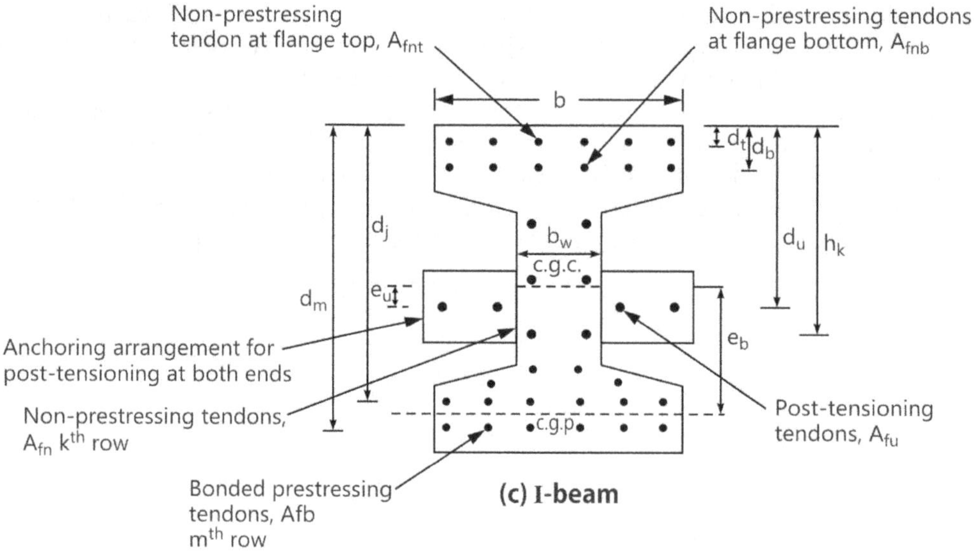

Figure 4.11. Typical cross-sections for bridge beams with bonded and unbonded tendons.

$$\text{where, } A = \frac{b d_m^2}{2} f_{fu}\left(1 - \frac{f_{pbmi}}{f_{fu}}\right)\frac{E_c}{E_f}$$

$$B = \left[F_{pi} d_m + \varepsilon_f d_m\left(\sum_{i=1}^{q} A_{fi}E_{fi} + \Omega_c A_{fu} E_{fp}\right)\right]$$

$$C = -\left[F_{pi} d_m - \varepsilon_f\left(\sum_{i=1}^{q} A_{fi}E_{fi}h_i + \Omega_c A_{fu} E_{fp} d_u\right)\right]$$

A_{fi} = cross-sectional area of bonded tendons in a particular layer (it is negative for tendons in compression zone),

b = flange width of beam,

d_m = distance from the extreme compression fiber to the centroid of the bottom prestressing tendons,

F_{pi} = total initial effective pre-tensioning and post-tensioning force,

E_{fi} = modulus of elasticity of bonded tendons in a particular layer,

h_i = distance of bonded tendons of an individual layer from the extreme compression fiber,

ε_f = difference in ultimate rupture strain and initial prestressing strain of bottom prestressing tendons,

q = number of layers of bonded tendon

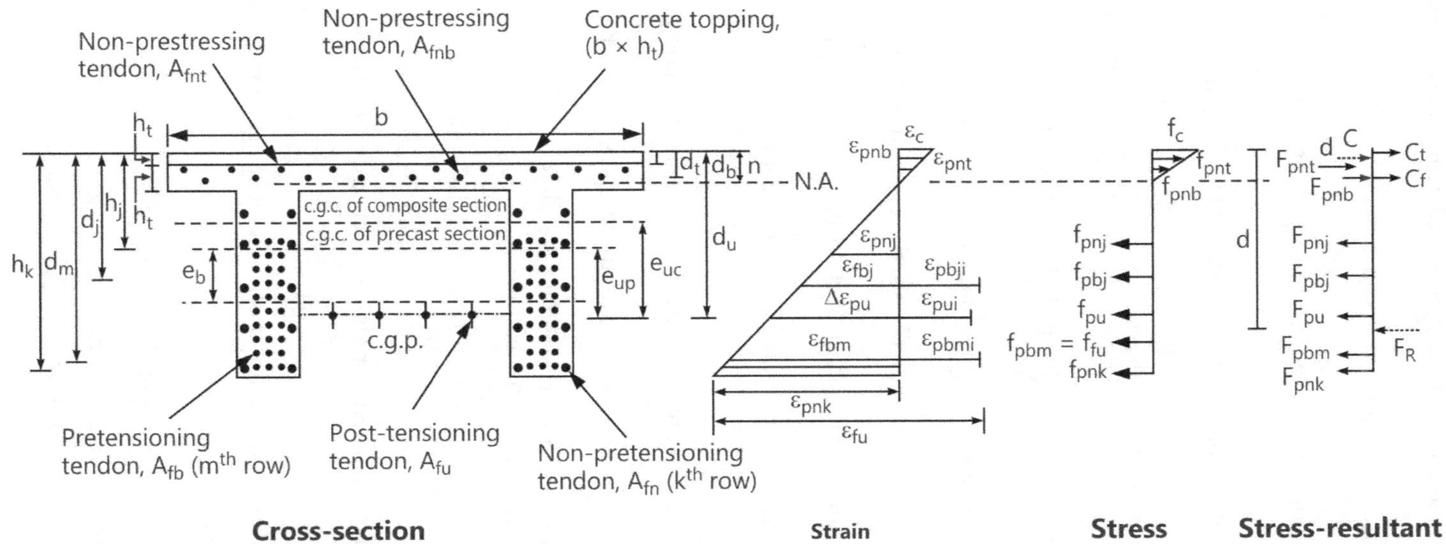

Figure 4.12. Strain and stress distributions for very under-reinforced beam at ultimate.

The following steps are taken to compute the moment carrying capacity of the beam:

1. Compute strains in bonded tendons and concrete:

Strain in prestressing tendons of an individual row, $\varepsilon_{pbj} = (\varepsilon_{fu} - \varepsilon_{pbmi})\dfrac{(d_j - n)}{(d_m - n)} + \varepsilon_{pbji}$ (for $j = 1, m$)

Strain in non-prestressing tendons of an individual row, $\varepsilon_{pnj} = (\varepsilon_{fu} - \varepsilon_{pbmi})\dfrac{(h_j - n)}{(d_m - n)}$ (for $j = 1, k$)

Strain in non-prestressing tendons of flange top, $\varepsilon_{pnt} = (\varepsilon_{fu} - \varepsilon_{pbmi})\dfrac{(n - d_t)}{(d_m - n)}$

Strain in non-prestressing tendons of flange bottom, $\varepsilon_{pnb} = (\varepsilon_{fu} - \varepsilon_{pbmi})\dfrac{(n - d_b)}{(d_m - n)}$

Strain in concrete at the extreme compression fiber, $\varepsilon_c = (\varepsilon_{fu} - \varepsilon_{pbmi})\dfrac{n}{(d_m - n)}$

where, d_j = depth from extreme compression fiber to the centroid of prestressing tendons of an individual row,

h_j = depth from extreme compression fiber to the centroid of non-prestressing tendons of an individual row,

h_k = depth from extreme compression fiber to the centroid of bottom non-prestressing tendons,

d_b = depth from extreme compression fiber to the centroid of non-prestressing tendons at the bottom of flange,

d_t = depth from extreme compression fiber to the centroid of non-prestressing tendons at the top of flange,

n = depth to the neutral axis from the extreme compression fiber.

2. Compute strain in unbonded post-tensioning tendons:

Strain in the unbonded post-tensioning tendons, $\varepsilon_{pu} = \varepsilon_{pui} + \Delta\varepsilon_{pu}$

$$\Delta\varepsilon_{pu} = \Omega_c \frac{(\varepsilon_{fu} - \varepsilon_{pbmi})(d_u - n)}{(d_m - n)}$$

where, Ω_c = bond reduction co-efficient for elastic cracked section (Grace and Singh, 2003)

$= \Omega \dfrac{I_{cr}}{I_{tr}}$

Ω = bond reduction co-efficient for elastic uncracked section (Grace and Singh, 2003)

$= \dfrac{1}{2}$ for one-point loading

$= \dfrac{1}{2}$ for two-point loading or uniform loading

I_{tr} = gross transformed moment of inertia of cross-section

I_{cr} = gross transformed moment of inertia of cracked section

It is to be noted that bond reduction co-efficients are introduced to take into account the lower strain in unbonded tendons with respect to equivalent bonded tendons. Also, it is assumed that the section is not cracked under the service load condition.

3. Compute stresses in tendons:

Stress in bonded prestressing tendons of an individual row, $f_{pbj} = E_f \times \varepsilon_{pbj} \leq f_{fu}$

Stress in unbonded post-tensioning tendons, $f_{pu} = E_{fp} \times \varepsilon_{pu} \leq f_{fup}$

Stress in non-prestressing tendons of an individual row, $f_{pnj} = E_{fn} \times \varepsilon_{pnj} \leq f_{fun}$

Stress in non-prestressing tendons at flange top, $f_{pnt} = E_f \times \varepsilon_{pnt} \leq f_{fu}$

Stress in non-prestressing tendons at flange bottom, $f_{pnb} = E_f \times \varepsilon_{pnb} \leq f_{fu}$

4. Compute resultant forces in tendons:

Resultant force in bonded prestressing tendons of each row, $F_{pbj} = f_{pbj} \times A_{fb}$

Resultant force in unbonded post-tensioning tendons, $F_{pu} = f_{pu} \times A_{fu}$

Resultant force in non-prestressing tendons of each row, $F_{pnj} = f_{pnj} \times A_{fn}$

Resultant force in non-prestressing tendons at flange top, $F_{pnt} = f_{pnt} \times A_{fnt}$

Resultant force in non-prestressing tendons at flange bottom, $F_{pnb} = f_{pnb} \times A_{fnb}$

where, A_{fb} = total cross-sectional area of bonded prestressing tendons in each row,
A_{fn} = total cross-sectional area of non-prestressing tendons in each row,
A_{fnt} = total cross-sectional area of compression non-prestressing tendons at flange top,
A_{fnb} = total cross-sectional area of compression non-prestressing tendons at flange bottom,
A_{fu} = total cross-sectional area of unbonded tendons,

5. Compute ultimate moment carrying capacity:

Let the centroid of the resultant (F_R) of tensile forces F_{pbj} (for $J = 1, m$), F_{pnj} (for $J = 1, k$) and F_{pu} lies at a distance, d and the centroid of resultant compression force (C) lies at a distance $\bar{d}$ from the extreme compression fiber. Nominal moment of resistance (M_n) is expressed as below:

$$M_n = F_R(d - \bar{d}) \tag{4.82}$$

where

$$F_R = \sum_{j=1}^{m} F_{pbj} + \sum_{j=1}^{k} F_{pnj} + F_{pu} \tag{4.83}$$

Design moment capacity, M_u, is expressed as follow:

$$M_u = \phi M_n \tag{4.84}$$

where, ϕ = strength reduction factor; $\phi = 0.85$ for CFRP tendons (Grace and Singh, 2003).

6. Compare design moment capacity and required moment capacity:

$$M_u \geq M_{\text{required}}, \text{ i.e., } \phi M_n \geq M_{\text{required}} \tag{4.85}$$

7. Check for stresses in concrete:

$$f_c = f_{fu}\left(1 - \frac{f_{pbmi}}{f_{fu}}\right)\left(\frac{n}{d_m - n}\right)\frac{E_c}{E_f} \tag{4.86}$$

E_c = modulus of elasticity of concrete
f_{pbmi} = initial effective prestress in the bottom tendons

$$f_c = \varepsilon_c \times E_c \leq 0.40 f_c' \tag{4.87}$$

where E_c can be taken as equal to $4730\sqrt{f_c'}$ MPa in the absence of experimental results. If $f_c' \geq 0.40 f_c'$, then moment capacity of beam should be calculated as per the procedure outlined here for under-reinforced beam.

B. *Under-reinforced beams:* The reinforcement ratio (ρ) of under-reinforced beams lies between $0.5\rho_b$ and ρ_b. In this class of beams, the rupture of bottom prestressing tendons governs failure. However, unlike significantly under-reinforced beams, the stress in the concrete at failure of under-reinforced beams will be within the non-linear range. The concrete stress distribution can be approximated by Whitney's rectangular stress block (Fig. 4.13). The strain distribution will be similar to that for the significantly under-reinforced beam. The depth to the neutral axis ($n = k_u d_m$) of under-reinforced beam can obtained using coefficient (k_u) defined by Eq. (4.88).

$$k_u = \frac{-B + \sqrt{B^2 - 4AC}}{2A} \tag{4.88}$$

where, $A = 0.85 f_c' b \beta_1 d_m^2$

$$B = -\left[A + F_{pi} d_m + \varepsilon_f d_m\left(\sum_{i=1}^{q} A_{fi} E_{fi} + \Omega_u A_{fu} E_{fp}\right)\right]$$

$$C = \left[F_{pi} d_m + \varepsilon_f\left(\sum_{i=1}^{q} A_{fi} E_{fi} h_i + \Omega_u A_{fu} E_{fp} d_u\right)\right]$$

1. Compute strains and stresses: The strains, stresses and forces in bonded tendons are calculated in the same manner as for significantly under-reinforced beams. However, computation of strains and stresses in the unbonded tendon is based on the ultimate bond reduction co-efficient (Ω_u), due to non-linearity of stress and strain relationship in the concrete. The resultant compressive force in concrete is computed using Whitney's equivalent rectangular stress block. The compression force resultants, C_t and C_f, are expressed as follows:

$$C_t = 0.85 f_c' \frac{E_{ct}}{E_c} b h_t \tag{4.89}$$

$$C_f = 0.85 f_c' b(\beta_1 n - h_t) \tag{4.90}$$

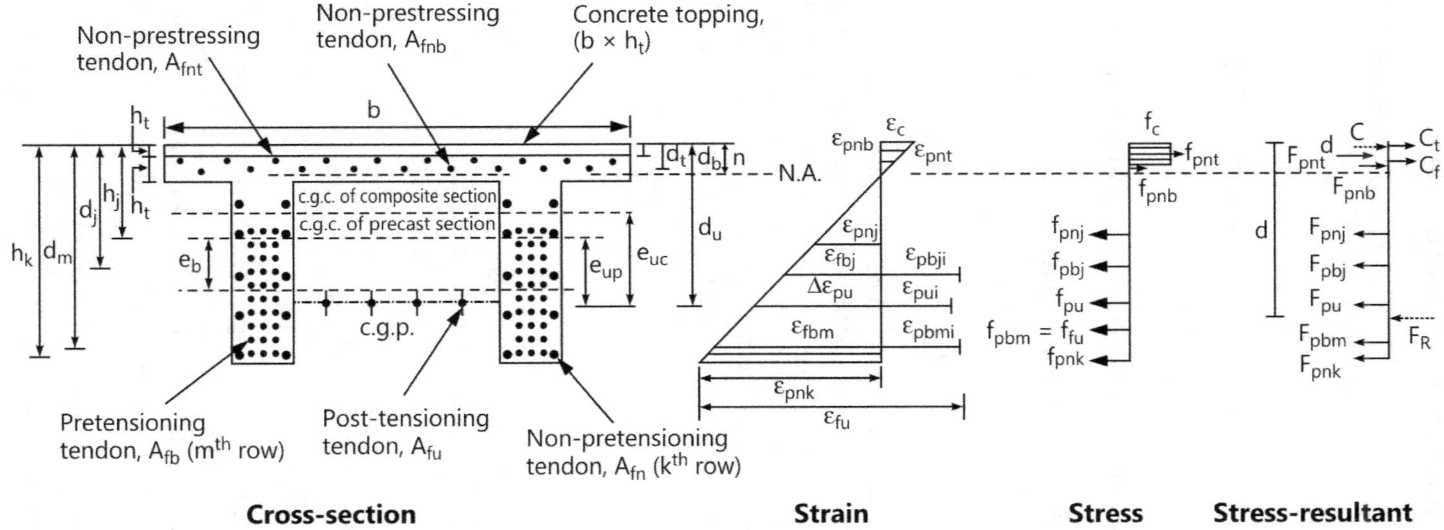

Figure 4.13. Strain and stress distributions for under-reinforced beam at ultimate.

The ultimate bond reduction coefficient is expressed as follows:

$$\Omega_u = \frac{2.6}{\left(\dfrac{L_u}{d_u}\right)} \quad \text{for one-point loading}$$

$$= \frac{5.4}{\left(\dfrac{L_u}{d_u}\right)} \quad \text{for two-point loading or uniform loading}$$

where, L_u = Horizontal distance between the ends of the post-tensioning strands.

2. Compute nominal and design moment capacities:

$$M_n = F_R(d - \bar{d}) \tag{4.91}$$

$$M_u = \phi M_n \geq M_{\text{required}} \tag{4.92}$$

C. *Over-reinforced beams:* For over-reinforced beams, reinforcement ratio ρ is greater than balanced ratio ρ_b. The failure of over-reinforced beams is governed by the crushing of concrete in the compression zone. The stress in the concrete at failure will be in the non-linear range and hence, the stress distribution can be approximated by Whitney's rectangular stress block (Fig. 4.14). In over-reinforced sections, the depth to the neutral axis ($n = k_u d_m$) can be determined using the co-efficient k_u defined in Eq. (4.93). This co-efficient is derived using equilibrium and compatibility equations for the section.

$$k_u = \frac{A + \sqrt{A^2 + 4B}}{2} \tag{4.93}$$

$$\text{where, } A = \frac{\left[\left(\sum_{j=1}^{m} A_{fb} E_f \varepsilon_{pbji} + A_{fu} \varepsilon_{pui} E_{fp}\right) - \varepsilon_{cu} \left(\sum_{j=1}^{q} A_{fj} E_{fj} + \Omega_u A_{fu} E_{fp}\right)\right]}{0.85 f_c' b \beta_1 d_m}$$

$$B = \frac{\varepsilon_{cu}\left(\sum_{j=1}^{q} A_{fj} E_{fj} h_j + \Omega_u A_{fu} E_{fp} d_u\right)}{0.85 f_c' b \beta_1 d_m^2}$$

A_{fj} = cross-sectional area of prestressing or non-prestressing bonded tendons in an individual row
E_{fj} = modulus of elasticity of bonded tendons of an individual row
h_j = distance of bonded tendons from the extreme compression fiber
q = total number of layers of bonded prestressing and non-prestressing tendons

1. Compute strains in tendons:

Strain in bonded prestressing tendons, $\varepsilon_{pbj} = \dfrac{0.003}{n}(d_j - n) + \varepsilon_{pbji}$ (for $j = 1, m$)

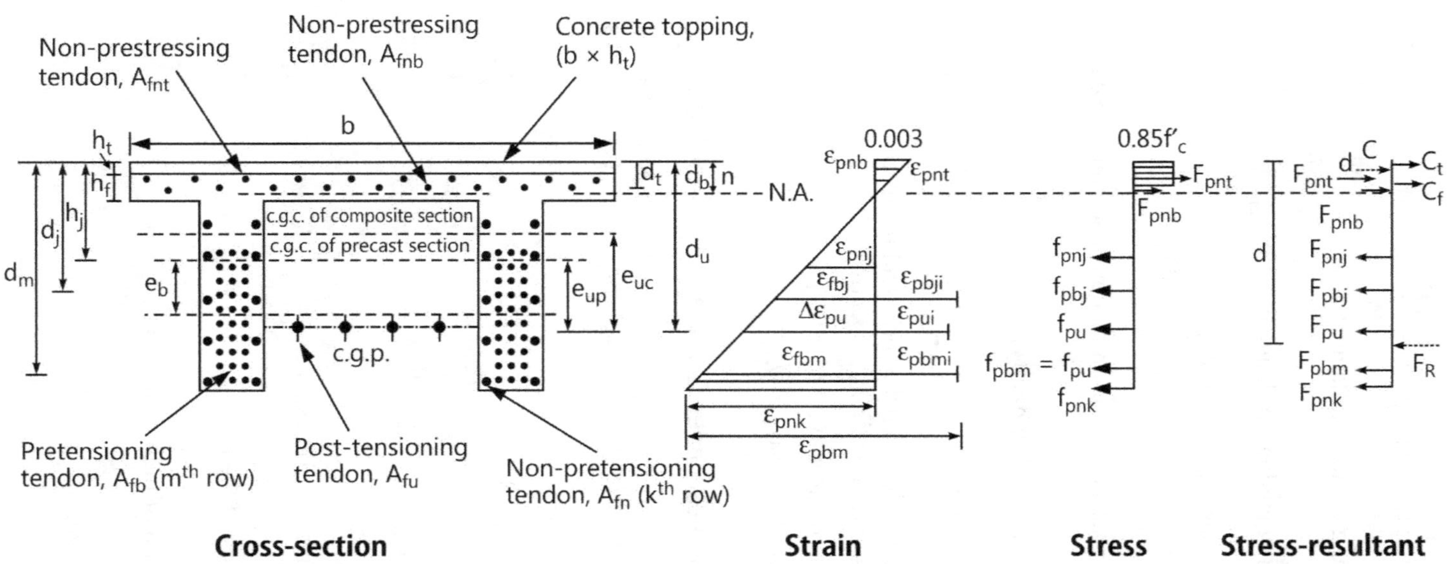

Figure 4.14. Strain and stress distributions for over-reinforced beam at ultimate.

Strain in non-prestressing tendons, $\varepsilon_{pnj} = \frac{0.003}{n}(h_j - n)$ (for $j = 1, k$)

Strain in non-prestressing tendons at flange top, $\varepsilon_{pnt} = \frac{0.003}{n}(n - d_t)$

Strain in non-prestressing tendons at flange bottom, $\varepsilon_{pnb} = \frac{0.003}{n}(n - d_b)$

Strain in unbonded tendons, $\varepsilon_{pu} = \varepsilon_{pui} + \Delta\varepsilon_{pu}$

$$= \varepsilon_{pui} + \Omega_u \frac{0.003}{n}(d_u - n)$$

It is to be noted that the stresses and forces in tendons and concrete can be calculated, as per equations for under-reinforced beams.

2. Compute the nominal moment capacity:

Resultant of the total tensile force, $F_R = \sum_{j=1}^{m} F_{pbj} + \sum_{j=1}^{k} F_{pnj} + F_{pu}$

Let the centroids of the resultant tensile force and resultant compression force lie at distances of d and $\bar{d}$ from the extreme compression fiber, respectively.

Ultimate nominal moment capacity, $M_n = F_R(d - \bar{d})$

Design flexural capacity, $M_u = \phi M_n \geq M_{\text{required}}$

4.4.2 Deflection and Stesses under Service Load Condition

Assuming that the service live load is applied through two-point loading system, the maximum beam deflection and tendon stresses prior to cracking load can be computed using the expressions given in the following:

Deflection due to applied load, $\delta_a = \frac{M_L L_1^2}{8E_c I_c}\left[\frac{8}{3} + 4\left(\frac{L_2}{L_1}\right) + \left(\frac{L_2}{L_1}\right)^2\right] \downarrow$ (4.94)

Deflection due to dead load, $\delta_d = \frac{5}{384}\frac{W_d L^4}{E_c I_c} \downarrow$ (4.95)

Deflection due to prestressing forces, $\delta_p = \frac{L^2}{8E_c I_c}\left[F_{\text{pre}} e_b + F_{\text{post}} e_u \left(\frac{L_u}{L}\right)^2\right] \uparrow$ (4.96)

Net downward deflection of the beam, $\delta = \delta_a + \delta_d - \delta_p \downarrow$ (4.97)

Stress in a bonded tendon of individual layer, $f_{pbj} = E_f \varepsilon_{pbji} + \frac{E_f}{E_c}\frac{M(d_j - y_{tc})}{I_c}$ (4.98)

Stress in unbonded tendon, $f_{pu} = E_{fp}\varepsilon_{pui} + \frac{E_{fp}}{E_c}\frac{(M - M_D)e_{uc}}{I_c}\Omega$ (4.99)

Stress in non-prestressing tendons, $f_{pn} = \frac{E_{fn}}{E_c}\frac{M(h_j - y_{tc})}{I_c}$ (4.100)

where, L_1 = distance between support and nearest load point
L_2 = longitudinal distance between two load points
L = effective span of the beam
M_L = service live load moment
e_b = eccentricity of resultant pretensioning force from the centroid of the section
e_u = eccentricity of unbonded tendons from the centroid of section

4.4.3. Nonlinear Response

The response of the beam under service load condition and before cracking can be determined using simple linear elastic beam theory. However, to predict the overall response of the beam from the onset of cracking of the section to the ultimate failure, a non-linear analysis is required. The non-linear stress and strain relationship for concrete can be modeled by the expression given in Eq. (4.101). Similar parabolic stress strain relationship for concrete was assumed by Tan *et al.* (2001), except, that they have taken ε_{cu} equal to 0.002. However, in the present investigation ε_{cu} has been taken as 0.003.

$$\frac{f_c}{f_c'} = 2.0\frac{\varepsilon_c}{\varepsilon_{cu}} - \left(\frac{\varepsilon_c}{\varepsilon_{cu}}\right)^2 \tag{4.101}$$

where, f_c = stress in concrete corresponding to strain ε_c.

The resultant compressive force in concrete based on non-linear stress-strain relation can be computed, using equivalent rectangular stress block factors at any load stage. The stress block factors can be determined, by equating the resultant compression force and its location obtained from non-linear stress-strain relation to that obtained by equivalent stress block. Equation (4.102) expresses the resultant compression force, while Eq. (4.103) can be used to locate the centroid of resultant compression force.

$$\int_0^n f_c\, b\, dy = \alpha f_c' \beta n b \tag{4.102}$$

$$\bar{y}_c = \frac{\int_0^n f_c\, by\, dy}{\int_0^n f_c\, b\, dy} \tag{4.103}$$

Using Eqs. (4.101), (4.102) and (4.102), the stress block factors for rectangular and flanged sections can be obtained using following expressions:

Rectangular section: For rectangular section, stress block factors (α_1, β_1) are evaluated using Eqs. (4.104a) and (4.104b).

$$\alpha_1\beta_1 = \frac{\varepsilon_t}{\varepsilon_{cu}} - \frac{1.0}{3.0}\left(\frac{\varepsilon_t}{\varepsilon_c}\right)^2 \tag{4.104a}$$

$$\beta_1 = \frac{4.0 - \dfrac{\varepsilon_t}{\varepsilon_{cu}}}{6.0 - \dfrac{2\varepsilon_t}{\varepsilon_{cu}}} \tag{4.104b}$$

where, ε_t = strain at the extreme compression fiber at a specific stage of loading

Flanged section: For a flanged section, the corresponding stress block factors (α_2, β_2) can be calculated using the following expressions:

$$\alpha_2 = \frac{A_1}{d_f b + (\beta_2 n - d_f) b_w} \tag{4.105a}$$

where, $$A_1 = \frac{\varepsilon_t}{n\varepsilon_{cu}}\left[(n-d_f)^2(b_w - b) + bn^2\right] - \frac{1}{3}\left(\frac{\varepsilon_t}{n\varepsilon_{cu}}\right)^2\left[(n-d_f)^3(b_w - b) + bn^3\right]$$

and β_2 can be evaluated by solving the following quadratic equation:

$$\beta_2^2 \frac{b_w n^2}{2} - \beta_2\left(b_w - n^2 - \frac{B_1}{A_1} b_w n\right) - d_f\left[\left(n - \frac{d_f}{2} - \frac{B_1}{A_1}\right)(b - b_w)\right] \tag{4.105b}$$

where, $$B_1 = \frac{2}{3}\frac{\varepsilon_t}{n\varepsilon_{cu}}\left[(n-d_f)^3(b_w - b) + bn^3\right] - \frac{1}{4}\left(\frac{\varepsilon_t}{n\varepsilon_{cu}}\right)^2\left[(n-d_f)^4(b_w - b) + bn^4\right]$$

Using linear elastic theory for computation of linear response and non-linear stress-strain relation of concrete for non-linear response, a computer program can be developed for predicting the overall load versus deflections, strains, stresses, forces in bonded and unbonded tendons and moment curvature relationships. The flow chart for computing the non-linear response of FRP prestressed concrete structure is given in Fig. 4.15.

E4.1. Design Example 1

A simply supported, normal weight concrete beam with f'_c equal to 27.6 MPa is needed in a medical facility to support MRI unit. The beam which is an interior beam is to be designed to carry a service live load of w_{LL} = 5.8 kN/m (20% sustained load) and a superimposed service dead load of w_{SDL} = 3.0 kN/m over a span of l = 3.35 m. The beam deflection should not exceed l/240, which is limitation for long-term deflection. Due to construction restriction, the depth of the member should not exceed 356 mm. GFRP reinforcing bars are selected to reinforce the beam;—material properties of the bars (as reported by the bar manufacturer) are shown in Table E4.1. Assume any other data. Use ACI 440.1R-06 standards.

Table E4.1. Manufacturer's reported GFRP bar properties.

Tensile strength, f^*_{fu}	620.6 MPa
Rupture strain, ε^*_{fu}	0.014
Modulus of elasticity, E_f	44 800 MPa

Design Procedure

The design procedure is presented in the following: Note that this procedure is equally applicable to other CFRP and AFRP bars.

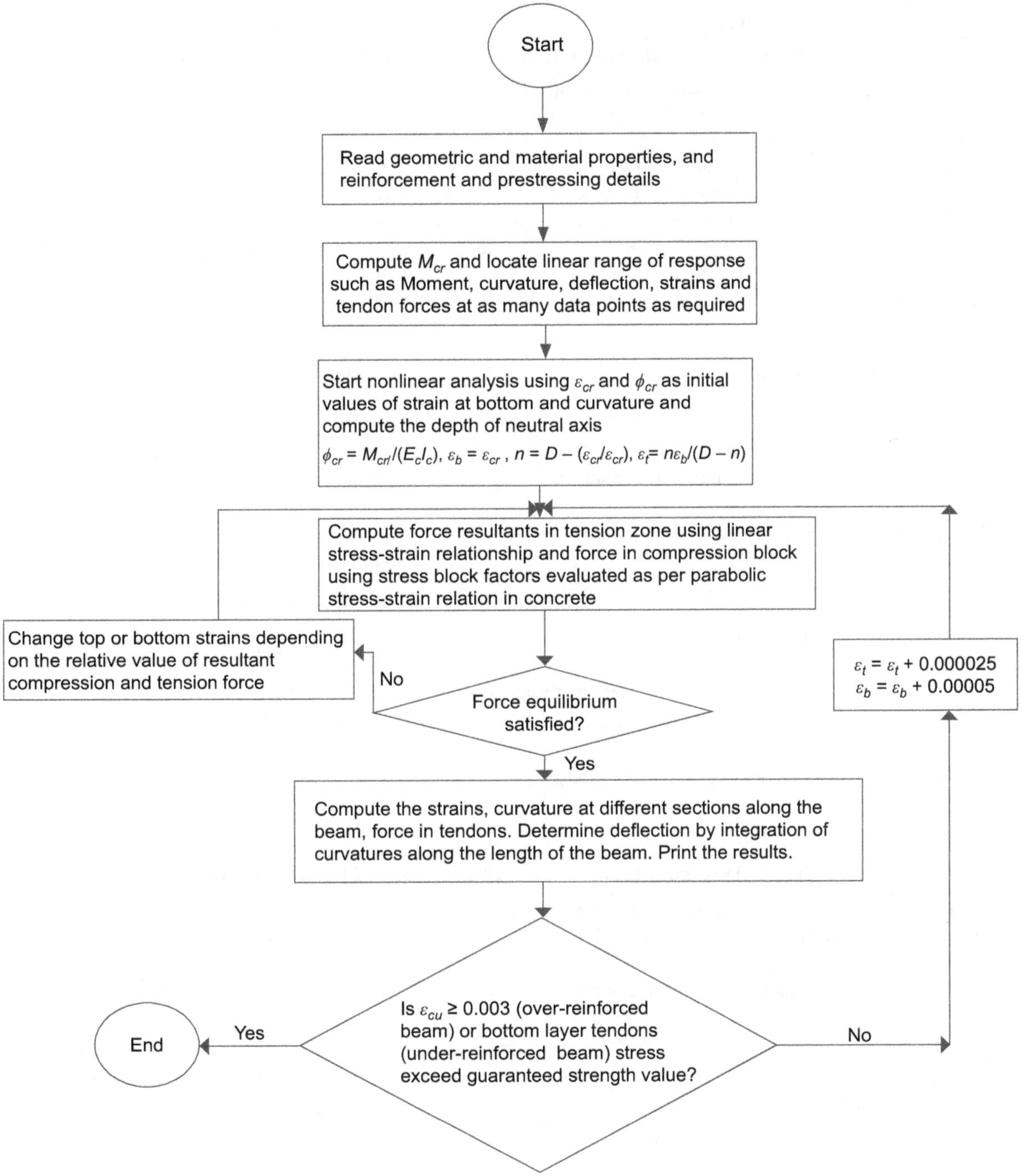

Figure 4.15. Flow chart for computation of linear and nonlinear response of prestressed beam.

Procedure	Calculation
Step 1: Estimate the appropriate cross-sectional dimensions of the beam	
An initial value for the depth of a simply supported reinforced concrete can be estimated from Table 8.2 of the ACI 318-95 Building Code. $h = l/16$ Recognizing that the lower stiffness of GFRP reinforcing bars will require greater depth than steel-reinforced concrete for deflection control, a larger overall height is used. The depth of the member is limited to 356 mm. An effective depth of the section is estimated using 38 mm clear cover: Estimated, $d = h - \text{cover} - d_{b,\text{shear}} - \frac{d_b}{2}$	$h = \frac{3.35}{16} = 0.209\,\text{m}$ Try $h = 305$ mm < 356 mm Assuming 2 Ø16 mm bars main and Ø9.5 mm bars for shear Clear concrete cover = 38 mm A minimum width of approximately 0.178 m is required when using 2 Ø16 mm or 2 Ø19 mm (2#5 or 2#6) bars with Ø9.5 mm (#3) stirrups Try $b = 0.178$ m $d = 305 - 38 - 9.5 - \frac{16}{2} = 250\,\text{mm}$
Step 2: Compute the factored load	
The uniformly distributed dead load can be computed including the self-weight of the beam: $w_{DL} = w_{SDL} + w_{sw}$ Compute the factored uniform load and ultimate moment $w_u = 1.2\, w_{DL} + 1.6\, w_{LL}$ $M_u = \frac{w_u l^2}{8}$	$w_{DL} = (3.0\text{ kN/m}) + 0.178\text{ m} \times 0.305\text{ m }(24\text{ kN/m}^3)$ $= 4.3$ kN/m $w_u = 1.2\,(4.3\text{ kN/m}) + 1.6\,(5.8\text{ kN/m})$ $w_u = 14.44$ kN/m $M_u = \frac{(14.44\text{ kN/m})\,(3.35)^2}{8}$ $= 20.3\,\text{kN/m}$
Step 3: Compute the design rupture stress of the FRP bars	
The beam will be located in an interior conditioned space. Therefore, for glass FRP bars, an environmental reduction factor (C_E) of 0.8 is used as per Table 4.1. $f_{fu} = C_E f^*_{fu}$	$f_{fu} = (0.80)(620.6\text{ MPa}) = 496$ MPa
Step 4: Determine the area of GFRP bars required for flexural strength	
Find the reinforcement ratio required for flexural strength by trial and error using Eqs. (4.8), (4.13) and (4.14). Assume an initial amount of FRP reinforcement: $\rho_f = \frac{A_f}{bd}$ $f_f = \left(\sqrt{\frac{(E_f \varepsilon_{cu})^2}{4} + \frac{0.85\beta_1 f'_c}{\rho_f} E_f \varepsilon_{cu}} - 0.5 E_f \varepsilon_{cu}\right)$	Try 2 Ø16 bars $\rho_f = \frac{400\text{ mm}^2}{(178\text{ mm})\,(250\text{ mm})}$ $= 0.009$ $f_f = \sqrt{\frac{(44800 \times 0.003)^2}{4} + \frac{0.85 \times 0.85 \times 27.6}{0.009} \times 44800 \times 0.003}$ $= -0.5 \times 44800 \times 0.003$ $= 482.6$ MPa

Procedure	Calculation
$M_n = \rho_f f_f \left(1 - 0.59\dfrac{\rho_f f_f}{f_c'}\right) bd^2$	$M_n = 0.009 \times 482.6\left(1 - 0.59\dfrac{0.009 \times 482.6}{27.6}\right)178 \times 250^2$ $= 43.7$ kN-m
$\rho_{fb} = 0.85\beta_1 \dfrac{f_c'}{f_{fu}} \dfrac{E_f \varepsilon_{cu}}{E_f \varepsilon_{cu} + f_{fu}}$ Compute the strength reduction factor $\phi = 0.3 + 0.25\dfrac{\rho_f}{\rho_{fb}}$ for $\rho_{fb} < \rho_f < 1.4\rho_{fb}$ Check $\phi M_n \geq M_u$	$\rho_{fb} = 0.85 \times 0.85 \times \dfrac{27.6}{496} \times \dfrac{44\,800 \times 0.003}{44\,800 \times 0.003 + 496}$ $= 0.0086$ $\phi = 0.3 + 0.25 \times \dfrac{0.009}{0.0086} = 0.56$ $\phi M_n = 0.56 \times 43.7$ kN·m $= 24.5$ kN·m > 20.3 kN·m
Step 5: Check for crack width	
$M_{DL} = \dfrac{w_{DL} l^2}{8}$ $M_{LL} = \dfrac{w_{LL} l^2}{8}$ $M_{DL+LL} = M_{DL} + M_{LL}$ $n_f = \dfrac{E_f}{E_c} = \dfrac{E_f}{4750\sqrt{f_c'}}$ $k = \sqrt{2\rho_f n_f + (\rho_f n_f)^2} - \rho_f n_f$ $f_f = \dfrac{M_{DL+LL}}{A_f d(1 - k/3)}$ Define the effective area of concrete $\beta = \dfrac{h - kd}{d(1-k)}$ $d_c = h - d$ $A = \dfrac{2d_c b}{\text{No. of bars}}$ Crack width, $w = \dfrac{2.2}{E_f}\beta k_b f_f \sqrt[3]{d_c A}$ Thus, crack width limitation controls the design. Try larger diameter bars of FRP reinforcement without affecting the width of beam. Repeat the above procedure.	$M_{LL} = \dfrac{5.8 \times 3.35^2}{8} = 8.14$ kN·m $M_{DL+LL} = M_{DL} + M_{LL} = 14.17$ kN·m $n_f = \dfrac{44\,800}{4750\sqrt{27.6}} = 1.8$ $k = \sqrt{(0.009 \times 1.8)^2 + 2(0.009)} - 0.009(1.8)$ $= 0.164$ $f_f = \dfrac{14.17 \times 10^6}{400 \times 250 \times (1 - 0.164/3)} = 149.9$ MPa $\beta = \dfrac{305 - 0.164 \times 250}{250\ (1 - 0.164)} = 1.263$ $d_c = 305 - 250 = 55$ mm $A = \dfrac{2 \times 55 \times 178}{2} = 9790\,\text{mm}^2$ $w = \dfrac{2.2}{44\,800} \times 1.263 \times 1.2 \times 149.9 \times \sqrt[3]{55 \times 9790}$ $= 0.90$ mm > 0.71 mm n.g. Trying 2 Ø19 mm results in crack width, $w = 0.68$ mm < 0.71 mm, Modified, $d = 248$ mm $k = 0.192$ $A_f = 567$ mm^2

Procedure	Calculation
Step 6: Check the Long-term deflection of the beam	
Compute the gross moment inertia of the section: $I_g = \dfrac{bh^3}{12}$	$I_g = \dfrac{178 \times 305^3}{12} = 4.209 \times 10^8 \text{mm}^4$
Calculate the cracked section properties and cracking moment: $f_r = 0.62\sqrt{f'_c}$ (SI) $M_{cr} = \dfrac{2 \times f_r \times I_g}{h}$ $I_{cr} = \dfrac{bd^3}{3}k^3 + n_f A_f d^2 (1-k)^2$	$f_r = 0.62\sqrt{27.6} = 3.25$ MPa $M_{cr} = \dfrac{2 \times 3.25 \times 4.209 \times 10^8}{305} = 8.97$ kN-m $I_{cr} = \dfrac{178 \times 248^3}{3} \times 0.192^3 + 1.8 \times (567) \times 248^2 \times (1-0.192)^2 = 4.74 \times 10^7 \text{ mm}^4$
Compute the reduction coefficient for deflection using the recommended, $\alpha_b = 0.50$ $\beta_d = \dfrac{1}{5}\left(\dfrac{\rho_f}{\rho_{fb}}\right) \le 1.0$	$\beta_d = \dfrac{1}{5} \times \left(\dfrac{0.0127}{0.0086}\right) = 0.30$
Compute the deflection due to dead load plus live load $I_e = \left(\dfrac{M_{cr}}{M_a}\right)^3 \beta_d I_g + \left[1 - \left(\dfrac{M_{cr}}{M_a}\right)^3\right] I_{cr} \le I_g$ $(\Delta_i)_{DL+LL} = \dfrac{5 M_{DL+LL} l^2}{48 E_c I_e}$	$I_e = \left(\dfrac{8.97}{14.17}\right)^3 \times 0.30 \times 4.209 \times 10^8 + \left[1 - \left(\dfrac{8.97}{14.17}\right)^3\right] \times 4.74 \times 10^7$ $= 0.674 \times 10^8 \text{ mm}^4 < I_g$ O.K. $(\Delta_i)_{DL+LL} = \dfrac{5 \times 14.17 \times 10^6 \times 3350^2}{48 \times 2.49 \times 10^4 \times 0.67 \times 10^8} = 9.93$ mm
Compute the deflection due to dead load alone and live load alone $(\Delta_i)_{DL} = \dfrac{w_{DL}}{w_{DL+LL}}(\Delta_i)_{DL+LL}$ $(\Delta_i)_{LL} = \dfrac{w_{LL}}{w_{DL+LL}}(\Delta_i)_{DL+LL}$ Compute the multiplier for long-term deflection using a $\xi = 2.0$ (recommended by ACI 318 for a duration of more than 5 years) $\lambda = 0.60\xi$ Compute the long-term deflection for 20% sustained live load and compare to long-term deflection limitations $\Delta_{LT} = (\Delta i)_{LL} + \lambda[(\Delta i)_{DL} + 0.20(\Delta i)_{LL}]$ Check $\varnothing_{LT}$ $\Delta_{LT} \le \dfrac{l}{240}$	$(\Delta_i)_{DL} = \dfrac{4.3}{4.3+5.8}(9.93) = 4.2$ mm $(\Delta_i)_{LL} = \dfrac{5.8}{4.3+5.8}(9.93) = 5.70$ mm $\lambda = 0.60 \times 2 = 1.2$ $\Delta_{LT} = 5.7 + 1.2[4.2 + 0.20(5.7)] = 12.1$ mm $12.1 \text{ mm} < \dfrac{3350}{240} = 14\text{mm},$ O.K.

Procedure	Calculation
Step 7: Check the creep rupture stress limits	
$M_s = \frac{w_{DL} + 0.20 w_{LL}}{w_{DL} + w_{LL}} M_{DL+LL}$	$M_s = \frac{4.3 + 0.20(5.8)}{4.3 + 5.8}(14.17)$
Compute the sustained stress level in the FRP bars:	$= 7.66$ kN·m
$f_{f,s} = \frac{M_s}{A_f d(1 - k/3)}$	$f_{f,s} = \frac{7.66 \times 10^6}{567 \times 248(1 - 0.192/3)}$
Check the stress limits given in Table 4.2 for glass FRP bars $f_{f,s} \leq 0.20 f_{fu}$	$= 58.2$ MPa $58.2 \text{ MPa} < 0.2 \times 496 = 99.2$ MPa, O.K.
Step 8: Design for shear	
Determine the factored shear demand at a distance d from the support $V_u = \frac{w_u l}{2} - w_u d$	$V_u = \frac{15.88 \times 3.35}{2} - 15.88 \times 0.248 = 22.7$ kN
Compute the shear contribution of the concrete for an FRP reinforced member	$k = \sqrt{2 \times 0.0127 \times 1.8 + (0.0127 \times 1.8)^2} - 0.0127 \times 1.8$ $= 0.192$
Neutral axis depth, $c = kd$ $k = \sqrt{2\rho_f n_f + (\rho_f n_f)^2} - \rho_f n_f$	$c = 0.192 \times 248 = 47.7$
$V_c = \frac{2}{5}\sqrt{f'_c}\, b_w c$	$V_c = \frac{2}{5}\sqrt{27.6} \times 178 \times 0.192 \times 248 \times 10^{-3}$ $= 17.8$ kN
Compute the strength of bend in the stirrups, f_{fb} $f_{fb} = \left(0.05\frac{r_b}{d_b} + 0.3\right) f_{fu}$	$f_{fb} = [(0.05 \times 3 + 0.3) + 0.3]496 = 223.2$ MPa
The design stress of FRP stirrups is limited to $f_{fv} = 0.004 E_f \leq f_{fb}$	$f_{fv} = 0.004 \times 44\,800 = 179.2 \text{ MPa} < 223.2$ MPa O.K.
The required spacing of the FRP stirrups can be computed by Eq. (4.42) $s = \frac{\phi A_f f_{fv} d}{(V_u - \phi V_{c,f})}$	$s = \frac{0.85 \times 142 \times 179.2 \times 248}{(22\,700 - 0.85 \times 17\,800)} = 708.6$ mm
Equation (4.46) for minimum amount of reinforcement can be rearranged as $s \leq \frac{A_{fv} f_{fv}}{0.35 b_w}$	$s \leq \frac{142 \times 179.2}{0.35 \times 178} = 408$ mm $d/2 = 248/2 = 124$ mm
Check maximum spacing limit $= d/2$	Thus, use No. 3 stirrups spaced at 120 mm on center.

E4.2. Design Example 2

A simply supported rectangular beam with effective span of 4.6 m (Fig. E4.2) has cross-section of 250 mm wide by 350 mm deep with 50 mm clear concrete cover to the main reinforcement. The beam is subjected to imposed load of 10 kN/m and exposed to earth and weather conditions. The CFRP reinforcement at tension is arranged in two layers. Each layer consists of three 10 mm bars. The two legged CFRP stirrups are provided at 100 mm c/c spacing. The c/c distance between the two layers of reinforcement is 60 mm. Using ACI 440.1R-06, evaluate the moment of resistance of section and check whether the beam cross-section is appropriate to carry the imposed load. The properties of concrete and CFRP materials are given in the following:

Concrete

Cylinder strength, $f_c' = 20$ N/mm^2, Cube Strength, $f_{ck} = 25$ N/mm^2

CFRP

Tensile strength, f_{fu}^*	1200 MPa
Rupture strain, ε_{fu}^*	0.015 mm/mm
Modulus of elasticity, E_f	150 GPa

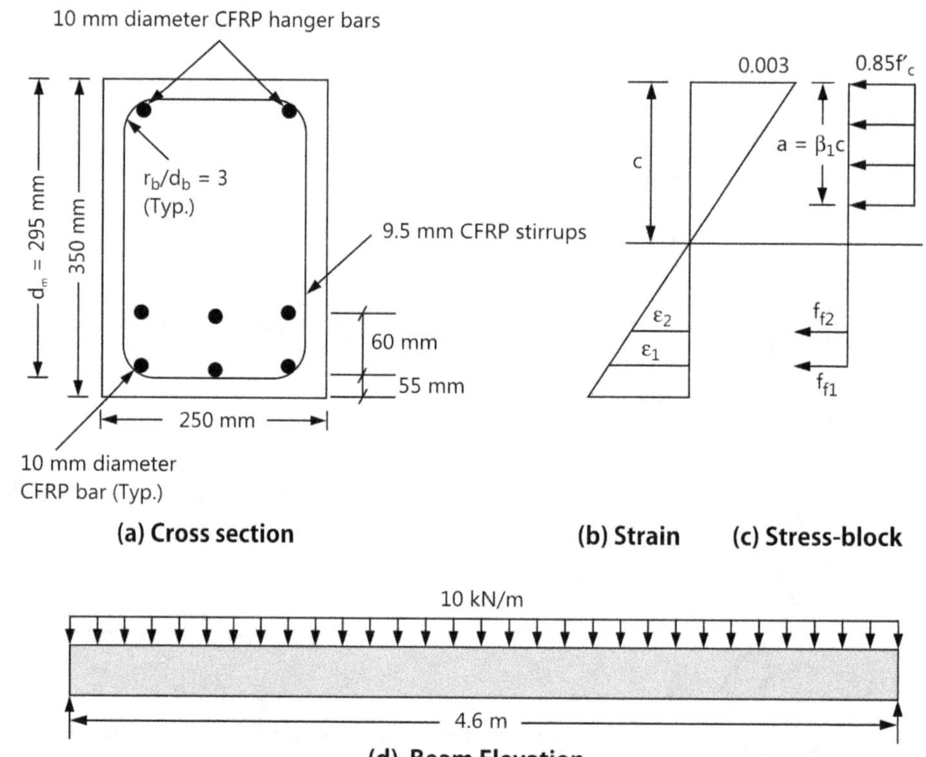

(a) Cross section (b) Strain (c) Stress-block

(d) Beam Elevation

Figure E4.2. Configuration cross-section and loading of simply supported beam.

Solution

Strength reduction factor for CFRP = 0.9 (Table 4.1)
Design strength, $f_{fu} = 0.9 \times 1200 = 1080$ MPa
Design strain, $\varepsilon_{fu} = 0.9 \times 0.015 = 0.0135$ mm/mm
Assume beam fails by crushing of concrete,
Stress block parameters, $\alpha = 0.85,\ \beta_1 = 0.85$
Total compression force, $C = 0.85\ f_c' ba$

$$= 0.85\ f_c'\beta_1 bc$$

$$= 0.85 \times 20 \times 0.85 \times 250 \times c = 3612.5c$$

Using compatibility equation, we have:

Strain in bottom layer of reinforcements, $\varepsilon_{f1} = \dfrac{0.003}{c} \times (295 - c)$

Strain in upper layer of reinforcements, $\varepsilon_{f2} = \dfrac{0.003}{c} \times (235 - c)$

Area of reinforcements in each layer, $A_{f1} = A_{f2} = 3 \times 78.5 = 235.6\ \text{mm}^2$

Force in bottom layer reinforcements, $F_1 = E_f \varepsilon_{f1} A_{f1}$

$$= 150 \times 10^3 \times \frac{0.003}{c}(295 - c) \times 235.6$$

$$= \frac{106028.75}{c}(295 - c)$$

Force in upper layer reinforcements, $F_2 = E_f\, \varepsilon_{f2} A_{f2}$

$$= 150 \times 10^3 \times \frac{0.003}{c}(235 - c) \times 235.6$$

$$= \frac{106028.75}{c}(235 - c)$$

From equilibrium

$$C = F_1 + F_2$$

$$3612.5c = \frac{106028.75}{c}(295 - c) + \frac{106028.75}{c}(235 - c)$$

$$c^2 + 58.70c - 1555.78 = 0$$

$$c = \frac{-58.70 + \sqrt{58.70^2 + 4 \times 1555.78}}{2} = 98.8\ \text{mm}$$

Check for CFRP strain limits at assumed failure mode

$$\varepsilon_{f1} = \frac{0.003}{c} \times (295 - c)$$

$$= \frac{0.003}{98.8} \times (295 - 98.8)$$

$= 0.006 < 0.013$ (i.e., ε_{fu}), O.K.

$$\varepsilon_{f2} = \frac{0.003}{c} \times (235 - c)$$

$$= \frac{0.003}{98.8} \times (235 - 98.8)$$

$$= 0.0041 < 0.013 \text{ (i.e., } \varepsilon_{fu}) \qquad \text{O.K.}$$

Since, the strains in both the layers are below the design ultimate tensile strain of CFRP reinforcements, this confirms the assumed compression mode of failure. This will also be verified later by comparing the actual reinforcement ratio and the corresponding balanced ratio for typical section with multiple layer reinforcements in tension.

Stress in bottom layer reinforcements, $f_{f1} = E_f \varepsilon_{f1}$

$$= 150 \times 10^3 \times 0.006;$$

$$= 900 \text{ MPa} < 1080 \text{ MPa (i.e., } f_{fu}); \qquad \text{O.K.}$$

Stress in upper layer reinforcements, $f_{f2} = E_f \varepsilon_{f2}$

$$= 150 \times 10^3 \times 0.0041;$$

$$= 615 \text{ MPa} < 1080 \text{ MPa (i.e., } f_{fu}); \qquad \text{O.K.}$$

Resultant force in bottom layer reinforcement, $F_1 = f_{f1} \times A_{f1}$

$$= 900 \times 235.6/1000$$

$$= 212.040 \text{ kN}$$

Resultant force in upper layer reinforcement, $F_2 = f_{f2} \times A_{f2}$

$$= 615 \times 235.6/1000$$

$$= 144.894 \text{ kN}$$

Depth stress block (Fig. E4.2c), $a = \beta_1 c = 0.85 \times 98.8 = 83.98 \text{ mm}$

Nominal moment carrying capacity of beam section,

$$M_n = F_1\left(d_1 - \frac{a}{2}\right) + F_2\left(d_2 - \frac{a}{2}\right)$$

$$= 212.040\left(298 - \frac{83.98}{2}\right) + 144.894\left(235 - \frac{83.98}{2}\right)$$

$$= 53648.2404 + 27965.9909$$

$$= 81614.231 \text{ kN-m} = 81.614 \text{ kN-m}$$

To compute the actual reinforcement ratio, ρ and corresponding balanced ratio, ρ_b of a section with two layers of tension reinforcement, let us use the equation of force resultant equilibrium in conjunction with strain compatibility condition for the balanced section.

Assuming section to be balanced, the force equilibrium equation is given as:

$$A_{f1} E_f \frac{\varepsilon_{cu}}{c_b} \times (d_1 - c_b) + A_{f2} E_f \frac{\varepsilon_{cu}}{c_b} \times (d_2 - c_b) = 0.85 f_c' \beta_1 b c_b \qquad \text{(E4.2a)}$$

Here, ε_{cu} is crushing strain of concrete.

At balanced condition, the bottom most layer of reinforcement will reach the failure strain, ε_{fu} first and hence,

$$\frac{\varepsilon_{cu}}{c_b} \times (d_1 - c_b) = \varepsilon_{fu} \qquad \text{(E4.2b)}$$

From Eq. (E4.2b), we can express c_b by Eq. (E4.2c),

$$c_b = \frac{\varepsilon_{cu}\, d_1}{\varepsilon_{cu} + \varepsilon_{fu}} \tag{E4.2c}$$

Using Eqs. (E4.2b) and (E4.2c), we can express Eq. (E4.2a) as:

$$A_{f_2} f_{fu} + A_{f_2}\left(f_{f_2}\right)_b = 0.85 f_c' \beta_1 \left(\frac{\varepsilon_{cu}}{\varepsilon_{cu} + \varepsilon_{fu}}\right) b d_1 \tag{E4.2d}$$

Rearranging the terms by diving the both sides by $f_{fu}(bd_1)$, we have:

$$\frac{A_{f1} + A_{f2}\dfrac{(f_{f2})_b}{f_{fu}}}{bd_1} = 0.85\,\beta_1 \frac{f_c'}{f_{fu}}\left(\frac{\varepsilon_{cu}}{\varepsilon_{cu} + \varepsilon_{fu}}\right) \tag{E4.2e}$$

Further, by multiplying the numerator and denominator of the terms in the bracket on the right hand side by E_f and substituting $E_f \varepsilon_f = f_{fu}$, we have:

$$\frac{A_{f1} + A_{f2}\dfrac{(f_{f2})_b}{f_{fu}}}{bd_1} = 0.85\beta_1 \frac{f_c'}{f_{fu}}\left(\frac{E_f \varepsilon_{cu}}{E_f \varepsilon_{cu} + f_{fu}}\right) \tag{E4.2f}$$

Thus, the left hand side of the equation actually represents the reinforcement ratio, ρ, which in this particular case is equal to balanced reinforcement ratio, ρ_b.

Thus, balanced ratio for the section is give as:

$$\rho_b = 0.85\beta_1 \frac{f_c'}{f_{fu}}\left(\frac{E_f \varepsilon_{cu}}{E_f \varepsilon_{cu} + f_{fu}}\right) \tag{E4.2g}$$

To find the actual reinforcement ratio for section with two layers of reinforcements, the left side expression of Eq. (E4.2f) can be used:

$$\rho = \frac{A_{f1} + A_{f2}\dfrac{(f_{f2})_b}{f_{fu}}}{bd_1} \tag{E4.2h}$$

Where $(f_{f2})_b$ represents the stress in upper layer bars at balanced condition; A_{f1} and A_{f2} are the total areas of reinforcement in bottom and upper layer, respectively.

For the example problem (E4.2h), we can use the afforementioned expression to show that failure mode is compression and hence, governs the design.

The stress in the upper layer bars at balanced condition can be determined using compatibility condition, as bottom layer bars will be at design stress, f_{fu}.

From Eq. (E4.2c), the depth of neutral axis at the balanced failure condition is given by:

$$c_b = \frac{\varepsilon_{cu} d_1}{\varepsilon_{cu} + \varepsilon_{fu}} = \frac{0.003 \times 295}{0.003 + 0.0135} = 53.64\,\text{mm}$$

$$\left(f_{f2}\right)_b = \left(\frac{f_{fu}}{295 - 53.64}\right) \times (235 - 53.64) = 0.75\, f_{fu}$$

So, actual reinforcement ratio of the section is determined using Eq. (E4.2h).

$$M_D = \frac{2.19 \times 4.6^2}{8} = 5.79\,\text{kN-m}$$

$$\rho = \frac{A_{f1} + A_{f2}\dfrac{(f_{f2})_b}{f_{fu}}}{bd_1} = \frac{235.6 + 0.75 \times 235.6}{250 \times 295} = 5.59 \times 10^{-3}$$

The balanced ratio is obtained using Eq. (E4.2g),

$$\rho_b = 0.85\,\beta_1 \frac{f'_c}{f_{fu}}\left(\frac{E_f \varepsilon_{cu}}{E_f \varepsilon_{cu} + f_{fu}}\right) = 0.85 \times 0.85 \times \frac{20}{1080} \times \left(\frac{0.003 \times 150 \times 10^3}{0.003 \times 150 \times 10^3 + 1080}\right) = 3.94 \times 10^{-3}$$

So $\rho > \rho_b$, which confirms that the compression mode of failure that is beams, failure–occurs due to crushing of concrete.

For $\dfrac{\rho}{\rho_b} = \dfrac{5.59 \times 10^{-3}}{3.94 \times 10^{-3}} = 1.42$, we have strength reduction factor, $\phi = 0.65$.

Design moment capacity of section = $\phi M_n = 0.65 \times 81.64 = 53.07$ kN-m.

Check for appropriateness of the beam to carry imposed load safely

Now to check the appropriateness of the beam to carry imposed load safely, let us compare the ultimate load carrying capacity of the beam section as per ACI 440.1R-06 document, with design moment capacity.

Ultimate factor moment, $M_u = 1.2\,M_D + 1.6M_L$

where, M_D is dead load moment and M_L is live load moment under service condition; and are given as:

$$M_L = \frac{10 \times 4.6^2}{8} = 26.45\,\text{kN-m}$$

Hence, $M_u = 1.2 \times 5.79 + 1.6 \times 26.45 = 49.27\,\text{kN-m} < \phi M_n$ (i.e., 53.07 kN-m)

Hence, the beam is safe in strength to carry the imposed load. The above approach can be applied for multilayered reinforcement also.

E4.3. Design Example 3

A simply supported rectangular beam with effective span of 5 m has precast concrete cross-section of 200 mm wide by 300 mm deep (see Fig. E4.3). The beam is subjected to service dead load (inclusive of self-weight) of 20 kN/m and service imposed load of 20 kN/m. The FRP reinforcement at tension side consists of three CFRP 10.0 mm dia. bars. As per ISIS Canada Design Guidelines, find the moment of resistance of section and check whether the beam cross-section is appropriate to carry the imposed load. Assume load factor of 1.5 for dead and imposed loads. Properties of concrete and CFRP materials are given in the following:

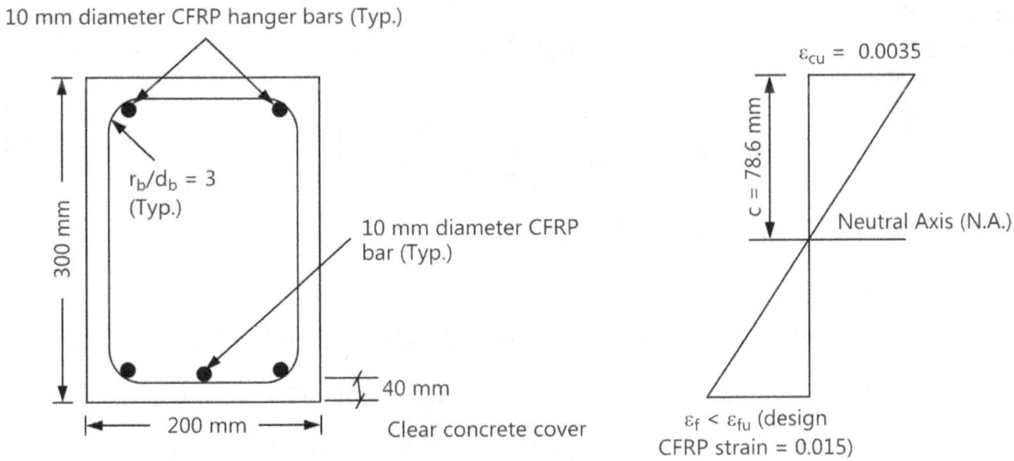

Figure E4.3. Cross-sectional details of simply supported beam and strain distribution at ultimate.

Concrete

$$f_c' = 30\,\text{N/mm}^2$$

CFRP

$$f_{fu}^* = 2000\,\text{N/mm}^2;\ \varepsilon_{fu}^* = 0.015\,\text{mm/mm};\ E_f = 150\,\text{GPa}$$

Solution

$W_{DL} = 20$ kN/m (including self-weight)

$W_{LL} = 20$ kN/m

$\phi_{frp} = 0.8$ for carbon fiber; $\phi_c = 0.65$ for concrete;

Cross-sectional area of reinforcement, $A_{frp} = 235.6\ \text{mm}^2$

Required concrete cover

The concrete cover shall be taken as $2.5d_b$ (where, d_b is diameter of reinforcing bar) or 40 mm, whichever is greater.

Here $2.5d_b = 2.5 \times 10 = 25$ mm

So 40 mm concrete cover governs.

Effective depth of beam cross-section $= h - \text{cover} - d_b/2 = 300 - 40 - 10/2 = 255$ mm

CFRP reinforcement ratio

Actual reinforcement ratio of the section is calculated as follows:

$$\rho_{frp} = \frac{A_{frp}}{bd} = \frac{235.6}{200 \times 255} = 0.0046 \tag{E4.3a}$$

Balanced ratio

As per ISIS Canada design guidelines, the balanced ratio of beam-cross section is given as:

$$\rho_{frpb} = \alpha_1 \beta_1 \frac{\phi_c}{\phi_{frp}} \frac{f_c'}{f_{frpu}} \left[\frac{\varepsilon_{cu}}{\varepsilon_{cu} + \varepsilon_{frpu}} \right] \quad \text{(E4.3b)}$$

$$\alpha_1 = 0.85 - 0.0015 f_c' = 0.85 - 0.0015 \times 30 = 0.81 > 0.67$$

So, consider $\alpha_1 = 0.81$. Similarly:

$$\beta_1 = 0.97 - 0.0025 f_c' = 0.97 - 0.0025 \times 30 = 0.9 > 0.67$$

So, consider $\beta_1 = 0.9$. Substituting the value of α_1 and β_1 in Eq. (E4.3b), we have

$$\rho_{frpb} = (0.81)(0.9)\left(\frac{0.65}{0.80}\right)\left(\frac{30}{2000}\right)\left[\frac{0.0035}{0.0035 + 0.015}\right]$$

$$= 0.0017$$

So, balanced ratio of the section is equal to 0.0017. This ratio will be compared with the actual reinforcement ratio, to decide whether section failure occurs due to crushing of concrete or rupture of CFRP reinforcements.

Mode of failure

Since, $\rho_{frp} = 0.0046 > \rho_{frpb} = 0.0017$, the section failure is governed by crushing of concrete, i.e., Compression Failure Mode. The stresses in tensile CFRP reinforcements will be lower than its ultimate design strength.

Computation of tensile stress in CFRP reinforcements at ultimate failure condition

The tensile stress in the CFRP reinforcement is given by Eq. (E4.3c),

$$f_{frp} = 0.5 E_{frp} \varepsilon_{cu} \left[\left(1.0 + \frac{4 \alpha_1 \beta_1 \phi_c f_c'}{\rho_{frp} \phi_{frp} E_{frp} \varepsilon_{cu}} \right)^{\frac{1}{2}} - 1.0 \right] \quad \text{(E4.3c)}$$

$$f_{frp} = (0.5)(150000)(0.0035)\left[\left(1.0 + \frac{4 \times (0.81)(0.90)(0.65)(30)}{(0.0046)(0.8)(150000)(0.0035)} \right)^{\frac{1}{2}} - 1.0 \right]$$

$$= 1185.6 \text{ MPa} < 2000 \text{ MPa} \quad \text{(O.K.)}$$

E4.4. Design Example 4: A Case Study Problem

Evaluate the shear resistance of two box-beams prestressed with 6 un-bonded and 7 bonded CFRP prestressing tendons. Each beam is provided with 4 non-prestressing tendons at bottom and 7 non-prestressing tendons at the top. The cross-section details of both beams are the same and are

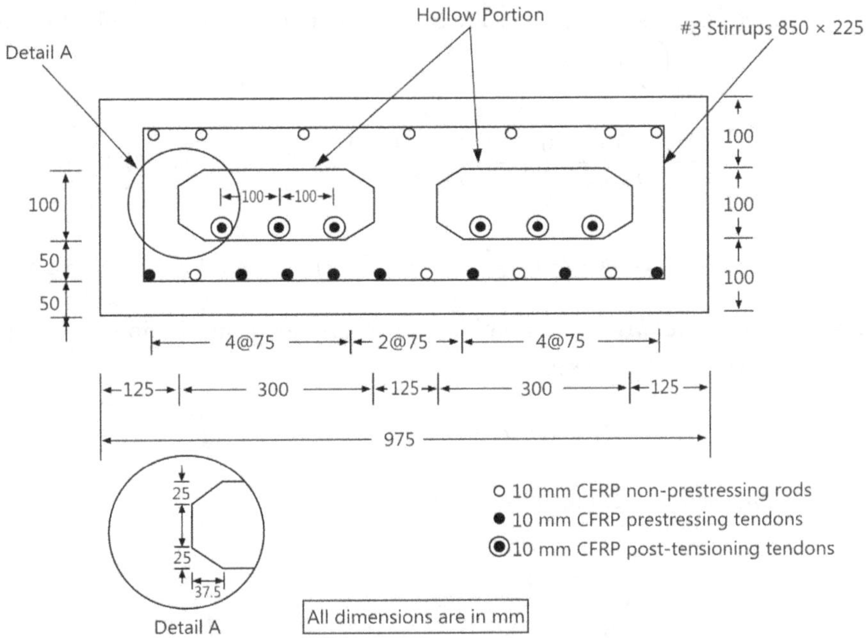

Figure E4.4. Cross-sectional details of typical CFRP prestressed concrete Box-beam.

shown in Fig. E4.4. The effective span of each beam is 4575 mm. Note that Beam 1 is provided with 10 mm CFRP stirrups at center-to-center spacing of $d/2$ (d is effective depth of the beam), while Beam 2 is provided with similar stirrups at center-to-center spacing of $d/3$. If the experimental central load carrying capacities of the control beam—beam without shear reinforcement, Beam 1 and Beam 2 are 355.1 kN, 516.65 kN and 535.8 kN, respectively, then, compute the percentage difference in theoretical and experimental shear strengths of beams. The following are the specified characteristics of stirrups and tendons:

Modulus of elasticity of CFRP stirrups, $E_f = 108\,900$ MPa
Ultimate breaking strength of stirrup bars, $f_{fu} = 937.7$ MPa
Diameter of stirrup bar, $d_b = 10$ mm
Diameter of each tendon, $d_p = 10$ mm
Total effective prestressing force in each tendon = 92.5 kN

Solution

The following steps are taken to evaluate the nominal and design shear capacity of the beams:

Step 1: Compute the nominal shear strength contribution of concrete

Nominal shear strength contribution of concrete is based on two failure modes.

(a) ***Web shear cracking:*** The nominal shear strength contribution of concrete for this failure mode is given by, V_{cw} where,

$$V_{cw} = b_w d(0.291\sqrt{f_c'} + 0.3 f_{cc}) + V_p \quad \text{(SI Units)}$$

Here, $V_p = 0$, since prestressing tendons are straight and normal to the applied shear force.

$$f_{cc} = \frac{F_{\text{pre}}}{A} + \frac{F_{\text{post}}}{A}$$

$$= \frac{7 \times 92.5 \times 1000}{225000} + \frac{6 \times 92.5 \times 1000}{225\,000} = 2.88 + 2.47 = 5.35 \text{ MPa}$$

Thus,

$$V_{cw} = 350 \times 245\ (0.291\sqrt{51} + 0.3 \times 5.35) = 315830 \text{ N} = 315.83 \text{ kN}$$

(b) ***Flexural-shear cracking:*** The nominal shear strength contribution of concrete for this failure mode is given by, V_{ci} where,

$$V_{ci} = 0.05\sqrt{f_c'}b_w d + V_o + \frac{V_i}{M_{\max}} M_{cr} \geq 0.14\sqrt{f_c'}b_w d$$

For the third-point loading, let the applied central load is P.

$$V_i = \frac{P}{2}$$

$$M_{\max} = \frac{P}{2} \times \frac{L}{3} = \frac{PL}{6}$$

$$\frac{V_i}{M_{\max}} = \frac{3}{L}$$

$$M_{cr} = \frac{I}{c_2}(0.5\sqrt{f_c'} + f_{2p} - f_o) \quad \text{(SI Units)}$$

Assuming self-weight of the beam as 24 kN/m^3

Cross-sectional area of section, $A = 950 \times 300 - 2 \times 300 \times 100 = 225\,000$ mm^2

Dead load of the beam/unit length, $W_D = 24 \times 225\,000 \times 10^{-6} \times 1 = 5.4$ kN/m

Dead load moment at third-point section $= \dfrac{W_D L^2}{9} = \dfrac{5.4 \times 4.575^2}{9} = 12.56$ kN-m

Moment of inertia of concrete cross-section, $I = 20\,875 \times 10^5$ mm^4

$$f_o = \frac{12.56 \times 10^6 \times 150}{20875 \times 10^5} = 0.903 \text{ MPa}$$

$$f_{2p} = \frac{F_{\text{pre}}}{A} + \frac{F_{\text{pre}} e c_2}{I} + \frac{F_{\text{post}}}{A} + \frac{F_{\text{post}} e c_2}{I}$$

$$= \frac{7 \times 92.5 \times 1000}{225\,000} + \frac{7 \times 92.5 \times 1000 \times 95 \times 150}{20875 \times 10^5} + \frac{6 \times 92.5 \times 1000}{225000} + \frac{6 \times 92.5 \times 1000 \times 45 \times 150}{225000}$$

$$= 2.88 + 4.42 + 2.47 + 1.79 = 11.56 \text{ MPa}$$

$$c_2 = 150 \text{ mm}$$

$$M_{cr} = \frac{20875 \times 10^5}{150}(0.5\sqrt{51} + 11.56 - 0.903) = 198 \times 10^6 \text{ N-mm} = 198 \text{ kN-m}$$

$$V_o = \frac{W_D L}{6} = \frac{5.4 \times 4.575}{6} = 4.12 \text{ kN}$$

$$0.14\sqrt{f_c'}b_w d = 0.14 \times \sqrt{51} \times 350 \times 245 = 85733 \text{ N} = 85.73 \text{ kN}$$

$$V_{ci} = 0.05\sqrt{f_c'}b_w d + V_o + \frac{V_i}{M_{\max}} M_{cr} \geq 0.14\sqrt{f_c'}b_w d$$

$$V_{ci} = \frac{0.05 \times \sqrt{51} \times 350 \times 245}{1000} + 4.12 + \frac{3 \times 198}{4.575} = 30.62 + 4.12 + 129.84$$

$$= 164.58 \text{ kN} > 85.73 \text{ kN} \qquad \text{O.K.}$$

Since, V_{ci} is smaller than V_{cw}, hence, $V_c = V_{ci} = 164.58$ kN.

Experimental value of the shear strength contribution of concrete = 0.5 × 355.1 = 177.6 kN. % difference in the theoretical and experimental values of the nominal shear strength of concrete

$$= \frac{(177.6 - 164.58)}{177.6} \times 100 = 7.3\%$$

Thus, design equations reasonably estimate the nominal shear strength contribution of concrete.

Step 2: Compute the nominal shear strength contribution of stirrups

$$\text{Strength of stirrups at bend, } f_{fb} = \left(0.05\frac{r_b}{d_b} + 0.3\right) f_{fu}$$

$$= (0.05 \times 3.81 + 0.3) \times 937.7 = 459.9 \text{ MPa}$$

$$f_{fv} = 0.002 \times 108\,900 = 217.8 \text{ MPa} < 459.9 \text{ MPa} \qquad \text{O.K.}$$

(a) *Shear strength contribution of stirrups provided at S = d/2:*

$$V_{sd2} = \frac{A_{fv} f_{fv} d}{S} = \frac{156.8 \times 217.8 \times d}{d/2} = 68302 \text{ N} = 68.3 \text{ kN}$$

(b) *Shear strength contribution of stirrups provided at S = d/3:*

$$V_{sd3} = \frac{A_{fv} f_{fv} d}{S} = \frac{156.8 \times 217.8 \times d}{d/3} = 102\,453 \text{ N} = 102.5 \text{ kN}$$

Step 3: Compute the nominal shear strength of beams

The nominal shear strength of Beam-1 and Beam-2 are computed as follows:

(a) *Nominal shear strength of Beam-1:*

$$V_n = V_c + V_{sd2} = 164.58 + 68.3 = 232.88 \text{ kN}$$

$$\text{Experimental shear capacity of Beam-1} = \frac{516.65}{2} = 258.3 \text{ kN}$$

$$\% \text{ difference} = \frac{(258.3 - 232.88)}{258.3} \times 100 = 9.8\%$$

(b) ***Nominal shear strength of Beam-2:***

$$V_n = V_c + V_{sd3} = 164.58 + 102.5 = 267.08 \text{ kN}$$

$$\text{Experimental shear capacity of Beam-2} = \frac{535.8}{2} = 267.9 \text{ kN}$$

$$\% \text{ difference} = \frac{(267.9 - 267.08)}{267.9} \times 100 = 0.3\%$$

Step 4: Compare the design and experimental shear strength values

(a) ***Design shear strength of Beam-1:***

Design shear strength of Beam-1, $V_d = \phi V_n = 0.85 \times 232.88 = 197.9$ kN

$$\% \text{ difference in design and experimental shear strengths} = \frac{(258.3 - 197.9)}{258.3} \times 100 = 23.4\%$$

(b) ***Design shear strength of Beam-2:***

Design shear strength of Beam-2, $V_d = \phi V_n = 0.85 \times 267.08 = 227.1$ kN

$$\% \text{ difference in design and experimental shear strengths} = \frac{(267.9 - 227.1)}{267.9} \times 100 = 15.3\%$$

Thus, on average design shear strength of beams is 19.4% lower than the actual strength value, which is consistent with strength reduction factor (ϕ) of 0.85.

E4.5. Design Example 5: Case Study of CFRP Prestressed Concrete Double-T Beam

Evaluate the load carrying capacity of a Double-tee (DT) concrete beam prestressed using a total of sixty bonded carbon fiber reinforced polymer (CFRP) Leadline pretensioning tendons and four unbonded carbon fiber composite cable (CFCC) strands. As shown in Fig. E4.5, the thirty of the sixty bonded pretensioning tendons are provided in one web of the DT-beam, while remaining thirty tendons are provided in the other web. The diameter of each Leadline tendon is 10 mm, while each unbonded CFCC post-tensioning strand is of 40 mm diameter. In addition, six layers of 12.5 mm diameter CFCC strands are used to reinforce each web of the DT-beam. Each of the top five layers of non-prestressing rods consists of two CFCC strands, while the bottom layer consists of four CFCC strands in each web. Two layers of non-prestressing 10 mm diameter Leadline tendons are provided in the flange of the DT-beam. The top layer of non-prestressing Leadline tendons has ten tendons, while the bottom layer has nine tendons. The DT-beam consists of a precast section and a 75 mm thick concrete topping. The pre-tensioning forces are applied only on the precast section, while 60% of the design post-tensioning forces are applied on the precast section and remaining 40% is applied on the composite section. CFRP NEFMAC sheets are used to reinforce the concrete topping. The material properties of CFRP Leadline tendons CFCC strands, NEFMAC sheets, and concrete are presented in Table E4.2, E4.3 and E4.4, respectively. The span of the DT-beam is 20.422 m. The dead weight of DT-beam is 31 kN/m length. The beam is loaded through 4-point loading system consisting of two pairs of load points. Each pair of load points is across the width of the DT-beam. The longitudinal

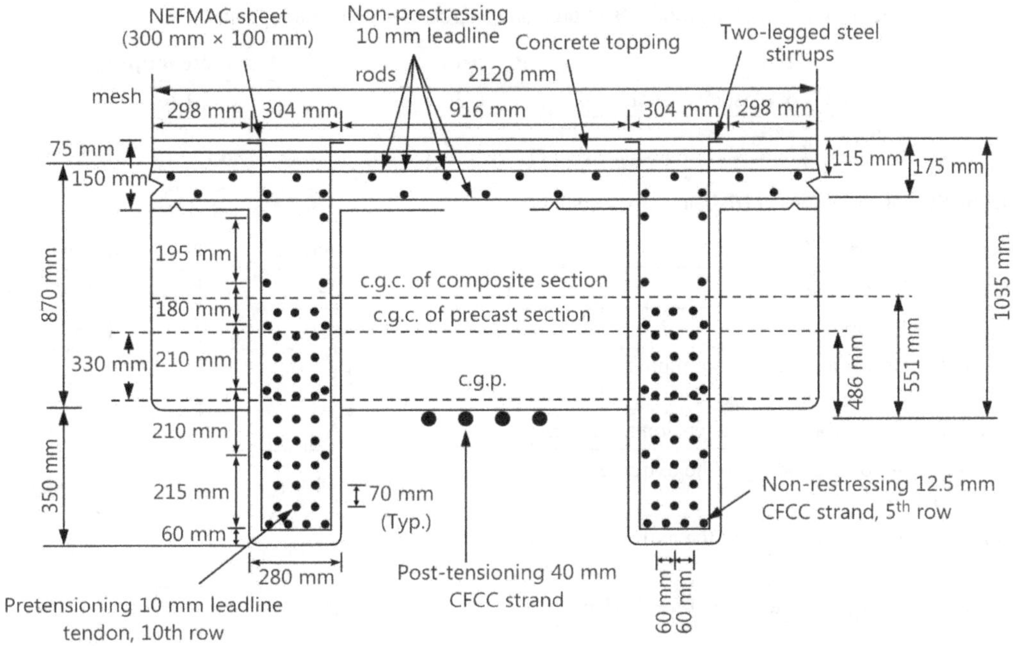

Figure E4.5. Cross-section of DT-beam at midspan.

Table E4.2. Material properties of CFRP tendons/CFCC strands.

Properties	Leadline^{TM} (MCC)	CFCC 1 × 7	CFCC 1 × 37
Nominal diameter, mm	10	12.5	40
Effective cross-sectional area, mm^2	71.6	76.0	752.6
Guaranteed tensile strength, kN/mm^2	2.26	1.87	1.41
Specified tensile strength, MPa	2860	2100	1870
Young's modulus of elasticity, kN/mm^2	147	137	127
Elongation, percent	1.9	1.5	1.5
Guaranteed breaking load, kN	162	142	1070
Ultimate breaking load, kN	204.7	160	1410

Tablle E4.3 Material properties of NEFMAC^{TM2} sheets

Modulus of elasticity, GPa	86.5
Ultimate strength, MPa	1 500
Ultimate strain, percent	1.8

Table E4.4. Material properties of precast concrete and concrete topping

Properties	Precast concrete	Concrete topping
Modulus of elasticity, GPa	36.7	31.6
Strength, MPa	53.8	39.3

Table E4.5. Section properties of DT-beam cross-section

Sectional properties	Precast section	Composite section
Cross-sectional area, mm^2	943 224	1 079 353
Moment of inertia, mm^4	1.3673×10^{11}	1.6808×10^{11}
Distance of centroid from extreme compression fiber, mm	474.5	484.6
Distance of centroid from extreme tension fiber, mm	745.5	810.3
Section modulus corresponding to the extreme tension fiber, mm^3	1.8340×10^8	2.0744×10^8
Section modulus corresponding to the extreme compression fiber, mm^3	2.8817×10^8	3.4683×10^8
Eccentricity of resultant pretensioned force, mm	330.2	N/A*
Eccentricity of unbonded post-tensioning strands, mm	485.6	550.4

*N/A refers to a non-applicable value for calculating stresses and/or moment due to pretensioning forces because pre-tensioning forces was not applied on the composite section.

distance between the two pairs of load points is 3.658 m. Distance between the centers of each support to the nearest load point is 8.382 m. If the maximum live load to be carried by DT-beam is 464 kN, compare the nominal moment capacity of the DT-beam and required moment capacity. Also, compute the design moment capacity of the DT-beam. Total pre-tensioning force in each of the Leadline tendon of five bottom layers is 86.8 kN, whereas, pre-tensioning force in Leadline tendons of top five layers is 82.3 kN. Post-tensioning force in each unbonded CFCC strand is 443 kN. Assume 25% loss of prestress in pre-tensioning tendons. Horizontal distance between anchorages of unbonded post-tensioning tendons is equal to 15 670 mm.

Solution

The following steps illustrate the procedure (Grace and Singh, 2003) to predict the behavior of the full-scale pretensioned and post-tensioned tested DT beam. It should be noted that the cross-sectional area of NEFMAC grids has not been included in the analysis presented below:

Procedure	Calculation in SI units
Step 1: Estimate the appropriate cross-sectional dimensions of the DT-beam.	
The depth of the DT-beam can be taken as about $L/16$, where, L is the span of the beam. However, a slightly higher value is considered, in view of the lower stiffness of CFRP bars in comparison to steel. Other dimensions are appropriately chosen in accordance with DT-beam sections, as shown in Fig. E4.5. Here, width of the flange section has been taken about 1.6 times the depth of beam.	$L/16 = 20422/16 = 1276$ mm. Take total depth of beam, $D = 1295$ mm Total width of flange section $b = 2120$ mm

<table>
<tr><th>Procedure</th><th>Calculation in SI units</th></tr>
<tr><td colspan="2">Step 2: Calculate required moment capacity</td></tr>
<tr><td>If W_d, M_D and M_L are Self-weight of the beam per unit length, Dead load moment and live load moment, respectively. L is the effective span of the beam.
Dead load moment at midspan,
$$M_D = \frac{W_d L^2}{8}$$
Let W be the total midspan load applied through 4-point loading system. The distance between the center of each support to the nearest load point is L_1 .
Design service live load $= P$
Service live load moment at midspan,
$$M_L = \frac{PL_1}{2}$$
Compute required moment capacity of the section,
$$M_{\text{required}} = 1.2M_D + 1.6M_L$$</td><td>Dead load/unit length, $W_d = 31$ kN/m
Dead load moment at midspan,
$$M_D = \frac{31 \times 20.422^2}{8} = 1{,}616 \text{ kN-m}$$
Design service live load = 464 kN
Service live load moment at midspan,
$$M_L = \frac{264 \times 8.382}{2} = 1944.6 \text{ kN-m}$$
Required moment capacity of the section,
$$M_{\text{required}} = 1.2M_D + 1.6M_L = 5\,050.6 \text{ kN-m}$$</td></tr>
<tr><td colspan="2">Step 3: Calculate balanced ratio (ρ_b)</td></tr>
<tr><td>Balanced ratio, $\rho_b = 0.85\beta_1 \dfrac{f'_c}{f_{fu}} \dfrac{\varepsilon_{cu}}{\varepsilon_{cu} + \varepsilon_{fu} - \varepsilon_{pbmi}}$
where, f'_c, f_{fu}, ε_{cu} and ε_{fu} are strength of precast concrete, specified strength of bonded pretensioning tendons, ultimate crushing strain of concrete and specified ultimate strain of bonded pretensioning tendons, respectively.
Based on experimental results, 25% loss in the prestressing force in the bonded tendons is considered. The initial effective strain in bottom pre-tensioning tendons, ε_{pbmi} can be calculated .</td><td>$f'_c = 53.8$ MPa (7810 psi) (see Table E4.4)
$f_{fu} = 2860$ MPa (see TableE4.2)
$\varepsilon_{cu} = 0.003$
$\varepsilon_{fu} = 0.019$ (see Table E4.2)
$$\beta_1 = 0.85 - \frac{(7810 - 4000)}{1000} \times 0.05 = 0.66$$
$$\varepsilon_{pbmi} = \frac{86.8 \times 0.75}{71.6 \times 147} = 0.0062$$
Balanced ratio,
$$\rho_b = 0.85 \times 0.66 \times \frac{53.8}{2860} \frac{0.003}{0.003 + 0.019 - 0.0062} = 0.002$$</td></tr>
<tr><td colspan="2">Step 4: Calculate cracking moment (M_{cr}) and cracking load (P_{cr})</td></tr>
<tr><td>Modulus of rupture of concrete,
$$f_r = 0.498\sqrt{f'_c}$$
Various sectional properties of the DT-beam section (see Fig. E4.5) are presented in Table E4.5.
Stress at the bottom fiber of the section due to pre-tensioning force,
$$(\sigma_c)_{b_1} = \frac{F_{\text{pre}}}{A_p} - \frac{F_{\text{pre}} e_b}{S_{bp}}$$</td><td>$$f_r = 0.498\sqrt{53.8} = 3.653 \text{ MPa}$$
$$F_{\text{pre}} = 5073 \times 0.75 = 3{,}804.8 \text{ kN}$$
$$F_{\text{post}} = 1772 \text{ kN}$$
From Table 4, $A_p = 943\,224$ mm^2 ,
$S_{bp} = 1.8340 \times 10^8$ mm^3 and
$e_b = 330.2$ mm
$$(\sigma_c)_{b1} = -\frac{3804.8}{943224} - \frac{3804.8 \times 330.2}{1.8340 \times 10^6} = -10.88 \text{ MPa}$$</td></tr>
</table>

<table>
<tr><th>Procedure</th><th>Calculation in SI units</th></tr>
<tr><td>The stress at the bottom fiber of section due to post-tensioning force, $(\sigma_c)_{b2}$
$$= -\frac{0.6\times F_{\text{post}}}{A_p} - \frac{0.6\times F_{\text{post}} e_{up}}{S_{bp}} - \frac{0.4\times F_{\text{post}}}{A_c} - \frac{0.4\times F_{\text{post}} \times e_{uc}}{S_{bc}}$$</td><td>$$(\sigma_c)_{b2} = -\frac{0.6\times 1772}{943224} - \frac{0.6\times 1772}{1.8340\times 10^8} - \frac{0.4\times 1772}{1079353}$$
$$= -\frac{0.4\times 1772\times 550.4}{2.0744\times 10^8} = -64.8 \text{ MPa}$$</td></tr>
<tr><td>Cracking moment can be obtained from the following expression
$$(\sigma_c)_{b1} + (\sigma_c)_{b2} + \frac{M_{cr}}{S_{bc}} = f_r$$
$$P_{cr} = \frac{(M_{cr} - M_D)\times 2}{L_1}$$</td><td>$M_{cr} = (10.88 + 64.8 + 3.653) \times 2.0744 \times 10^8$
$= 4358.94$ kN-m
$$P_{cr} = \frac{(4358.94 - 1616)\times 2}{8.382}$$
$= 1944.6$ kN > service load (464 kN)
Experimental value of cracking load = 643.9 kN
Percent difference in theoretical and experimental cracking loads
$$= \frac{(654.5 - 643.9)}{643.9} \times 100 = 1.6\%$$</td></tr>
<tr><th colspan="2">Step 5: Compute flexural moment capacity</th></tr>
<tr><td>Ultimate bond reduction coefficient for external strands,
$$\Omega_u = \frac{5.4}{\dfrac{L_u}{d_u}}$$
Reinforcement ratio,
$$\rho = \frac{\sum_{j=1}^{m} F_{pi} + f_{pbb}A_{pb} + f_{pnbb}A_{pn} + F_{pui} + f_{pub}A_{fu} - f_{pnfb}A_{pnf}}{b\times d_m \times f_{fu}}$$
Since, the section is over-reinforced, failure of the beam will occur due to the crushing of concrete topping. The steps to calculate the moment capacity of the DT-beam are given below.</td><td>$L_u = 15670$ mm
$d_u = 1035$ mm
$$\Omega_u = \frac{5.4}{\dfrac{L_u}{d_u}} = \frac{5.4}{\dfrac{15670}{1035}} = 0.36$$
$d_m = 1194$ mm
$b = 2120$ mm
$\Sigma F_{pi} = 5576.8$ kN
$F_{pui} = 1772$ kN
$f_{pbb} = 130$ MPa
$f_{pnbb} = 1050.8$ MPa
$f_{pub} = 497.8$ MPa
$f_{pnf} = 275.4$ MPa
$A_{pb} = 4{,}296.8 \text{ mm}^2$
$A_{pn} = 2{,}131.6 \text{ mm}^2$
$A_{fu} = 3019.3 \text{ mm}^2$
$A_{pnf} = 1360.6 \text{ mm}^2$
$\rho =$
$$\frac{5576.8 + 1301\times 4296.8 + 1050.8\times 2131.6 + 1772 + 497.8\times 3019.3 - 275.4\times 1360.6}{2120\times 1194\times 2860}$$
$= 0.0023 > 0.002(\rho_b)$
$A_{fb} = 429.6 \text{ mm}^2$
$A_{fu} = 3101.4 \text{ mm}^2$</td></tr>
</table>

Procedure	Calculation in SI units
	$b = 2\,120$ mm $E_f = 147\,000$ MPa $E_{fp} = 127\,000$ MPa $E_{fn} = 137\,000$ MPa
a. Compute neutral axis depth of the section at ultimate failure	
In an over-reinforced section, the depth to the neutral axis ($c = k_u d_m$) can be computed using coefficient k_u (Grace and Singh, 2003) as expressed below: $k_u = \dfrac{A+\sqrt{A^2+4B}}{2}$ $A = \dfrac{\left[\sum_{j=1}^{m} A_{fb}E_f\varepsilon_{pbji} + A_{fu}\varepsilon_{pui}E_{fp} - \varepsilon_{cu}\left(\sum_{j=1}^{q} A_{fj}E_{fj} + \Omega_u A_{fu}E_{fp}\right)\right]}{0.85 f_c' b\beta_1 d_m}$ $B = \dfrac{\varepsilon_{cu}\left(\sum_{j=1}^{q} A_{fj}E_{fj}h_j + \Omega_u A_{fu}E_{fp}d_u\right)}{0.85 f_c' b\beta_1 d_m^2}$ $E_{fj} = E_f$ (for Leadline tendons); E_{fp} (for post-tensioning CFCC strands); and E_{fn} (for non-prestressing CFCC strands). Calculate various terms of the parameters A and B and then substituting in the expression of A and B, we can get A and B. Further, substitute the values of A and B in the expression for k_u The depth to the neutral axis, $c = k_u\, d_m$	$\varepsilon_{pbji} = 0.0059$ (for row number 1–5) $= 0.0062$ (for row number 6–10) $\varepsilon_{pui} = 0.0046$ $A_{fj} = 429.6$ mm^2 (for Leadline pretensioning tendons of rows 1 through 10) $= 304$ mm^2 (for bonded CFCC strands of rows 1 through 5) and 608 mm^2 (for 6th row, i.e., bottom row of CFCC strands) $= 716$ and 644.4 mm^2 (for non-prestressing Leadline tendons provided at top and bottom of flange respectively.) $h_j = 564$, 634, 704, 774, 844, 914, 984, 1054, 1124, 1194 mm for Leadline pretensioning tendons of rows 1 to 10, respectively, $h_j = 225$, 420, 600, 810, 1020, 1235 mm for CFCC strands of rows 1 to 6, respectively. $= 115$ and 175 mm for Leadline tendons at the top and bottom of the flange, respectively $\sum_{j=1}^{m} A_{fb}E_f\varepsilon_{pbji} = 5 \times 429.6 \times 147\,000 \times 0.0059 + 5 \times 429.6 \times 0.0062 = 3\,820\,647.6$ $A_{fu}\varepsilon_{pui}E_{fp} = 3101.4 \times 0.0046 \times 127\,000 = 1\,758\,675.68$ $\varepsilon_{cu}\left(\sum_{j=1}^{q} A_{fj}E_{fj} + \Omega_u A_{fu}E_{fp}\right) = 0.003(631\,512\,000 + 291\,536\,000 + 199\,978\,800 + 137\,635\,488) = 3\,781\,986.864$ $0.85 f_c' b\beta_1 d_m = 0.85 \times 53.8 \times 2120 \times 0.66 \times 1194 = 76\,398\,586.7$ $0.85 f_c' b\beta_1 d_m^2 = 0.85 \times 53.8 \times 2120 \times 0.66 \times 1194^2 = 9.122 \times 10^{10}$ $A = 0.0235$ and $B = 0.03148$, $k_u = 0.1896$ $c = 0.1896 \times 1194 = 226.4$ mm

Procedure	Calculation in SI units
b. Compute strains in tendons/strands	
Strain in bonded prestressing tendons, $\varepsilon_{pbji} = \frac{0.003}{c}(d_j - c) + \varepsilon_{pbji}$ (for $j = 1, 10$)	$\varepsilon_{pbji} = 0.0059$ for layer number, $j = 1$ to 5 $= 0.0062$ for layer number, $j = 6$ to 10
Strain in non-prestressing tendons, $\varepsilon_{pnj} = \frac{0.003}{c}(h_j - c)$ (for $j = 1, 6$)	$d_1 = 564$ mm; $d_2 = 634$ mm; $d_3 = 704$ mm; $d_4 = 774$ mm; $d_5 = 844$ mm; $d_6 = 914$ mm;
Strain in non-prestressing tendons at flange top, $\varepsilon_{pnt} = \frac{0.003}{c}(c - d_t)$	$d_7 = 984$ mm; $d_8 = 1054$ mm; $d_9 = 1124$ mm; $d_{10} = 1194$ mm.
Strain in non-prestressing tendons at flange bottom, $\varepsilon_{pnb} = \frac{0.003}{c}(c - d_b)$	$h_1 = 225$ mm; $h_2 = 420$ mm; $h_3 = 600$ mm; $h_4 = 10$ mm; $h_5 = 844$ mm and h_6 are 225, 420, 600, 810, 1020 and 1235 mm
Strain in unbonded tendons, $\varepsilon_{pu} = \varepsilon_{pui} + \Delta e_{pu}$ $= e_{pui} + \Omega_u \frac{0.003}{c}(d_u - c)$	$d_t = 115$ mm; $d_b = 175$ mm; $d_u = 1035$ mm.
Using the above equations, strain in tendons/strands of different layer can be calculated by substituting the values and then using stress-strain relation and force-stress relations, stresses and forces in tendons/strands can be computed. For example,	$\varepsilon_{pb1} = \frac{0.003}{226.4}(564 - 226.4) + 0.0059$ $= 0.01037$ $f_{pb1} = 147{,}000 \times 0.01037$ $= 1524.4$ MPa < 2860 MPa (f_{fu}) O.K.
$\varepsilon_{pb1} = \frac{0.003}{c}(d_1 - c) + \varepsilon_{pb1i}$ $f_{pb1} = E_f \times \varepsilon_{pb1}$ $F_{pb1} = f_{pb1} \times$ cross-sectional area of each tendon	$F_{pb1} = 1524.4 \times 71.6 = 109.1$ kN The computed values of strains, stresses and forces for tendons and strands of various layers are presented in Table E4.6, E4.7 and E4.8, respectively,
c. Compute resultant forces in tendons/strands and concrete	
Compressive force in concrete = $C_t + C_f$ $= 0.85 f'_c \frac{E_{ct}}{E_c} b h_t + 0.85 f'_c b(\beta_1 c - h_t)$	$C_t + C_f = 0.85 \times 53.8 \times 2120\ [0.86 \times 75 + (0.66 \times 226.4 \times 75)] = 13\,468.3$ kN $C = 13\,468.3 + 10 \times 15.8 + 9 \times 7.4$ $= 13692.9 \times 13693$ kN
Total compressive force, $C = C_t + C_f + F_{pnt} + F_{pnb}$ Total resultant tension force, $F_R = \sum_{j=1}^{10} F_{pbj} + \sum_{j=1}^{6} F_{pnj} + F_{pu}$	$F_R = 14\,692$ kN $F_R = C = \frac{14692 + 13693}{2} = 14192.5$ kN $C_t \frac{h_t}{2} = 6215.12 \times 0.5 \times 75 = 233\,067$
F_R can be obtained by substituting the forces of each tendon (Table E4.8) in the above equation. The difference in the resultant compressive force and tensile forces can be balanced by taking their average value.	$C_f\left(h_t + \frac{a - h_t}{2}\right) = 715.2 \times \left(75 + \frac{149.24 - 75}{2}\right) = 809\,632$ $F_{pnt} d_t + F_{pnb} d_n = 158 \times 115 + 66.6 \times 175 = 29\,825$

Procedure	Calculation in SI units
Distance of the center of gravity of resultant compression force from the extreme compression fiber, $\bar{d} = \dfrac{C_t \frac{h_t}{2} + C_f(h_t + \frac{a-h_t}{2}) + F_{pnt}d_t + F_{pnb}\,d_b}{C}$	$\bar{d} = \dfrac{(233067 + 809632 + 29825}{13693} = 78.43$ mm
Nominal moment capacity, $M_n = F_R(d - \bar{d})$	$M_n = 14\,192.5 \times (959.9 - 78.43)$ $= 12\,506$ kN-m
Design moment capacity, $M_u = \phi M_n > (M_{required})$	$M_u = 0.85 \times 12506$ $= 10\,630.1$ kN-m $> 5\,050.6$ kN-m ($M_{required}$) O.K.

d. Compare the theoretical and experimental values of moment capacity and post-tensioning tendon force

Experimental ultimate failure load of DT-beam = 2443 kN

Experimental value of the moment capacity of DT-beam

$$\frac{2443 \times 8.382}{2} + 1616 = 11855 \text{ kN-m}$$

Theoretical nominal moment capacity of DT-beam = 12 506 kN-m

Percent difference in experimental and theoretical moment capacities

$$\frac{(12506 - 11\,855)}{11\,855} \times 100 = 5.5\%$$

Experimental value of post-tensioning force in each post-tensioning strand = 806.8 kN

Theoretical value of post-tensioning force in each tendon = 808 kN

Percent difference in experimental and theoretical values of post-tensioning force

$$= \frac{(808 - 806.8)}{806.8} \times 100 = 0.1\%$$

e. Compute stresses due to service loads:

Moment due to service loads, M = 1616 + 1944.6

= 3560.6 kN-m < 4358.94 kN-m (cracking moment)

Since, moment due to service loads is less than the cracking moment, the section will remain uncracked at the service load. The maximum stresses can be calculated as follows:

Maximum compressive stress in concrete at the extreme compression fiber,

$$f_{ct} = \frac{F_{pre}}{A_p} - \frac{F_{pre}e_b}{S_{tp}} + \frac{0.6 \times F_{post}}{A_p} - \frac{0.6 \times F_{post}e_{up}}{S_{tp}} + \frac{0.4 \times F_{post}}{A_c} - \frac{0.4 \times F_{post} \times e_{uc}}{S_{tc}} + \frac{M}{S_{tc}} \quad \text{(E4.4)}$$

Substituting the values of sectional properties (Table E4.5) and other parametric values in Eq. E4.4, we get:

$f_{ct} = 8.8$ MPa < 32.3 MPa$(0.6f'c)$ O.K.

Maximum concrete stress due to applied load $= \dfrac{1944.6 \times 1000}{3.4683 \times 10^8} = 5.6$ MPa

Experimental value of concrete stress = 6.9 MPa, which is higher than theoretical value and is attributed to the change in the strain condition, because of changing of support condition of the DT-test beam during various construction stages, and due to different properties of concrete topping and precast concrete.

Stress in bottom prestressing tendons,

$$f_{pb10} = E_f \varepsilon_{pb10i} + \frac{E_f}{E_c} \frac{M \; (d_{10} - y_{tc})}{I_c} \tag{E4.5}$$

$$= 147000 \times 0.0062 + 4.0 \times \frac{3560.6 \times 10^6 \times (1194 - 484.6)}{1.6808 \times 10^{11}}$$

$$= 911.4 + 60.11 = 971.5 \text{ MPa} < 2860 \text{ MPa} \qquad \text{O.K.}$$

Stress in bottom prestressing tendons due to applied load

$$= 4.0 \times \frac{1944.6 \times 10^6 \times (1194 - 484.6)}{1.6808 \times 10^{11}}$$

$$= 32.8 \text{ MPa} \cong 32.6 \text{ MPa (Experimental value)}$$

Stress in post-tensioning strands,

$$f_{pu} = E_{fp}\, \varepsilon_{pui} + \frac{E_{fp}}{E_c} \frac{(M - M_D) e_{uc}}{I_c} \Omega$$

$$= 127\,000 \times 0.0046 + 3.46 \times \frac{(3560.6 - 1616) \times 550.4}{1.6808 \times 10^{11}} \times \frac{2}{3}$$

$$= 598.9 \text{ MPa} < 1870 \text{ MPa}$$

Force in a CFCC strand at service load = 598.9 × 752.6 = 450.7 kN

Experimental value of post-tensioning force = 450 kN

Percent difference in theoretical and experimental values of post-tensioning force at service load

$$= \frac{450.7 - 450}{450} \times 100 = 0.2\%$$

f. Compute maximum delection under applied service load (No cracking occurs under service load)

$$M_L = M - M_D = 1{,}944.6 \text{ kN}$$

Distance between the support and nearest load point, $L_1 = 8{,}382$ mm

Longitudinal distance between load points, $L_2 = 3{,}658$ mm

Deflection due to applied load,

$$\delta_a = \frac{M_L L_1^2}{8E_c I_c}\left[\frac{8}{3} + 4.0\left(\frac{L_2}{L_1}\right) + \left(\frac{L_2}{L_1}\right)^2\right] \tag{E4.6}$$

$$= \frac{1944.6\times10^6\times(8382)^2}{8\times36700\times1.6808\times10^{11}}\left[\frac{8}{3} + 4.0\left(\frac{3658}{8382}\right) + \left(\frac{3658}{8382}\right)^2\right]$$

$$= 12.7 \text{ mm} \downarrow \cong 12.8 \text{ mm} \quad \text{(Experimental value)}$$

Assuming that the beam is supported at two end supports, the deflection due to dead load is given by,

$$\delta_d = \frac{5}{384}\frac{W_d L^4}{E_c I_c} = \frac{5}{384}\frac{31\times20422^4}{36\,700\times1.6808\times10^{11}} = 11.4 \text{ mm} \downarrow$$

Assuming that the loss of pre-tensioning forces up to the instant of initial post-tensioning be negligible,

Effective pretensioning force = 5073 kN

Total effective post-tensioning force = 1772 kN

Deflection due to pre-tensioning force at the instant of initial post-tensioning

$$= \frac{F_{\text{pre}} e_b L^2}{8E_c I_p} \uparrow = \frac{5073\times1000\times330.2\times(20\,422)^2}{8\times36700\times1.3673\times10^{11}} = 17.4 \text{ mm} \uparrow$$

Horizontal distance between the two end anchor points of post-tensioning tendons,

$$L_u = 15\,670 \text{ mm}$$

Deflection due initial post-tensioning

$$= \frac{F_{\text{post}} e_{up} L_u^2}{8E_c I_p} = \frac{0.6\times1772\times1000\times485.6\times15670^2}{8\times36\,700\times1.3673\times10^{11}} = 3.2 \text{ mm} \uparrow$$

Increase in deflection due to final post-tensioning

$$= \frac{F_{\text{post}} e_{uc} L_u^2}{8E_c I_c} \uparrow = \frac{0.4\times1772\times1000\times550.4\times15670^2}{8\times36\,700\times1.6808\times10^{11}} = 1.9 \text{ mm} \uparrow$$

Total deflection due to pre-stressing, $\delta_p = 17.4 + 3.2 + 1.9 = 22.5$ mm ↑

Net upward camber prior to loading = $\delta_p - \delta_d = 22.5 - 11.4 = 11.1$ mm ↑

Experimental value of the camber after final post-tensioning = 14.2 mm ↑

Total deflection of the DT-beam under service load,

$$\delta = \delta_{a} + \delta_{d} - \delta_{p} = 12.7 + 11.4 - 22.5 = 1.6 \text{ mm} \uparrow$$

Note 1: the difference in the theoretical and experimental value of the beam deflection, is due to changes in support condition from initial post-tensioning to final post-tensioning and due to transportation of beam from the precast plant to the testing facility.

Note 2: A simple analytical procedure is presented herein to theoretically evaluate the strength, failure mode, deflection, stresses and strains in tendons and concrete and post-tensioning forces in unbonded tendons. The presented procedure precisely estimates the moment capacity and the failure mode, cracking load and forces in the unbonded post-tensioning strands of the DT-test beam. Overall, theoretical results are closer to the corresponding experimental results, especially under the service load condition. The theoretical nominal moment capacity of the DT-test beam is about 5.5% higher in comparison to the corresponding experimental value, whereas, the theoretical post-tensioning forces at the service and ultimate conditions are almost equal to their corresponding experimental values. Thus, unified design approach (Grace and Singh, 2003) can reasonably predict ultimate load behavior of CFRP reinforced and prestressed concrete beams, using internally bonded pre-tensioning and externally unbonded post-tensioning tendons/strands arranged vertically in distributed layers along with non-prestressing strands/tendons with different material combinations.

Table E4.6. Strains in tendons of different layers at ultimate load condition of DT-beam.

	Strain				
	Tendon type				
Layer No.	**Bonded pretensioning tendons**	**Non-prestressing tendons located in webs**	**Unbonded post-tensioning tendons**	**Non-prestressing tendons located at flange top**	**Non-prestressing tendons located at flange bottom**
1	0.01037	–0.0000186*	0.008457	0.0015	0.00007
2	0.0113	0.0049510			
3	0.0122	0.002565			
4	0.01316	0.004951			
5	0.01408	0.007700			
6	0.01530	0.01336	N/A	N/A	N/A
7	0.01620				
8	0.01720	N/A			
9	0.01810				
10	0.01900				

* This strain value is very small due to the neutral axis located close to the layer 1 and negative (-ve) value refers to the compressive strain. N/A refers to not-applicable.

Table E4.7. Stresses in tendons of different layers at ultimate load condition of DT-beam.

Strain (MPa)*					
Layer No.	Tendon type				
	Bonded pretensioning tendons	Non-prestressing tendons located in webs	Unbonded post-tensioning tendons	Non-prestressing tendons located at flange top	Non-prestressing tendons located at flange bottom
1	1524.4	–2.5	1074	220.5	11.2
2	1661.1	351.5	N/A	N/A	N/A
3	1793.4	678.3			
4	1933.9	1054.9			
5	2069.8	1438.5			
6	2249.1	1830.3			
7	2381.4	N/A			
8	2528.4				
9	2660.7				
10	2793.0				

* Stresses in tendon/strands are less than specified strength of corresponding tendons or strands and hence, applicable; N/A refers to not-applicable.

Table E4.8. Resultant forces in tendons of different layers at ultimate load condition of DT-beam.

Force (kN)					
Layer No.	Tendon type				
	Bonded pretensioning tendons	Non-prestressing tendons located in webs	Unbonded post-tensioning tendons	Non-prestressing tendons located at flange top	Non-prestressing tendons located at flange bottom
1	109.1	–0.2*	808.0	15.8	7.4
2	118.9	26.7	N/A	N/A	N/A
3	128.4	51.6			
4	138.5	80.2			
5	148.2	109.3			
6	161.0	139.1			
7	170.5	N/A			
8	181.0				
9	190.5				
10	200.0				

N/A refers to not-applicable. * Negative (-ve) value refers to the compression force.

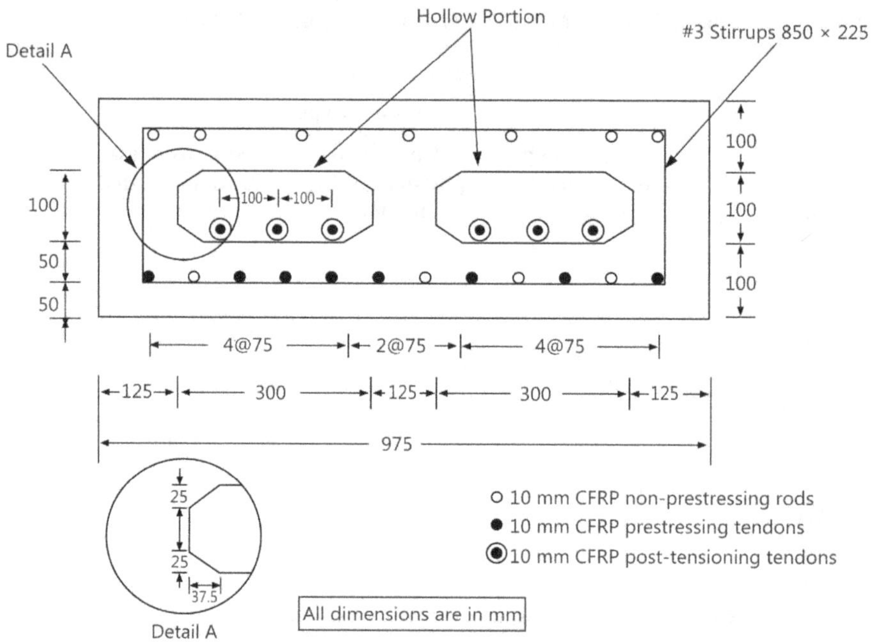

Figure E4.6. Cross-section Configuration of Box Beam.

E4.6. Design Example 6: Case Study of CFRP Prestressed Concrete Box-Beam

Problem Statement

Determine the flexure capacity of a box-beam, prestressed with 6 unbonded and 7 bonded CFRP tendons, and reinforced with 2 non-prestressing bars at bottom and 7 non-prestressing bars at the top. The cross-section details of the beam are shown in Fig. E4.6. The effective span of the beam is 4.575 m. The beam is reinforced and prestressed with DFI bars and tendons having f_{fu} = 1390 MPa. All tendons and bars have a diameter (d) of 9.5 mm. All other properties of tendons and concrete are listed in the following.

Material properties

$$f_{fu} = 1390 \text{ MPa}$$

$$E_f = 131 \text{ GPa}$$

$$\varepsilon_{fu} = 0.0147$$

$$f_c' = 48.3 \text{ MPa}$$

$$E_c = 32.9 \text{ GPa}$$

$$f_r = 3.45 \text{ MPa}$$

Section properties

Bar and tendons diameter, $d = 9.5$ mm

Area of bars and tendons, $A = 71\ \text{mm}^2$

Width of box beam, $b = 950$ mm

Moment of inertia of section, $I = 2080 \times 10^6\ \text{mm}^4$

$d_m = 250$ mm

Prestressign forces

Force per tendon = 93.5 kN

$f_i = 1316.3$ MPa

$\varepsilon_{pbmi} = 0.01$

Total pre-tensioning force on section = 656.4 kN

Total post-tensioning force on section = 549.6 kN

Solution

Balancec section

$$\rho_b = 0.85\beta_1 \frac{f_c'}{f_{fu}} \frac{\varepsilon_{cu}}{\varepsilon_{cu} + \varepsilon_{fu} - \varepsilon_{pbmi}}$$

$\beta_1 = 0.7$

$f'_c = 48.3$ MPa

$f_{fu} = 1930$ MPa

$\varepsilon_{cu} = 0.003$

$\varepsilon_{fu} = 0.0147$

$\varepsilon_{pbmi} = 0.01$

$$\rho_b = 0.85 \times 0.7 \times \frac{48.3}{1930} \times \frac{0.003}{0.003 + 0.0147 - 0.01} = 0.0058$$

$$\rho = \frac{\sum_{i=1}^{p} A_{fi}\alpha_i}{b\, d_m}$$

$b = 950$ mm

$d_m = 250$ mm

Figure E4.7 shows the balanced section and the strains induced in the tendons and bars. The stresses in the tendons and bars are calculated accordingly and are shown as follows:

Stress at non-prestressing tendons = 614.3 MPa

Stress at prestressing tendons = 1930 MPa

Stress at post-tensioning tendons = 1730 MPa

Stress at compression non-prestressing tendons = 191.7 MPa

$$\therefore \sum_{i=1}^{p} A_{fi}\alpha_i = 7\times 71\times 1 + 2\times 71\times \frac{614.3}{1930} + 6\times 71\times \frac{1730}{1930} - 7\times 71\times \frac{191.7}{1930} = 874.7$$

$$\therefore \rho = \frac{847.7}{950\times 250} = 0.0037$$

ρ is less than ρ_b and greater than $0.5\rho_b$, thus the section is under reinforced.

Cracking moment

$$M_{cr} = (f_r + \Sigma\sigma_{bp})S_b$$

$$\sum \sigma_{bp} = \frac{549.6\times 1000}{225000} + \frac{549.6\times 1000\times 20\times 150}{2088\times 10^6} + \frac{656.4\times 1000}{225000} + \frac{656.4\times 1000\times 100\times 150}{2088\times 10^6}$$

$$= 12.05 \text{ MPa}$$

$$M_{cr} = (3.45 + 12.05) \times 1392 \times 10^4 = 215.76 \times 10^6 \text{ N-mm.}$$

$$M_{cr} = 215.76 \text{ kN-m}$$

Flexural capacity

$$n = k_u\, d_m$$

$$k_u = \frac{-B \pm \sqrt{B^2 - 4AC}}{2A}$$

$$A = 0.85\ f_c' b \beta_1 d_m^2$$

$$= 0.85 \times 48.3 \times 950 \times 0.7 \times 250^2 = 17063.5 \times 10^5$$

$$B = -\left[A + F_{pi} d_m + \varepsilon_f d_m\left(\sum_{i=1}^{q} A_{fi}E_{fi} + \Omega_u A_{fu} E_{fp}\right)\right]$$

$$F_{pi} = 1206 \text{ kN}$$

$$d_m = 250 \text{ mm}$$

$$\varepsilon_f = 0.0047$$

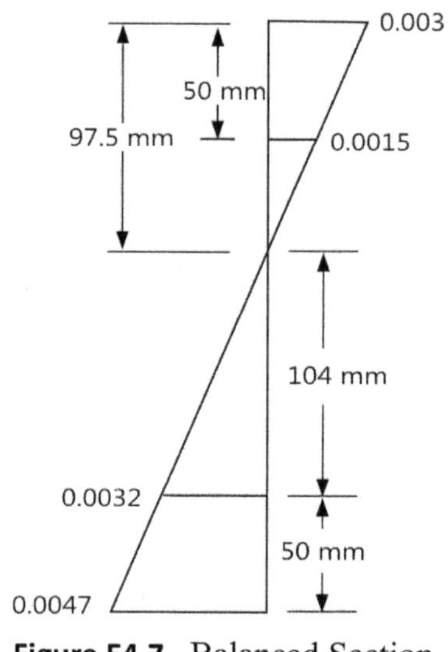

Figure E4.7. Balanced Section.

$$\sum_{i=1}^{q} A_{fi}E_{fi} = 7 \times 71 \times 131 \times 10^3 + 2 \times 71 \times 131 \times 10^3 - 7 \times 71 \times 131 \times 10^3 = 1860.2 \times 10^4$$

$$\Omega_u = \frac{5.4}{\left(\dfrac{L_u}{d_u}\right)} \qquad (L_u = 4800 \text{ mm}, d_u = 200 \text{ mm})$$

$$\therefore \Omega_u = \frac{5.4}{\left(\dfrac{4800}{200}\right)} = 0.225$$

$$A_{fu} = 6 \times 71 = 426 \text{ mm}^2$$

$$E_{fp} = 131 \times 10^3 \text{ MPa}$$

$$\therefore B = -\begin{bmatrix} 17063.5\times10^5 + 1206\times10^3\times250 + 0.0047\times250\times \\ \left(1860.2\times10^4 + 1255.635\times10^4\right) \end{bmatrix}$$

$$= -20444.61\times10^5$$

$$C = \left[F_{pi}d_m + \varepsilon_f\left(\sum_{i=1}^{q} A_{fi}E_{fi}h_i + \Omega_u A_{fu}E_{fp}d_u\right)\right]$$

$$F_{pi}\, d_m = 3015 \times 10^5$$

$$\varepsilon_f = 0.0047$$

$$\sum_{i=1}^{q} A_{fi}E_{fi}h_i = 497 \times 131 \times 10^3 \times 250 + 142 \times 131 \times 10^3 \times 250 - 491 \times 131 \times 10^3 \times 50$$

$$= 1.77112 \times 10^{10}$$

$$\therefore k_u = 0.24$$

$$\Omega_u A_{fu} E_{fp} \times d_u = 0.225 \times 426 \times 131 \times 10^3 \times 200 = 251\,127 \times 10^4$$

$$C = 3015 \times 10^5 + 0.0047\,(177\,112 \times 10^{10} + 251\,127 \times 10^4) = 396\,545\,609$$

$$\therefore k_u = 0.24$$

$$\therefore n = k_u\, d_m = 0.24 \times 250 = 60 \text{ mm}$$

Strain in prestressing tendons

$$\varepsilon_{pbj} = (0.0147 - 0.01)\frac{(250-60)}{(250-60)} + 0.01 = 0.0147$$

Strain in non-prestressing tendons

$$\varepsilon_{pnj} = (0.0147 - 0.01)\frac{(250-60)}{(250-60)} = 0.0047$$

Strain in concrete at the extreme compression fiber

$$\varepsilon_c = (0.0147 - 0.01)\frac{60}{(250-60)} = 0.00148$$

Strain in non-prestressing tendons in compression zone,

$$\varepsilon_{ct} = (0.0147 - 0.01)\frac{(60-50)}{(250-60)} = 0.000247$$

$$\Delta\varepsilon_{pu} = 0.225\frac{(0.0147-0.01)(200-60)}{(250-60)} = 0.00078$$

∴ Strain in the unbonded prestressing tendons

$$(\varepsilon_{pu}) = 0.01 + 0.00078 = 0.01078$$

Stress in bonded prestressing tendons,

$$f_{pbj} = 131\,000 \times 0.0147 = 1925.7 \text{ MPa} \leq 1930 \text{ MPa} \qquad \text{O.K.}$$

Stress in non-prestressing tendons,

$$f_{pnj} = 131\,000 \times 0.0047 = 615.7 \text{ MPa} \leq 1930 \text{ MPa} \qquad \text{O.K.}$$

Stress in non-prestressing tendons in compression,

$$f_{pnt} = 131\,000 \times 0.000247 = 32.41 \text{ MPa} \leq 1930 \text{ MPa} \qquad \text{O.K.}$$

Stress in unbonded post-tensioning tendons,

$$f_{pu} = 131\,000 \times 0.01078 = 1412.2 \text{ MPa} \qquad \text{O.K.}$$

Resultant force in bonded prestressing tendons,

$$F_{pbj} = 1925.7 \times 497/1000 = 957.1 \text{ kN}$$

Resultant force in non-prestressing tendons,

$$F_{pnb} = 615.7 \times 142/1000 = 87.4 \text{ kN}$$

Resultant force in unbonded post-tensioning tendons,

$$F_{pu} = 1412.2 \times 426/1000 = 601.6 \text{ kN}$$

Resultant force in non-prestressing tendons,

$$f_{pnt} = 32.41 \times 497/1000 = 16.11 \text{ kN}$$

Resultant compression force in concrete,

$$C_c = 0.85 f_c' b\beta_1 c = 0.85 \times 0.7 \times 48.3 \times 950 \times 60/1000 = 1638.1 \text{ kN}$$

Distance of resultant compression force from extreme compression fiber,

$$\bar{d} = \frac{(1638.1 \times 0.7 \times 60/2 + 16.11 \times 50}{1638.1 + 16.11} = 21.28$$

Distance of centroid of resultant compression force from extreme compression fiber = 21.28 mm.

Ultimate moment carrying capacity

$$M_n = 957.1 \times (250 - 21.28) + 87.4 \times (250 - 21.28) + 601.6 \times (200 - 21.28)$$
$$= 346.416 \times 10^3 = 346.42 \text{ kN-m}$$

E4.7. Design Example

A simply supported rectangular beam of 5.25 m overall length is resting on 250 mm wide masonry walls at its ends. The cross-section of beam is 300 mm wide by 400 mm deep with 50 mm clear concrete cover. The CFRP reinforcement at tension side consists of three 10.5 mm CFRP bars and 2-legged CFRP stirrups at 100 mm c/c spacing. Assume cast in situ beam. Properties of concrete and CFRP materials are given as follows:

Concrete

$$f_c' = 20 \text{ N/mm}^2,\ f_{ck} = 25 \text{ N/mm}^2$$

CFRP

$$f_{fu}^* = 600 \text{ N/mm}^2,\ \varepsilon_{fu}^* = 600 \text{ mm/mm},\ E_f = 45000 \text{ MPa}$$

Determine the flexural capacity of Beam as per ISIS Canada Guidelines.

Solutions

Step 1: Preliminary calcuations

$\phi_{frp} = 0.8$ for CFRP; $\phi = 0.6$ for Cast-in-situ Concrete

$$A_{frp} = 3 \times \frac{\pi}{4} \times 10.5^2 = 259.8 \text{mm}^2$$

Effective depth, $d = 400 - 50 = 350$ mm

Step 2: CFRP reinforcement ratio

$$\rho_{frp} = \frac{A_{frp}}{b \times d} = \frac{259.8}{300 \times 350} = 0.0025$$

Step 3: Balanced ration

$$\alpha_1 = 0.85 - 0.0015 f_c' \geq 0.67$$
$$= 0.85 - 0.0015 \times 20 = 0.82$$

$$\beta_1 = 0.97 - 0.0025 f_c' \geq 0.67$$
$$= 0.97 - 0.0025 \times 20 = 0.92$$

$$\rho_{frpb} = \alpha_1 \beta_1 \frac{\varphi_c}{\varphi_{frp}} \frac{f_c'}{f_{rpu}} \left[\frac{\varepsilon_{cu}}{\varepsilon_{cu} + \varepsilon_{frpu}} \right] = 0.82 \times 0.92 \times \left(\frac{0.60}{0.80} \right) \times \left(\frac{20}{600} \right) \left[\frac{0.0035}{0.0035 + 0.014} \right]$$
$$= 0.0038$$

Step 4: Check wheter the section will fail due to tension failure or compressin failure

Since $\rho_{frp} = 0.0025 < \rho_{frpb} = 0.0038$, the failure of beam will be governed by tension failure. The strain distribution is shown in Fig. E4.8.

Step 5: Perform iterative strain analysis for neutral axis depth, *c*

Let the depth to the neutral axis, $c = 48$ mm. From the strain compatibility condition, we have,

$$\frac{\varepsilon_c}{c} = \frac{\varepsilon_{frpu}}{d - c} \Rightarrow \varepsilon_c = 48 \times \frac{0.014}{(350 - 48)} = 2.23 \times 10^{-3}$$

Tensile force,

$$T = \phi_{frp} f_{frpu} A_{frp} = \frac{0.8 \times 600 \times 259.8}{1000} = 124.7 \, \text{kN}$$

Using Figs. 4.9 and 4.10, for $f_c' = 20 \, \text{MPa}$, $\varepsilon_c = 2.23 \times 10^{-3}$

$$\alpha = 0.93 \text{ and } \beta = 0.78$$

Resultant Compression Force,

$$C = \alpha \phi_c f_c' \beta c b = 0.93 \times 0.6 \times 20 \times 0.78 \times 48 \times 300 \times 10^{-3} = 125.3 \, \text{kN} \cong T = 124.7 \, \text{kN}$$

So take $c = 48$ mm

Step 6: Calculate moment capacity of the section

$$M_r = \phi_{frp} A_{frp} f_{frpu} \left(d - \frac{\beta c}{2} \right) = 124.7 \left(350 - \frac{0.78 \times 48}{2} \right) \times 10^{-3} = 41.31 \text{ kN-m}$$

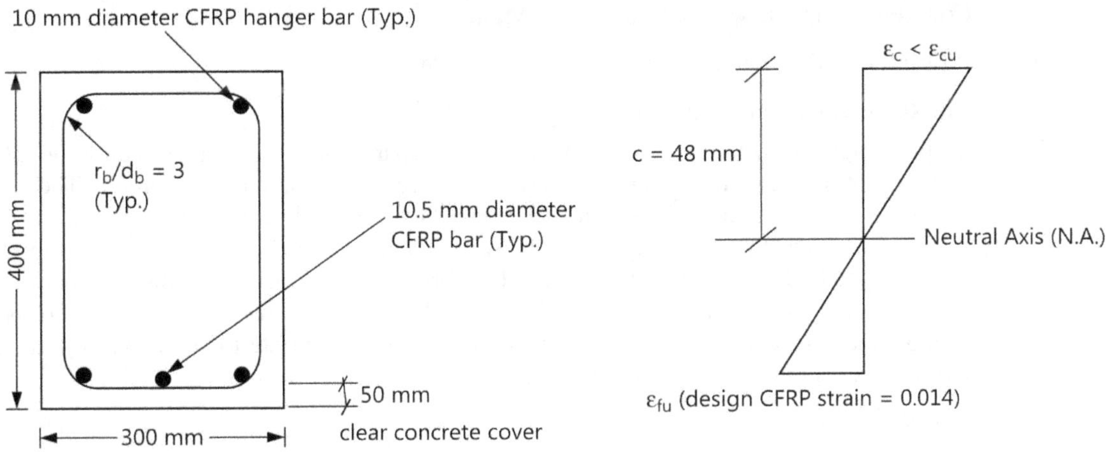

Figure E4.8. Cross-section details and strain distribution of CFRP Reinforced RC Beam.

Exercise Problems

1. A simply supported rectangular beam of 6.3 m overall length is resting on 300 mm wide masonry walls at its ends. The cross-section of beam is 350 mm wide by 550 mm deep with 50 mm clear concrete cover. The CFRP reinforcement at tension side consists of 3 Nos. 10 mm CFRP bars and 2-legged CFRP stirrups at 200 mm c/c spacing. Assume cast in situ beam. Properties of concrete and CFRP materials are given as follows:

   Concrete:

$$f'_{fu} = 1000 \text{ N/mm}^2, f_{ck} = 25 \text{ N/mm}^2$$

   CFRP:

$$f'_{fu} = 1000 \text{ N/mm}^2,\ \varepsilon_{fu} = 0.015 \text{ mm/mm}^2,\ E_f = 150 \text{ GPa}$$

   Determine the flexural capacity of Beam as per ACI 440.1R-06 and ISIS Canada Guidelines.
2. Using ISIS Canada Design guidelines, calculate the factored moment resistance, M_r, for a precast ($\phi_c = 0.65$) FRP-reinforced concrete section with the following dimensions:
   Width of beam, $b = 300$ mm, depth $h = 450$ mm
   The tensile reinforcement consists of eight 10.5 mm diameter GFRP ISOROD bars (bundled in pairs) in a single layer. Assume that the shear reinforcement consists of 6 mm diameter Leadline™ stirrups and that the beam has an interior exposure condition. The properties of materials are given as follows:

Concrete compressive strength, $f'_c = 35$ MPa;

ISOROD GFRP tensile strength, $f_{frpu} = 617$ MPa

ISOROD GFRP tensile modulus, $E_{frp} = 42$ GPa

3. A simply supported rectangular beam with effective span of 4.5 m has cross-section of 350 mm wide by 550 mm deep with 50 mm clear concrete cover to the main reinforcement. The beam is subjected to imposed load of 15 kN/m and exposed to earth and weather conditions. The CFRP reinforcement at tension is arranged in two layers. Each layer consists of 3 Nos. of 8 mm bars. The two legged CFRP stirrups are provided at 200 mm c/c spacing. The c/c distance between the two layers of reinforcement is 80 mm. Using ACI 440.1R-06, evaluate the moment of resistance of section and check whether the beam cross-section is appropriate to carry the imposed load. The properties of concrete and CFRP materials are given as follows:

   Concrete:

   Cylinder strength, $f'_c = 20$ N/mm^2, Cube Strength, $f_{ck} = 25$ N/mm^2

   CFRP:

Tensile strength, f^*_{fu}	1800 MPa
Rupture strain, ε^*_{fu}	0.015 mm/mm
Modulus of elasticity, E_f	120 GPa

4. Calculate the factored moment resistance, M_r, for a precast ($\phi_c = 0.65$) FRP-reinforced concrete section with the following dimensions:

   Section width, $b = 400$ mm; Section depth, $h = 600$ mm

   The tensile reinforcement consists of six #10 ISOROD CFRP bars in a single layer. Assume that the shear reinforcement consists of 5 mm diameter Leadline™ stirrups and the beam has interior exposure conditions. The material properties are given as follows:

   Concrete compressive strength, $f'_c = 35$ MPa.

   ISOROD CFRP tensile strength, $f_{frpu} = 1600$ MPa

   ISOROD CFRP tensile modulus, $E_{frp} = 110$ GPa

   The area of one #10 bar, $A_{bar} = 71$ mm^2

   The diameter of one #10 bar, $d_b = 9.3$ mm

5. As per ISIS Canada design guidelines, calculate the factored moment resistance, M_r, in positive bending, for the precast ($\phi_c = 0.65$) FRP-reinforced concrete section as shown in Fig. P4.1. Also calculate the moment of resistance as per ACI 440.1R-06 design standards. Assume that the beam has an interior exposure condition:

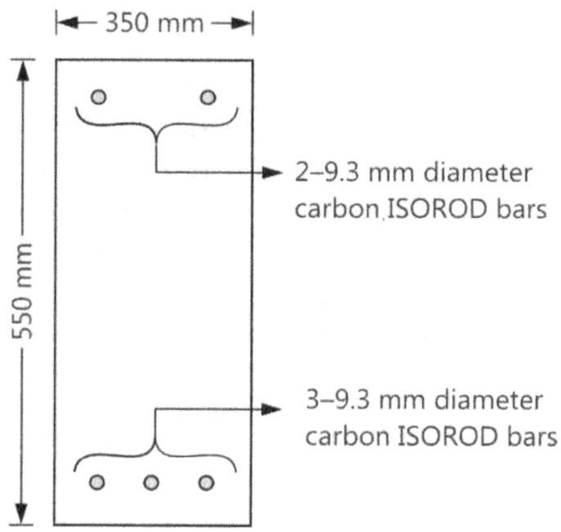

Figure P4.1.

Material Properties:

Concrete Compressive Strength, $f'_c = 45$ MPa;

FRP Ultimate Strength, $f_{frpu} = 1500$ MPa;

FRP Elastic Modulus, $E_{frp} = 100$ GPa;

Area of FRP Bars, $A_{bar} = 71\ \text{mm}^2$

6. Use ISIS Canada Design guidelines and calculate the factored moment resistance, M_r, in positive bending, for the precast ($\phi_c = 0.65$) FRP-reinforced concrete section shown in Fig. P4.2. Assume that the beam has an interior exposure condition:

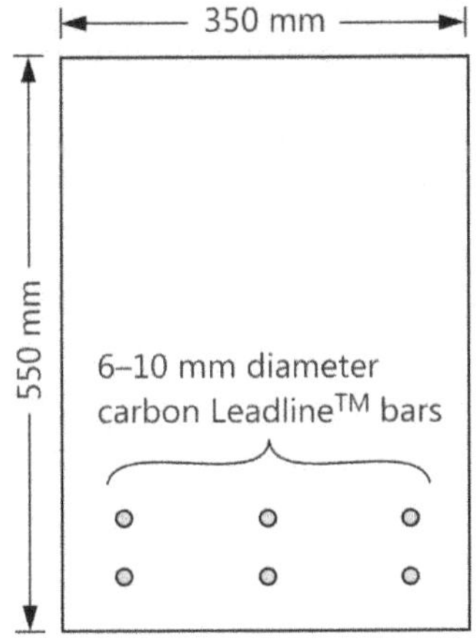

Figure P4.2.

Material Properties:

Concrete Compressive Strength, $f'_c = 40$ MPa

FRP Ultimate Strength, $f_{frpu} = 2255$ MPa

FRP Elastic Modulus, $E_{frp} = 147$ MPa

Area of FRP Bars, $A_{bar} = 113\ \text{mm}^2$

Maximum aggregate size, MAS = 14 mm

7. Determine the flexure capacity of a box-beam, prestressed with 6 unbonded and 7 bonded CFRP tendons and reinforced with 2 non-prestressing bars at bottom and 7 non-prestressing bars at the top. The cross-section details of the beam are shown in Fig. P4.3. The effective span of the beam is 5 m. The beam is reinforced and prestressed with DFI bars and tendons having $f_{fu} = 2000$ MPa. All tendons and bars have a diameter (d) of 10 mm. All other properties of tendons and concrete are listed as follows:

Material properties:

$f_{fu} = 2000$ MPa;

$E_f = 150$ GPa;

$f_c = 48$ MPa;

Modulus of rupture of concrete, $f_r = 3.5$ MPa

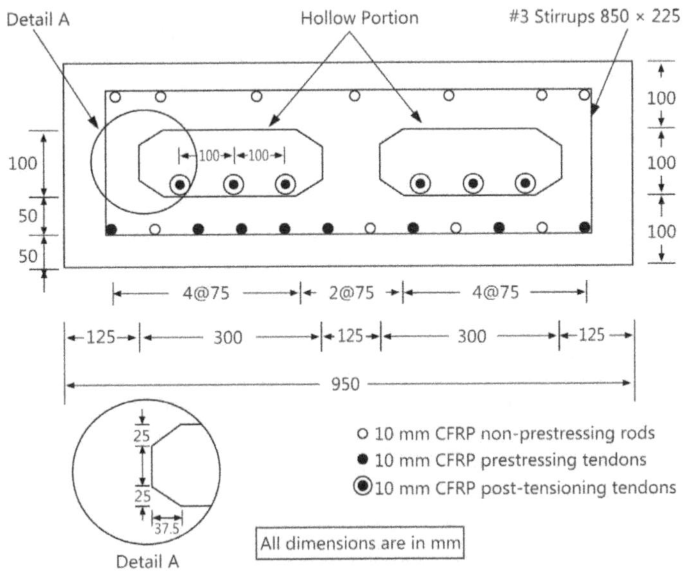

Figure P4.3. Cross-section dimensions of the box-beam.

CHAPTER 5

Design Philosophy for FRP External Strengthening Systems

5.1. Introduction

Recently, the bonding of FRP plates or sheets has become a popular method for the flexural strengthening of RC beams. The traditional methods used for the flexural and shear strengthening of structures using steel plates or external post-tensioning suffer from inherent disadvantages such as difficult application procedures and lack of durability. The flexural strengthening of a simply supported RC beam using FRP composites is generally achieved by bonding an FRP plate or sheet to the soffit of the beam after adequate surface preparation. The purpose of adequate surface preparation is to remove the weak surface layer of the concrete, expose the concrete aggregate to improve the bond with FRP and provide an even surface. The FRP plate may be pre-fabricated using pultruded manufacturing technique, in which case some preparation of the bonding surface of the FRP plate may be necessary. On the other hand, the FRP plate may be constructed on the site in a wet lay-up process where FRP fabric sheets are impregnated with resin and bonded to the prepared surface of the concrete surface. Brief descriptions of the variations in the basic procedure of FRP strengthening are given in the following subsections.

5.1.1. Non-Prestressed Soffit Plates

The bonding of unstressed FRP plates to the soffit of RC beams is the most common method and has received the greatest amount of theoretical and experimental research till date. There are three schemes for the adhesion of unstressed FRP plates to the soffit of an RC beam. These schemes are: (i) adhesive bonding of prefabricated FRP plates, (ii) wet lay-up, and (iii) resin infusion. In the first approach, prefabricated or pultruded FRP plates are cut to the required size and bonded to the soffit of an RC beam. The second approach is perhaps the most commonly used and gives the greatest flexibility for field use. It is also the cheapest. In this case, the adhesive also forms the matrix of the FRP and creates a strong bond with the RC beam. This method is, however, sensitive to unevenness of the RC beam soffit and such unevenness can lead to debonding. The third method, i.e., resin infusion method has similar characteristics to those of wet lay-up but is the least commonly used.

5.1.2. End Anchorage for Unstressed (Non-Prestressed) Plates

To prevent debonding at the ends of the soffit plate, mechanical anchors can be installed. Pre-fabricated steel or FRP U-strips of an end anchorage can be bonded or bolted. Moreover, FRP-end anchorage strips can also be formed by wet lay-up and they can be completely or partially wrapped around the RC beam near the ends of the plate. Experimental investigations have established that mechanical anchors delay, if not prevent, the onset of debonding although.

5.1.3. Prestressed Soffit Plates

A pultruded FRP plate can be bonded to the RC beam soffit with pressurizing. The main advantage of prestressing the FRP plate is that the bonded plate contributes to the load-carrying capacity before the additional loading is applied on the structure. The prestressing also contributes to the reduction of crack widths if this is an important issue. Since FRPs have high tensile strength, prestressing them improves their efficiency as a prestressed FRP plate is more likely to reach its ultimate tensile strength at failure. This technique is still in its infancy as there have been limited experimental studies or field applications.

5.2. Flexural Failure Modes and Typical Behavior

A schematic representation of typical failure modes observed in experimental tests is shown in Fig. 5.1. The failure modes are classified into seven main categories and are termed as: (a) flexural failure by FRP rupture, (b) flexural failure by crushing of compressive concrete, (c) shear failure, (d) concrete cover separation, (e) plate-end interfacial debonding, (f) intermediate flexural crack-induced interfacial debonding and (g) intermediate flexural shear crack-induced interfacial debonding. Collectively, failure modes (d) and (e) are termed as plate-end debonding failure, while failure modes (f) and (g) are referred to as intermediate crack-induced interfacial debonding failures. It must be noted that all the failure modes, except intermediate crack-induced interfacial debonding with a long process of debonding propagation, occur in a brittle manner. In particular, debonding at the plate ends occurs with little or no indication of failure.

5.2.1. Flexural Failure

If the ends of the plate are properly anchored, the ultimate flexural capacity of the beam is reached when either the FRP plate fails by tensile rupture (Fig. 5.1a) or the compressive concrete is crushed (Fig. 5.1b). This is very similar to the classical flexural failure of RC beams, except for small differences due to the brittleness of the bonded FRP plate. FRP rupture generally occurs following the yielding of the longitudinal steel bars, although steel yielding may not have been reached, if the steel bars are located quite far away from the tension face.

5.2.2. Shear Failure

As shown in Fig. 5.1c, the strengthened beam can fail in brittle mode in shear while the normal RC beams are designed to fail in flexure rather than in shear which is a brittle mode, the shear failure mode can be made critical by flexural strengthening. In such cases, the shear capacity of the RC beam alone dictates its strength in shear as the FRP soffit plate provides little contribution to shear resistance. In

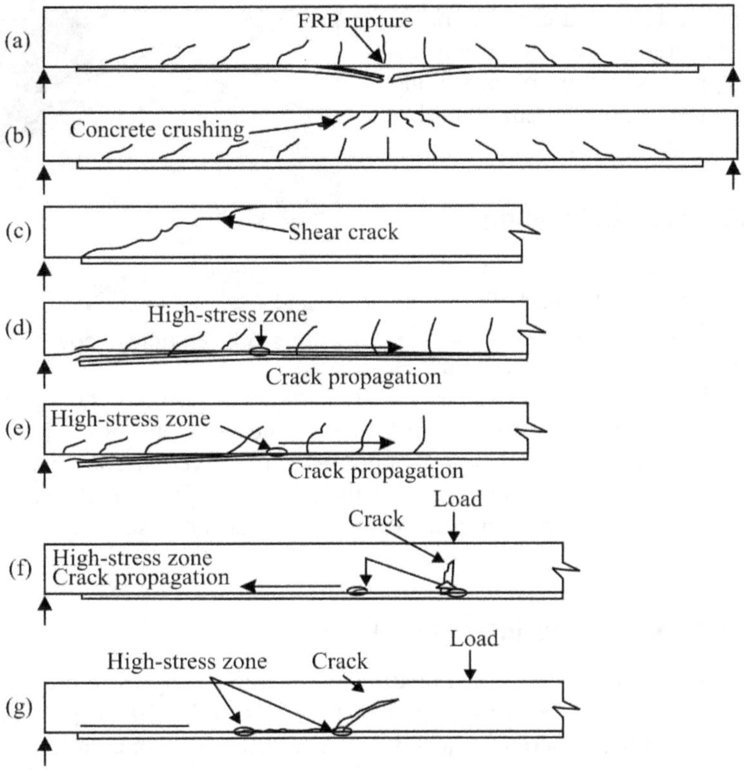

Figure 5.1. Failure modes of FRP-strengthened RC beams.

such situations, shear strengthening of the RC beam should be carried out simultaneously to ensure that the required flexural strength is not compromised by shear failure and the flexural failure still precedes shear failure. The flexural failure mode of a plated RC beam is desirable although brittle, and it is still more ductile than its shear failure mode.

5.2.3. Plate-End Debonding Failures

The most commonly reported mode of debonding in experimental studies has been the separation of the concrete cover (Fig. 5.1d) that initiates at or near one of the two ends of the soffit plate (referred to as the end of the soffit plate, the end of the plate, or the plate end). It has been observed that beam may fail owing to debonding before the ultimate flexural capacity. A less commonly observed debonding failure mode that also initiates at or near the plate end is the separation of the FRP plate from the RC beam soffit (Fig. 5.1e). In the real beams, the two failure modes may be mixed together. The plate-end debonding failures result in less ductility. The concrete cover separation mode of failure is most commonly reported premature failure mode. This failure mode has been referred to elsewhere as 'end-of-plate failure through concrete' or 'concrete rip-off failure', 'debond at rebar layer', 'local shear failure' and 'concrete cover delamination'. A typical picture of such failure is shown in Fig. 5.2. It is generally believed that failure of concrete cover is initiated by the formation of a crack at or near the plate-end, due to high interfacial shear and normal stresses caused by abrupt termination of the plate here. Once a crack forms in the concrete at or near the end of the plate, the crack propagates to the level of tension reinforcement and then progresses horizontally along the level of the steel tension reinforcement, resulting in the separation of the concrete cover.

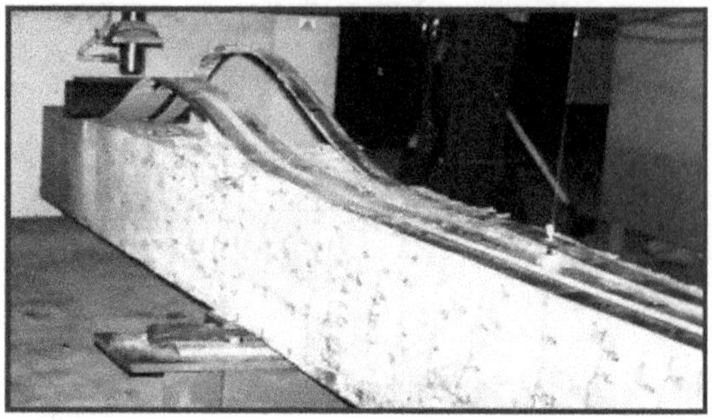

Figure 5.2. Concrete cover delamination.

5.2.4. Plate-End Interfacial Debonding

Debonding between the adhesive and the beam (Fig. 5.1e) which propagates from the end of the plate is referred to as plate-end interfacial debonding. The load-deflection response for this mode of failure is similar to that of failure by concrete cover separation. This mode of failure occurs primarily due to the combined action of high interfacial shear and normal stresses near the end of the plate that exceed the weakest element, generally the concrete. Upon debonding, a thin layer of concrete generally remains attached to the plate. This suggests that failure occurs in the concrete adjacent to the concrete-to-adhesive interface. However, the risk of plate-end interfacial debonding is increased by inferior adhesive and uneven concrete surfaces.

5.2.5. Intermediate Crack-Induced Interfacial Debonding

Debonding may initiate at a flexural or a mixed flexural shear crack away from the plate-ends and then propagates towards one of the plate-ends as shown in Figs.5.1f and 5.1g, respectively. In this case, debonding generally occurs in the concrete, adjacent to the adhesive-to-concrete interface and referred to herein as intermediate flexural crack-induced interfacial debonding or intermediate flexural shear crack-induced interfacial debonding. Intermediate crack-induced interfacial debonding failures appear to be more likely in shallow beams and more ductile than plate-end debonding failures.

5.2.5.1. Intermediate flexural crack-induced interfacial debonding

The mechanism of intermediate flexural crack-induced debonding may be explained by formation of the major crack and its effects on developing high interfacial stresses between concrete and FRP at the location of crack. When a major crack is formed in the concrete, the tensile stresses released by the cracked concrete are transferred to the FRP plate and result in high local interfacial stresses between the FRP plate and the concrete near the crack. As the applied loading increases further, the tensile stresses in the plate and hence the interfacial shear stresses between the FRP plate and the concrete near the crack also increase. When these stresses reach critical values, debonding initiates at the crack and then propagates towards one of the plate-ends generally the nearer end.

5.2.5.2. Intermediate flexural shear crack-induced interfacial debonding

In the case of debonding induced by a flexural crack, the widening of the crack is the driving force for the propagation of debonding. However, in an intermediate flexural shear crack-induced debonding failure, the relative vertical displacement between the two faces of the crack produces peeling stresses on the FRP plate-to-concrete interface. However, it is believed by many researchers that the widening of the crack is normally the more important factor, with the relative movement of crack faces being secondary.

5.2.6. Other Debonding Failures

In many experimental tests it has been observed that debonding failures involve a mixture of distinct modes, e.g., a combination of concrete cover separation and plate-end interfacial debonding, i.e., a mixed mode of failure where concrete cover separation occurred at an intermediate crack, but at a small distance away from the crack, interfacial debonding occurs between the adhesive and the concrete. Interlaminar shear failure within the FRP plate is another possible mode of failure but this mode does not appear to have been reported in test to date.

5.2.7. Some Additional Aspects of Debonding

The risk of debonding is increased by a number of factors associated with the quality of on-site application. These include poor workmanship and the use of inferior adhesives. The effects of these factors can be minimized if due care is exercised in the application process. The following sections describe some important aspects which could be taken care to minimize the chances of debonding.

5.2.7.1 Concrete surface preparation

The concrete surface should be properly prepared to avoid failure at the adhesive-to-concrete interface by following a suitable procedure. An important requirement for good concrete surface preparation is that uneven concrete surface is evened out. When a thin FRP plate bonded to an uneven concrete surface is subject to tension, it tries to straighten itself, which results in interfacial peeling stresses that can cause debonding in the concrete adjacent to the adhesive-to-concrete interface. FRPs formed in a wet lay-up process are more sensitive to the surface unevenness, as their profile follows the uneven surface.

5.2.7.2. Adhesive

Commercially strong adhesives are available for FRP plate bonding and their strength generally exceeds that of concrete so failure in the adhesive is rare. However, if sub-standard adhesives are used or if adhesives are not properly applied, adhesives failure may occur within the adhesive layer, adjacent to the adhesive-to-concrete interface, and adjacent to the FRP plate-to-adhesive interface.

5.2.7.3. FRP surface preparation

In the case of pultruded FRP plates, failure can also occur at the FRP-to-adhesive interface, if the surface of the FRP plate is not properly prepared, e.g., inadequate removal of impurities such as grease from the FRP plate surface.

5.2.7.4. U-strip-end anchorage

It has been observed that an experimentally installed mechanical anchors at the plate-ends delay or prevent plate-end debonding failures. A U-strip wrapped around the FRP soffit plate was found to be more effective than one placed above the soffit plate as mechanical anchor systems. In the following sections, flexural and shear strength design approaches followed as per ACI 440-2R-02 and ISIS Canada Guidelines have been presented with design examples.

5.3. Flexural Design Considerations

In this section, general design considerations are presented based on the traditional reinforced concrete design principles stated in the requirements of ACI 440-2R-02, ACI 318-99, BS8110 (1997) and the knowledge of the specific mechanical behavior of FRP reinforcement. The FRP strengthening systems should be designed to resist tensile forces while maintaining strain compatibility between the FRP and the concrete substrate. FRP reinforcement should not be relied upon to resist compressive forces. However, it is acceptable for FRP tension reinforcement to experience compression due to moment reversals or changes in the load pattern. It is recommended that compressive strength of the FRP reinforcement should be neglected.

5.3.1. Flexural Design Philosophy (ACI 440-2R-02)

These design recommendations are based on limit state design principles, which set acceptable levels of safety against the occurrence of both serviceability limit states (excessive deflections and cracking) and ultimate limit states (failure, stress rupture and fatigue). In assessing the nominal strength of a member, the possible failure modes and subsequent strains and stresses in each material should be assessed. For evaluating the serviceability of a member, engineering principles, such as modular ratios and transformed sections, can be used.

The FRP strengthening systems should be designed in accordance with ACI 318-99 strength and serviceability requirements using the load factors stated in ACI 318-99. The strength reduction factors required by ACI 318-99 should also be used. Moreover, additional reduction factors applied to the contribution of FRP reinforcement are recommended by ACI 440-2R-02 to reflect the lesser existing knowledge of FRP systems compared with reinforced and prestressed concrete. If needed, designers may incorporate more conservative strength reduction factors in the case of uncertainties regarding existing material strengths or substrate conditions greater than those recommended by ACI 440-2R-02 standards.

In the case of design of FRP systems for seismic retrofit of a structure, it may be appropriate to use capacity design principles (Paulay and Priestley, 1992) which assume that a structure should develop its full capacity and require that members be capable of resisting the associated required shear strengths. These FRP systems, particularly when used for columns, should be designed to provide seismic resistance through energy dissipation and deflection capacity at the code defined base shear levels. Unless additional performance objectives are specified by owner, life safety is the primary performance objectives of the seismic designs with an allowance for some level of structural damage to provide energy dissipation. Consequently, retrofitted members may require some level of repair or replacement following a seismic event. Caution should be exercised upon re-entering a seismically damaged structure especially during or after a subsequent fire.

5.3.2. Strengthening Limits

Strengthening limits are imposed to guard against collapse of the structure should bond or other failure of the FRP system occur due to fire, vandalism, or other causes. Some designers and system manufacturers have recommended that the unstrengthened structural members, without FRP reinforcement, should have sufficient strength to resist a certain level of load. Thus, using this philosophy, in the event that the FRP system is damaged, the structure will still be capable of resisting a reasonable level of load without collapse. As per ACI 440-2R-02 recommendation, the existing strength of the structure be sufficient to resist a level of load as described by Eq. (5.1) while more specific limits for structures requiring a fire endurance rating are given in the next section.

$$(\phi R_n)_{\text{existing}} \geq (1.2S_{DL} + 0.85S_{LL})_{\text{new}} \tag{5.1}$$

where, R_n = nominal strength of a member
S_{DL} = dead load effects
S_{LL} = live load effects
ϕ = strength reduction factor

5.3.2.1. Structural fire endurance

The level of strengthening that can be achieved through the use of externally bonded FRP reinforcement is often limited by the code required and fire-resistance rating of a structure. The polymer resins used in wet layup and prepreg FRP systems and polymer adhesives used in prepreg FRP systems lose structural integrity at temperatures exceeding the glass transition temperature of the polymer. While glass temperature can vary depending on the polymer chemistry, a typical range for field applied resins and adhesives is 60°C to 82°C. Due to the high temperatures associated with fire and low temperature resistance of the FRP system, the FRP system will not be capable of enduring a fire for any appreciable amount of time. Furthermore, it is most often not feasible to insulate the FRP system to substantially increase its fire endurance, because the amount of insulation that would be required to protect the FRP system is far more than can be realistically applied.

Although the FRP system itself has low fire endurance, the combination of the FRP system with an existing concrete structure may still have an adequate level of fire endurance. This is attributable to the inherent fire endurance of the existing concrete structure alone. To investigate the fire endurance of an FRP-strengthened concrete structure, it is important to recognize that the strength of traditional reinforced concrete structures is somewhat reduced during exposure to the high temperatures associated with a fire event as well. The yield strength of reinforcing steel is reduced and the compressive strength of concrete is reduced. As a result, the overall resistance of a reinforced concrete member to load effects is reduced. ACI 216R suggests limits that maintain a reasonable level of safety against complete collapse of the structure in the event of a fire. The existing strength of a structural member with a fire resistance rating should satisfy the conditions of Eq. (5.2), if it is to be strengthened with FRP system. The load effects, S_{DL} and S_{LL}, should be determined using the current load requirements for the structure. If the FRP system is meant to allow greater load carrying strength, such as an increase in live load, the load effects should be computed using these greater loads.

$$(R_{n\theta})_{\text{existing}} \geq (S_{DL} + S_{LL})_{\text{new}} \tag{5.2}$$

where, $R_{n\theta}$ = nominal resistance of a member at an elevated temperature
S_{DL} = dead load effects
S_{LL} = live load effects

5.4. Design Material Properties

The material properties reported by manufacturers, such as the ultimate tensile strength typically do not consider long term exposure to environmental conditions and should be considered as initial properties. Since the long term exposure to various types of environments can reduce the tensile properties, creep rupture and fatigue endurance of FRP laminates, the material properties should be based on the environmental exposure conditions.

The design of ultimate tensile properties should be determined using the environmental reduction factor as given in Table 5.1. Equations (5.3) to (5.5) give the design tensile strength, rupture strain and modulus of elasticity.

$$f_{fu} = C_E f_{fu}^* \tag{5.3}$$

$$\varepsilon_{fu} = C_E \varepsilon_{fu}^* \tag{5.4}$$

where C_E refers to environmental reduction factor; f_{fu}^* refers to specified/guaranteed strength of materials given by manufacturer; ε_{fu}^* refers to specified/guaranteed rupture strain specified by manufacturer.

$$E_f = \frac{f_{fu}}{\varepsilon_{fu}} \tag{5.5}$$

The expression for the modulus of elasticity is based on the fact that FRP materials are linearly elastic until failure and are typically unaffected by environmental conditions. The modulus given in Eq. (5.5) will be the same as the initial value reported by the manufacturer. It must be noted that the constituent materials, fibers and resins of an FRP system affect its durability and resistance to environmental exposure. The environmental-reduction factors given in Table 5.1 are conservative estimates based on the relative durability of each fiber type. As more research information is developed and becomes available, these values will be refined. However, the methodology regarding the use of these factors will remain unchanged. Durability test data for FRP systems with and without protective coatings may be obtained from the manufacturer of the FRP system under consideration.

As shown in Table 5.1, if the FRP system is located in a relatively benign environment such as indoor environment conditions, the reduction factor is close to unity. If the FRP system is located in an aggressive environment where prolonged exposure to high humidity, freeze thaw cycles, salt water, or alkalinity is expected, a lower reduction factor should be used. The reduction factor can reflect the use of a protective coating if the coating has been shown through testing to lessen the effects of environmental exposure and the coating is maintained for the life of the FRP system.

Table 5.1. Environmental reduction factor for various FRP systems and exposure conditions.

Exposure conditions	Fiber and resin type	Environmental reduction factor, C_E
Interior exposure	Carbon/epoxy	0.95
	Glass/epoxy	0.75
	Aramid/epoxy	0.85
Exterior exposure (bridges, piers and unenclosed parking garages)	Carbon/epoxy	0.85
	Glass/epoxy	0.65
	Aramid/epoxy	0.75
Aggressive environment (chemical plants and waste water treatment plants)	Carbon/epoxy	0.85
	Glass/epoxy	0.50
	Aramid/epoxy	0.70

5.5. General Considerations for Flexural Strengthening

This section gives the guidance on the calculation of the flexural strengthening effect of adding longitudinal FRP reinforcement to the tension face of a reinforced concrete member. A specific illustration of the concepts applied to the strengthening of the existing rectangular sections reinforced in the tension zone with non-prestressed steel is given. However, the general concepts outlined here can be extended to non-rectangular shapes (T- and I-sections) and to the members with compression steel reinforcement. In the case of prestressed members, strain compatibility, with respect to the state of strain in the stressed member should be used to evaluate the FRP contribution. Additional failure modes controlled by rupture of prestressing tendons should also be considered.

5.5.1. Assumptions

As per ACI 440.2R-02 standards, the following assumptions are made in calculating the flexural resistance of a section strengthened with an externally applied FRP system:

1. Design calculations are based on the actual dimensions, internal reinforcing steel arrangement and material properties of the existing member being strengthened.
2. The plane sections before loading remains plane after loading, i.e., the strains in the reinforcement and concrete are directly proportional to the distance from the neutral axis.
3. There is no relative slip between external FRP reinforcement and the concrete.
4. The shear deformation within the adhesive layer is neglected since the adhesive layer is very thin with slight variations in its thickness.
5. The maximum usable compressive strain in the concrete is 0.003.
6. The tensile strength of concrete is neglected.
7. The FRP reinforcement has a linear elastic stress–strain relationship to failure.

It must be noted that some of the above assumptions are necessary for the easy computation of the moment capacity of sections; however, the assumptions do not accurately reflect the true fundamental behavior of FRP flexural reinforcement. For example, there will be shear deformation in the adhesive layer causing relative slip between the FRP and the substrate. However, the inaccuracy of the assumptions will not significantly affect the computed flexural strength of an FRP strengthened member. An additional strength reduction factor is used to obtain design moment capacity will conservatively compensate for any such discrepancies.

5.5.2. Section Shear Strength

When FRP reinforcement is being used to increase the flexural strength of a member, it is important to verify that the member will be capable of resisting the shear forces associated with the increased flexural strength. The potential for shear failure of the section should be considered by comparing the design shear strength of the section to the required shear strength. If additional shear strength is required, FRP laminates oriented transversely to the section can be used to resist shear forces as described in the section entitled 'Shear Strength'.

5.5.3. Existing Substrate Strain

The initial strain level on the bonded substrate, ε_{bi}, due to sustained load (before FRP installation) can be determined from an elastic analysis of the existing member considering all loads that will be on the member, during the installation of the FRP system. It is recommended that the elastic analysis of the existing member should be based on cracked-section properties. These initial strains should be excluded from the FRP strain to get the effective FRP strains.

5.6. Nominal Strength, M_n

The nominal flexural strength of the FRP strengthened concrete member can be determined based on strain compatibility, internal force equilibrium and the controlling mode of failure. In the strength design approach, the design flexural strength (ϕM_n) of a member exceeds its required moment strength as indicated by Eq. (5.6). The design flexural strength refers to the nominal strength of the member multiplied by a strength reduction factor (ϕ) and the required moment strength, M_u refers to the load effects calculated from factored loads (e.g., $\alpha_{DL}M_{DL} + \alpha_{LL}M_{LL} + \ldots$). ACI 440.2R-02 standards recommends that required moment strength of a section be calculated by the use of load factors as required by ACI 318-99. Also, the use of ϕ should be based on ACI 318-99 with an additional strength reduction factor of 0.85 applied to the flexural contribution of the FRP reinforcement alone ($\psi_f = 0.85$). This additional reduction factor is meant to account for lower reliability of the FRP reinforcement, as compared with steel reinforcement.

$$\phi M_n \geq M_u \tag{5.6}$$

5.6.1. Controlling Failure Modes

The flexural strength of a section depends on the controlling failure mode. The following flexural modes should be investigated for an FRP strengthened section.

- Crushing of the concrete in compression before yielding of the reinforcing steel;
- Yielding of the steel in tension followed by rupture of the FRP laminate;
- Yielding of the steel in tension followed by concrete crushing;
- Shear or tension delamination of the concrete cover (cover delamination); and
- Debonding of the FRP from the concrete substrate (FRP debonding).

To predict these failure modes, it is assumed that concrete crushing occurs if the compressive strain in the concrete reaches its maximum usable strain ($\varepsilon_c = \varepsilon_{cu} = 0.003$). Similarly, rupture of the FRP laminate is assumed to occur if the strain in the FRP reaches its maximum design rupture strain.

5.6.1.1. Cover delamination

Cover delamination or FRP debonding can occur if the force in the FRP cannot be sustained by the susbtrate. To prevent debonding of the FRP laminate, a limitation should be placed on the strain level developed in the laminate. A bond dependent co-efficient, k_m, has been defined in Eq. (5.7) to provide the limiting value of strain ($k_m\varepsilon_{fu}$) in the FRP.

In SI Units:

$$k_m = \begin{cases} \dfrac{1}{60\varepsilon_{fu}}\left(1 - \dfrac{nE_f t_f}{360\,000}\right) \leq 0.90 \quad \text{for } nE_f t_f \leq 180\,000 \\ \dfrac{1}{60\varepsilon_{fu}}\left(\dfrac{90\,000}{nE_f t_f}\right) \leq 0.90 \quad \text{for } nE_f t_f > 180\,000 \end{cases} \tag{5.7}$$

where n is number of plies of FRP flexural reinforcement at the location along with the length of member, where the moment strength is being computed (this term recognizes that laminates with greater stiffnesses are more prone to delamination); t_f is thickness of each ply; ε_{fu} is design rupture strain of FRP; E_f is modulus of elasticity of FRP.

From Eq. (5.7), it is clear that as the laminate stiffness increases, the strain limitation becomes more severe. For laminates with a unit stiffness $nE_f t_f$ greater than 180 000 N/mm, k_m limits the force in the laminate as opposed to the strain level. This effectively places an upper bound on the total force that can be developed in an FRP laminate, regardless of number of plies. The width of the FRP laminate is not included in the calculation of the unit stiffness, $nE_f t_f$, because an increase in the width of the FRP results in a proportional increase in the bond area.

5.6.2. Strain Level in FRP Reinforcement

Since FRP materials are linearly elastic until failure, the level of strain in the FRP will dictate the level of stress developed in the FRP. The maximum strain level that can be achieved in the FRP reinforcement will be governed by either the strain level developed in the FRP at the point at which concrete crushes, the point at which the FRP ruptures, or the point at which the FRP debonds from the substrate. This maximum strain or the effective strain level in the FRP reinforcement at the ultimate limit state can be found from Eq. (5.8).

$$\varepsilon_{fe} = \varepsilon_{cu}\left(\frac{h-c}{c}\right) - \varepsilon_{bi} \leq k_m \varepsilon_{fu} \tag{5.8}$$

5.6.3. Stress Level in the FRP Reinforcement

The effective stress level in the FRP reinforcement is the maximum level of stress that can be developed in the FRP reinforcement before flexural failure of the section. This effective stress level can be found from the strain level in the FRP, assuming perfectly elastic behavior using Eq. (5.9).

$$f_{fe} = E_f \varepsilon_{fe} \tag{5.9}$$

5.7. Ductility

Since, the use of externally bonded FRP reinforcement for flexural strengthening will reduce the ductility of the original member, the strain level in steel at the ultimate limit state should be checked. Adequate ductility is achieved if the strain in steel at the point of concrete crushing or failure of the FRP, including delamination or debonding, is at least 0.005 as per the definition of tension controlled section of ACI 318-99. It is recommended by ACI 440 committee (ACI 440.2R-02) that a section with low ductility should compensate with a higher reserve of strength. The higher reserve of strength is achieved by applying a strength reduction factor of 0.70 to brittle sections as opposed to 0.9 for ductile sections. The strength reduction factor, ϕ, can be obtained using Eq. (5.10) or using Fig. 5.3. Equation (5.10) and Fig. 5.3 set the reduction factor at 0.90 for ductile sections and 0.70 for brittle sections where the steel does not yield and provides a linear transition for the reduction factor between these two extremes (Fig. 5.3).

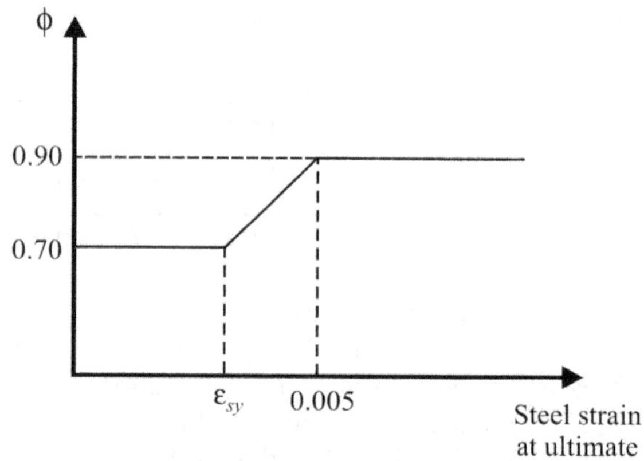

Figure 5.3. Graphical representation of the strength reduction factor as a function of ductility.

$$\phi = \begin{cases} 0.90 & \text{for } \varepsilon_s \geq 0.005 \\ 0.70 + \dfrac{0.20(\varepsilon_s - \varepsilon_{sy})}{0.005 - \varepsilon_{sy}} & \text{for } \varepsilon_{sy} < \varepsilon_s < 0.005 \\ 0.70 & \text{for } \varepsilon_s \leq \varepsilon_y \end{cases} \tag{5.10}$$

5.8. Serviceability

The effect of the FRP external reinforcement on the serviceability can be assessed using the transformed section analysis. To avoid inelastic deformations of the reinforced concrete members strengthened with an external FRP reinforcement, the existing internal steel reinforcement should be prevented from yielding under service load levels. The stress ($f_{s,s}$) in the steel under service load should be limited to 80% of the yield strength (f_y) as given by Eq. (5.11).

$$f_{s,s} \leq 0.80 f_y \tag{5.11}$$

5.9. Creep-Rupture and Fatigue Stress Limits

The stress levels in the FRP reinforcement under sustained and/or under cyclic service, loading situations should be checked to ensure that the FRP reinforcement does not fail due to creep-rupture under sustained loading or due to cyclic stresses under fatigue loading. Since these stress levels will be within elastic response of the member, these stresses can be computed by the use of an elastic analysis and an applied moment due to all sustained loads (dead loads and sustained portion of live load) plus the maximum moment induced in a fatigue loading cycle. The sustained and cyclic stress limit ($F_{f,s}$) as presented in Table 5.2 should exceed

Table 5.2. Sustained and cyclic service load stress limits ($F_{f,s}$) in FRP reinforcement.

Stress type	FRP type		
	GFRP	AFRP	CFRP
Sustained plus cyclic stress limit	$0.20f_{fu}$	$0.30f_{fu}$	$0.55f_{fu}$

the corresponding stresses ($f_{f,s}$) in the FRP under sustained and/or fatigue service loading conditions as expressed by Eq. (5.12).

$$F_{f,s} \geq f_{f,s} \tag{5.12}$$

5.10. Applications of Flexural Design Considerations to a Singly Reinforced Rectangular Section

In this section, the flexural design concepts for the FRP externally strengthened sections have been applied to an individual reinforced and non-prestressed rectangular section.

5.10.1. Ultimate Flexural Strength

Figure 5.4 shows the internal strain and stress distribution for a rectangular section under flexure at the ultimate limit state. The ultimate flexural strength should be calculated based on strain compatibility, force equilibrium and the governing failure mode. Several calculation procedures can be derived to satisfy these conditions. The calculation procedure described herein involves a trial and error method. The trial and error procedure involves the following steps:

- Selecting an assumed depth to the neutral axis, *c*;
- Calculating the strain level in each material using strain compatibility;
- Calculating the associated stress level in each material; and
- Checking internal force equilibrium;
- If the internal force equilibrium is not satisfied; then
- The depth to the neutral axis must be revised and the above procedure should be repeated.

For any assumed depth to the neutral axis, *c*, the strain level (ε_{fe}) in the FRP reinforcement can be computed using Eq. (5.8). This equation considers the governing mode of failure for the assumed neutral axis depth. If the first term in the equation controls, concrete crushing controls flexural failure of the section. If the second term controls, FRP failure (rupture or debonding) controls flexural failure of the section. The effective stress level (f_{fe}) in the FRP can be obtained using Eq. (5.9) assuming linear elastic behavior until failure of the FRP. Based on the strain level in the FRP reinforcement, the strain level in the non-prestressed tension steel can be found from Eq. (5.13) using strain compatibility (Fig. 5.4).

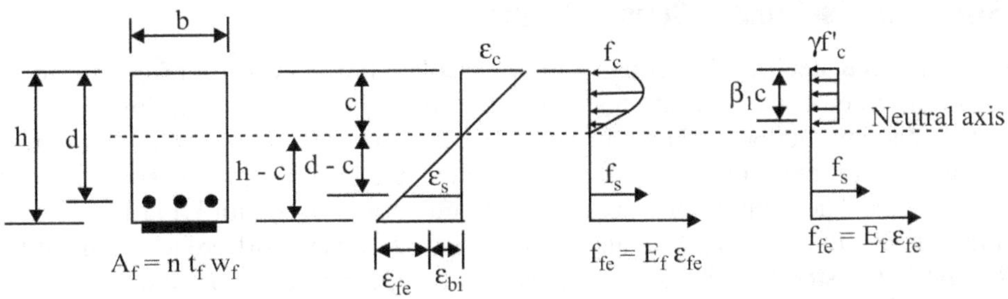

Figure 5.4. Internal strain and stress distribution for a rectangular section under flexure at ultimate.

$$\varepsilon_s = \left(\varepsilon_{fe} + \varepsilon_{bi}\right)\left(\frac{d-c}{h-c}\right) \tag{5.13}$$

where, d and h are the effective depth and overall depth of the rectangular cross-section.

The stress level in the steel is calculated from the strain level in the steel using Eq. (5.14), which is based on elastic plastic behavior of steel.

$$f_s = E_s\varepsilon_s \le f_y \tag{5.14}$$

The stress levels in the FRP and steel reinforcement determined for the assumed neutral axis depth can be used to check the internal force equilibrium as per Eq. (5.15).

$$c = \frac{A_s f_s + A_f f_{fe}}{\gamma f_c' \beta_1 b} \tag{5.15}$$

When the calculated neutral axis depth from the above equation is close to the assumed depth of the neutral axis, then the assumed depth of neutral axis is correct and satisfies the equilibrium condition and can be used in Eq. (5.16) to calculate the nominal flexural moment capacity of section. The actual depth to the neutral axis, c, is found by simultaneously, satisfying Eqs. (5.8), (5.9), (5.13), (5.14) and (5.15), thus, establishing internal force equilibrium and strain compatibility. In Eq. (5.15), A_s and A_f are the area of steel and FRP reinforcements. The terms γ and β_1 are parameters defining a rectangular stress block in the concrete equivalent to the actual non-linear distribution of stress. If concrete crushing is the controlling mode of failure (before or after steel yielding), γ and β_1 can be taken as the values associated with the Whitney's stress block $\gamma = 0.85$ and $\beta_1 = 0.85$ for $f'_c \le 27.6$ MPa. For $f'_c > 27.6$ MPa, the value of β_1 decreases by 0.05 for each 6.895 MPa increase in the concrete strength (f'_c). If FRP rupture, cover de-lamination, or FRP de-bonding control failure occurs, the Whitney stress block will give reasonably accurate results. A more accurate stress block for the actual strain level reached in the concrete at the ultimate limit state may be used as presented in the section entitled 'Design Procedure for Strengthening of RC beam using NSM Bars'. Moreover, methods considering a non-linear stress distribution in the concrete can also be used. In Eq. (5.16), an additional reduction factor ψ_f is applied to the flexural strength contribution of the FRP reinforcement. A factor, $\psi_f = 0.85$ is recommended by ACI 440.2R-02 standards.

$$M_n = A_s f_s\left(d - \frac{\beta_1 c}{2}\right) + \psi_f A_f f_{fe}\left(h - \frac{\beta_1 c}{2}\right) \tag{5.16}$$

5.10.2. Stress in Steel under Service Loads

The stress level in the steel reinforcement can be calculated based on a cracked elastic analysis of the strengthened reinforced concrete section as given by Eq. (5.17). The distribution of strain and stress in the reinforced concrete section is shown in Fig. 5.5. As in the case of conventional reinforced concrete, the depth to the neutral axis at service, kd, can be calculated by taking the first moment of areas of the transformed section. The transformed area of FRP is obtained by multiplying the area of FRP by the modular ratio of FRP to concrete. Although this method ignores the difference in the initial strain level of the FRP, the initial strain level does not greatly influence the depth to the neutral axis in the elastic response range of the member. The stress in the steel under service loads should be computed using Eq. (5.17) and should satisfy the limits imposed by Eq. (5.11).

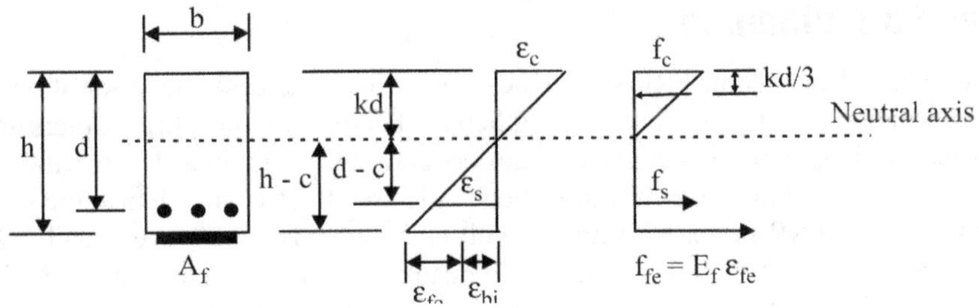

Figure 5.5. Elastic strain and stress distribution.

$$f_{s,s} = \frac{\left[M_s + \varepsilon_{bi} A_f E_f \left(h - \frac{kd}{3}\right)\right](d - kd) E_s}{A_s E_s \left(d - \frac{kd}{3}\right)(d - kd) + A_f E_f \left(h - \frac{kd}{3}\right)(h - kd)} \tag{5.17}$$

where, M_s is equal to the moment due to all sustained loads (dead loads and the sustained portion of the live load) plus the maximum moment induced in a fatigue loading cycle as shown in Fig. 5.6.

5.10.3. Stress in FRP under Service Loads

The stress level in the FRP reinforcement can be computed using Eq. (5.18) with $f_{s,s}$ from Eq. (5.17). The stress in FRP under service loads as computed from Eq. (5.18) should be compared against the limits presented in Table 5.2.

$$f_{f,s} = f_{s,s}\left(\frac{E_f}{E_s}\right)\frac{(h - kd)}{(d - kd)} - \varepsilon_{bi} E_f \tag{5.18}$$

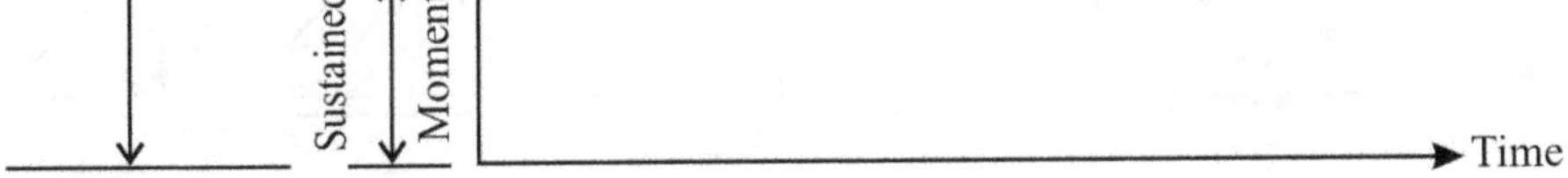

Figure 5.6. Illustration of the level of applied moment to be used to check the stress limits in the FRP reinforcement.

5.11. Shear Strengthening

Like flexural strength of the beam, FRP systems can be used to increase the shear strength of the existing concrete columns and beams. To externally strengthen the beams, columns, bridge piers and other structural components, such as slabs, bridge deck, bridge girders and metal sign boards, etc., the FRP systems with fibers oriented transverse to the axis or at a specified angle are effective to enhance the shear strength of the structural members. The FRP systems saturated with resin such as Epoxy are usually wrapped along the periphery of the members with all sides exposed. In beams, the FRP systems can be applied on three sides or on two sides of the beam to increase it shear strength. Figure 5.7 shows the three types of FRP wrapping schemes used for increasing the shear strength of prismatic rectangular beams or columns and/or bridge piers.

It may be noted that shear strengthening using external FRP can be provided at locations of plastic hinges or stress reversal and for enhancing the post-yield flexural response of members in moment resisting frames designed for seismic loads by completely wrapping the section. It is also worth noting that the center-to-center, distance between the FRP shear strips should not exceed the sum of $d/4$ plus the width of shear strips. The FRP system wrapped completely around the section of the structural members is considered to be the most efficient wrapping scheme. Complete wrapping is most commonly used in columns and/or bridge pier applications where all four sides of the sections are accessible for wrapping. In the case of beam applications, the shear strength can be improved by wrapping the FRP system around the three sides of the member usually called U-wrap or by bonding on two sides of the beam. All the four sides of the beams are not accessible because of integral slab which makes it impractical to completely wrap. Figure 5.8 shows the dimensional variables and strengthening pattern along the length of beams. Bonding of the FRP on two sides of the beam is least effective in enhancing the shear strength while three sided wrap is commonly used to strengthening scheme for beams where only three sides are accessible. In case, all the four sides are accessible, it will be most efficient to wrap on all the four sides. The FRP system can be installed continuously along the length of the member or can be placed as discrete strips. The use of continuous FRP reinforcement that completely encases the member is the most preferred choice which may prevent the migration of moisture.

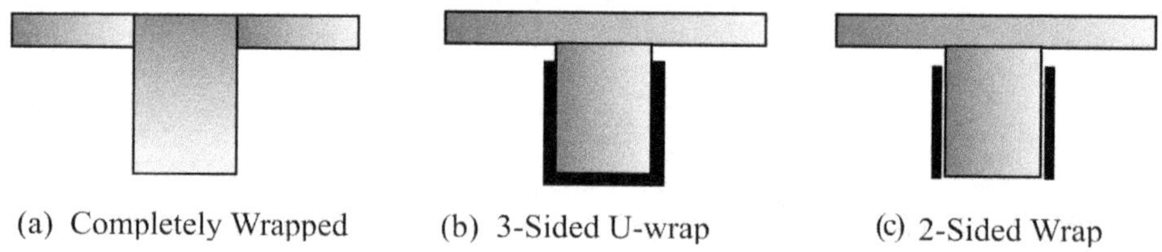

Figure 5.7. Typical strengthening schemes for shear strengthening using FRP laminates.

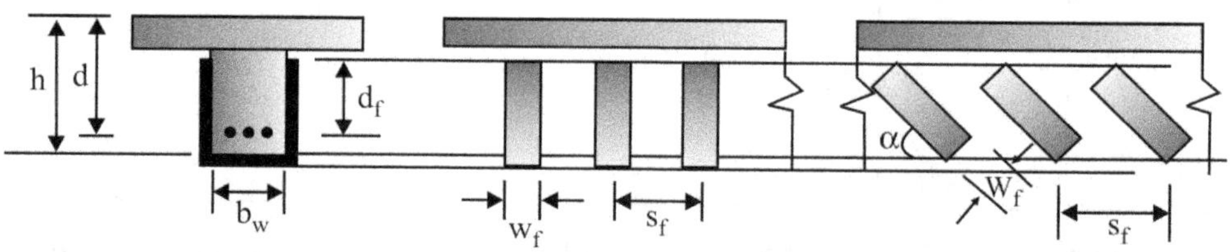

Figure 5.8. Schematics of dimensional variables for shear strengthening evaluations.

5.11.1. Nominal Shear Strength (ACI 440.2R-02)

In this section, evaluation of shear strength of FRP strengthened members based on ACI 440.2R-02 code has been presented. The design shear strength of a concrete member strengthened with an FRP system should be more than ultimate shear demand as given by Eq. (5.19).

$$\phi V_n \geq V_u \tag{5.19}$$

where, ϕ = strength reduction factor and is taken as equal to 0.85
V_n = nominal shear strength of member
V_u = ultimate shear demand or ultimate shear force

The nominal shear strength, V_n, of an FRP strengthened concrete member can be determined by summing the contributions of concrete, steel stirrups/ties/spirals and FRP reinforcements can be expressed by Eq. (5.20).

$$V_n = V_c + V_s + \psi_f V_f \tag{5.20}$$

where, V_c is shear strength contribution of concrete; V_s is shear strength contribution of steel stirrups/ties and/or spirals; V_f is shear strength contribution of FRP; the factor ψ_f is an additional reduction factor applied to the shear contribution of the FRP reinforcement. The recommended values of ψ_f for bond critical and contact critical shear reinforcement are given in Table 5.3.

Table 5.3. Recommended additional reduction factors, ψ_f, for FRP shear reinforcement.

Type of wrap	Type of shear criticality	ψ_f
Completely wrapped members	Contact-critical	0.95
Three-sides U-wraps or bonded face plies	Bond-critical	0.85

5.11.2. Shear Strength Contribution of FRP System

The shear strength contribution of FRP systems, such as FRP plates and/or fabrics depends on the fiber orientation and an assumed crack width and can be calculated using Eq. (5.21).

$$V_f = \frac{A_{fv} f_{fe} (\sin\alpha + \cos\alpha) d_f}{s_f} \tag{5.21}$$

where,

$$A_{fv} = 2nt_f w_f \tag{5.22}$$

$$f_{fe} = \varepsilon_{fe} E_f \tag{5.23}$$

In the aforementioned equations, d_f refers to the effective depth of FRP from its free end to the level of steel reinforcement; n refers to the number of FRP laminates; t_f refers to the thickness of each laminate; w_f is width of each laminate; A_{fv} is the cross-section of the FRP laminates resisting shear force; f_{fe} is the effective tensile stress developed in laminates; and ε_{fe} is the effective tensile strain developed in FRP laminates. Figure 5.8 shows the dimensional variables used in the above equations. Also, the effective strain defined herewith, is the maximum strain that can be achieved in the FRP plates or fabrics at the ultimate stage and is governed by the failure modes of the FRP system and of the strengthened reinforced concrete members. Thus, the effective strain is representative of the critical failure mode. The effective strain (ACI-440.2R-02) for different configurations of the FRP laminates used for shear strengthening of reinforced concrete members is provided in the following sections.

5.11.2.1. Completely wrapped members

In the case of reinforced concrete members such as columns and beams completely wrapped by the FRP system, the loss of concrete aggregate interlock occurs at fiber strains less than in the case of ultimate fiber strain. Hence, to avoid this mode of failure, the maximum strain used for design is limited to 0.004 for members completely wrapped around the cross-section of members. This limiting effective strain is expressed by Eq. (5.24) and is based on experimental observations and experience.

$$\varepsilon_{fe} = 0.0004 \leq 0.75\varepsilon_{fu} \tag{5.24}$$

5.11.2.2. Bonded U-wraps or bonded face plies

In the case of two- and three-sided wraps, the FRP systems have been observed to de-laminate from the concrete surface before the loss of concrete aggregate interlock of the section. Hence, the bond stresses become the critical factors to determine the level of effective strain that can be developed in the FRP laminates and/or fabrics. In this case, a bond reduction co-efficient (k_v) is employed to determine the effective strain as given by Eq. (5.25).

$$\varepsilon_{fe} = k_v \varepsilon_{fu} \leq 0.004 \tag{5.25}$$

It may be noted that the bond reduction co-efficient, k_v, is a function of the concrete strength and type of the wrapping (i.e., two-sided wrap or three-sided wrap) and the stiffness of the laminate. Equation (5.26) is used to determine the value of k_v in SI unit system.

$$k_v = \frac{k_1 k_2 L_e}{11900\varepsilon_{fu}} \leq 0.75 \tag{5.26}$$

where, L_e is the active bond length over which the majority of the bond stress is maintained and is expressed by Eq. (5.27) for SI unit system.

$$L_e = \frac{23300}{(nt_f E_f)^{0.58}} \tag{5.27}$$

In Eq. (5.26), the modification factors k_1 and k_2, respectively, account for the concrete strength and the type of wrapping scheme used for FRP shear strengthening. The modification factor, k_1, is expressed by Eq. (5.28) for SI unit system. Similarly, the modification factor, k_2 is expressed by Eq. (5.29) for U-wrap and two-sided wraps.

$$k_1 = \left(\frac{f'_c}{27}\right)^{2/3} \tag{5.28}$$

$$k_2 = \begin{cases} \dfrac{d_f - L_e}{d_f} & \text{for U-wraps} \\ \dfrac{d_f - 2L_e}{d_f} & \text{for two sides bonded wrap} \end{cases} \tag{5.29}$$

It may be noted that the equation for computing the bond reduction co-efficient, k_v, has been validated for simply supported members in regions of high shear and low moment. The above methodology has not been confirmed (ACI 440.2R-02) for shear strengthening in areas subjected to combined high flexural and shear stresses or in regions where web is primarily in compression, i.e., in negative moment regions. However, the value of bond reduction co-efficient, k_v, is found to be sufficiently conservative for regions of combined high flexural and shear stresses or for negative moment zones. It has also been reported in literature that mechanical anchorage should be used at termination points to develop larger tensile stresses. The effectiveness of such mechanical anchorages along with the level of tensile stresses they can develop should be substantiated through representative physical testing. However, it should be ensured that the effective strain in FRP laminates and/or FRP fabrics should not exceed 0.004.

5.12. Spacing of FRP Strips

As per ACI-440.2R-02 standards, the spaced FRP strips used for shear strengthening should be investigated to evaluate their contribution to the shear strength. Also, the spacing of FRP shear strips should be in accordance with the limits imposed by ACI 318-99 for internal steel shear reinforcements. The spacing of FRP shear strips is defined as the distance between the center line of the strips. Moreover, the structural testing should validate the use of discretely spaced FRP stirrups for shear strengthening.

5.13. Reinforcement Limits

The total shear reinforcement should be taken as the sum of the contribution of the FRP shear reinforcement and the steel shear reinforcement. The total shear reinforcement should be limited based on the criteria given for steel alone in ACI 318-99 section 11.5.6.9. The limits for total shear reinforcement are prescribed by the Eq. (5.30) for SI unit system.

$$V_s + V_f \leq 0.66\sqrt{f_c'}\,b_w d \tag{5.30}$$

5.14. Design Procedure for Strengthening of RC Beam Using NSM Bars

It has already been demonstrated that fiber-reinforced polymer (FRP) techniques for repair and strengthening of structures especially concrete structures have been gaining worldwide popularity. In spite of other methods of external strengthening using FRP plates or FRP fabric sheets, the near surface mounted (NSM) FRP reinforcement has been observed to perform efficiently as a flexural and shear strengthening technology for reinforced concrete and prestressed concrete members. It has been indicated by experimental test data that failure of the strengthened beams that may occur by the mechanisms accounted by the conventional RC theory as well as debonding of NSM rebars. This section presents a design procedure for flexural and shear strengthening of RC beams with NSM FRP reinforcement as well as for anchorage length. The design equations account for conventional failure mechanisms and debonding mechanisms which are critical for NSM reinforcement. In the following subsections, details of flexural and shear strengthening are explained.

Advantages of NSM FRP strengthening: The following are the main advantages of NSM FRP strengthening techniques:

a. The NSM FRP technique does not require any surface preparation work and after groove cutting requires minimum installation time compared to FRP laminate.
b. It provides feasibility of anchoring the bars into members adjacent to the one being strengthened.
c. This technique becomes particularly attractive for strengthening in the negative moment regions where external reinforcement would be subjected to mechanical and environmental damages and require protecting cover which could interfere with the presence of a floor finish.

5.14.1. Flexural Strengthening

The flexural capacity of a section depends on the controlling failure mode. The following flexural failure modes should be investigated for an FRP strengthened section:

- Crushing of concrete in compression before yielding of the reinforcing steel
- Yielding of steel in tension followed by concrete crushing
- Shear tension delamination of the concrete cover (cover delamination).

Many models have been proposed to predict the flexural strength of beams associated with debonding of FRP and/or steel plates used for external strengthening of beams. The so called 'Concrete Tooth Models' are based on the concept of concrete tooth between two adjacent cracks behaving like a cantilever under the horizontal shear stresses acting at the interface of the beam with reinforcement bonded to the tension face. The preliminary models for debonding strength of NSM FRP reinforced RC beams are presented here:

The minimum stabilized crack spacing, l_{min}, is given by Eq. (5.31).

$$\text{Minimum crack spacing, } l_{min} = \frac{A_e f_{ct}}{u_s \sum O_s + u_f \sum O_f} \tag{5.31}$$

where, A_e is the area of concrete in tension which is equal to the product of beam width and twice the distance from the centroid of the tension reinforcement to the base of RC beam; f_{ct} is the concrete tensile strength, u_s is average bond strength between concrete and steel reinforcing bars; ΣO_s is the total perimeter of steel bars; u_f is average bond strength between NSM FRP bars and surrounding material and ΣO_f is the total perimeter of the FRP bars.

It may be noted that Eq. (5.31) is an extension of the classical expression of the minimum crack spacing in reinforced concrete, based on the assumption of a uniform distribution of the bond stresses at both steel-to-concrete and NSM bar-to-concrete interfaces. The average bond strength, u_f, can be reasonably approximated by the local bond strength for estimating the typical values of minimum crack spacing. As expressed by Eq. (5.32), the maximum stabilized crack, l_{max}, spacing is taken as twice the minimum crack spacing.

$$l_{max} = 2l_{min} \tag{5.32}$$

The values of u_s and f_{ct} can be respectively evaluated by Eqs. (5.33) and (5.34).

$$u_s = 0.28\sqrt{f_{cu}} \quad \text{(in MPa)} \tag{5.33}$$

$$f_{ct} = 0.36\sqrt{f_{cu}} \quad \text{(in MPa)} \tag{5.34}$$

where, f_{cu} is concrete cube compressive strength.

The concrete tooth model assumes that the failure of the concrete tooth between two adjacent cracks occurs when the stress at Point A (Fig. 5.9a) exceeds the concrete tensile strength. The stress at Point A can be determined as follows:

$$\sigma_A = \frac{M_A}{I_A}\left(\frac{l}{2}\right) \tag{5.35}$$

where M_A refers to the moment at the base of the tooth (see Eq. (5.36)); I_A refers to the moment of inertia of the tooth (see Eq. (5.37)) and l refers to the crack spacing (minimum or maximum). In Eq. (5.36), h' refers to the distance from the base of the steel tension reinforcement to the centroid of the NSM FRP reinforcement; τ is the shear stress at the interface between NSM FRP bars and surrounding material; n is the number of NSM FRP bars; d_b is diameter of NSM bars; and b is width of beam.

$$M_A = \tau n \pi d_b l h' \tag{5.36}$$

$$I_A = \frac{bl^3}{12} \tag{5.37}$$

Substituting M_A and I_A in Eq. (5.35) and assuming that at the instant of debonding, $\sigma_A = f_{ct}$. Thus, the value of τ at which delamination of the concrete covers occurs is given by Eq. (5.38).

$$\tau_{del} = \frac{f_{ct} l}{6h'} \frac{b}{n\pi d_b} \tag{5.38}$$

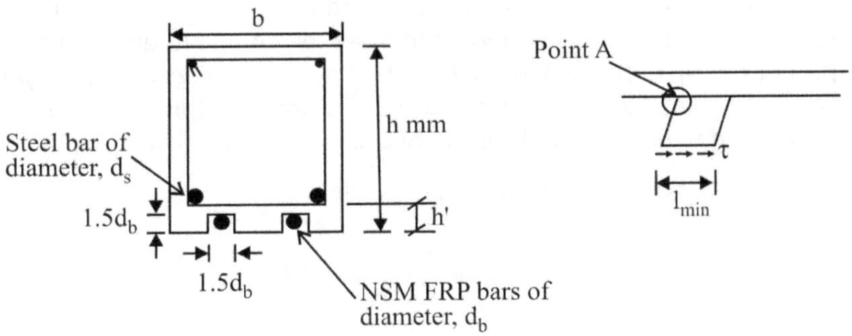

Figure 5.9a. Concrete tooth model.

For a beam under four-point bending, the shear stress, τ, within shear span of the beam is balanced by the axial stress, σ_f, in the NSM bars. At the location under the point load at the delamination, the value of σ_f is given by Eq. (5.39).

$$\sigma_{fdel} = \frac{4}{d_b} L_p \tau_{del} = \frac{2bL_p l}{3n\pi d_b^2 h'} f_{ct} \tag{5.39}$$

where L_P is the effective length of the NSM bars in the shear span over which equivalent shear stresses at the interface of the bars with the surrounding material may be assumed to remain constant. The value of L_P (Eq. (5.40)) is smaller of the plate length in the shear span, L_{P1} and an equivalent length, L_{P2}, as function of l_{min}.

$$L_P = \text{smaller of } (L_{P1}, L_{P2}) \tag{5.40}$$

$$\begin{aligned} L_{P2} &= 1.86 l_{min}^2 - 127 l_{min} + 2436 \quad && \text{if } l_{min} \le 50\,\text{mm} \\ &= 736\,\text{mm} && \text{if } l_{min} > 50\,\text{mm} \end{aligned} \tag{5.41}$$

Thus, the minimum and maximum stress in the FRP required to cause flexural crcking and failure of a tooth can be determined from Eq. (5.39) with l taken as l_{min} or l_{max}, respectively.

5.14.2. Design Procedure for Flexural Strengthening Using NSM FRP Rebars

To compute the moment capacity of an RC cross-section strengthened in bending with NSM FRP rebars, the following design equations and procedures are proposed.

1. Obtain local bond strength (u_f) of NSM bars from literature data or by running bond tests with short bonded lengths with the same type of bar, concrete strength, groove-filling materials and groove depth to bar diameter ratio to be used in the beams.
2. Compute l_{min} from Eq. (5.31) and l_{max} from Eq. (5.32).
3. Compute $\sigma_{fdel\,max}$ using Eq. (5.39) by taking $l = l_{max}$.
4. Compute the effective strain in NSM FRP system

$$\sigma_{fueff} = \text{Min}\left(\sigma_{fu}, \sigma_{fdelmax}, f_{fec}\right) \tag{5.42a}$$

where, σ_{fu} is the tensile strength of the FRP bars and f_{fec} stress in the FRP bars corresponding to the crushing mode of concrete and determined based on compatibility of strains and constitutive relationships. For example, for an assumed depth of neutral axis, c, effective strain and stresses in FRP bars can be calculated and then by force equilibrium condition, the assumed depth of neutral axis could be verified to be close to the exact one as described below:

Effective strain in FRP based on crushing of concrete,

$$\varepsilon_{fe} = \varepsilon_{cu}\left(\frac{d_f - c}{c}\right) - \varepsilon_{bi} \le k_m \varepsilon_{fu} \tag{5.42b}$$

Effective stress in the FRP bars based on the concrete crushing mode of failure,

Effective stress in the FRP bars based on concrete crushing,

$$f_{fec} = E_f \varepsilon_{fe} \tag{5.42c}$$

The term k_m is bond reduction co-efficient and is based on general trend. The term k_m for NSM systems can be taken equal to 0.50, 0.60 or 0.70 depending upon the surface configuration of the reinforcement, which is very influential on the bond behavior. The smallest value 0.50 should be used for smooth, sand blasted or sanded bars, the value 0.60 should be used for ribbed bars with high rib protrusion, while the largest value 0.70 should be used for ribbed bars with low rib protrusion, spirally wound bars and rectangular bars with large b_b/a_b ratio. The term, d_f, is distance of NSM bars from extreme compression fiber, whereas b_b/a_b refers to depth to width ratio of NSM bar.

It may be noted that if $\varepsilon_{fe} < k_m\varepsilon_{fu}$ or $f_{fec} < \sigma_{f\text{del max}}$, then concrete crushing governs the failure of the beam. The strain in steel bars could be estimated by Eq. (5.42d) as follows:

$$\varepsilon_s = \frac{(\varepsilon_{fe} + \varepsilon_{bi})}{d_f - c}(d - c) \tag{5.42d}$$

The stress in steel can be estimated by Eq. (5.42e) as follows:

$$f_s = E_s \varepsilon_s \leq f_y \tag{5.42e}$$

From the force equilibrium equation, depth of neutral axis at failure can be computed and verified using Eq. (5.42f).

$$c = \frac{A_s f_s + A_f \sigma_{fuef}}{\alpha f_c' b} \tag{5.42f}$$

where value of α depends on the actual strain in the concrete. The value of c has to be obtained by trial and error. If $\varepsilon_c \geq 0.003$, α could be taken as 0.85. However, the actual value of α is based on the Hognestad's stress–strain curve for concrete presented as follows:

$$\alpha = 1 + \frac{\varepsilon_c}{\varepsilon_o}\left(1 - \frac{\varepsilon_c}{3\varepsilon_o} - \frac{\varepsilon_o^2}{\varepsilon_c^2}\right) - \left(\frac{0.15}{0.004 - \varepsilon_o}\right)\left(\frac{\varepsilon_c}{2} - \varepsilon_o\right) - \left(\frac{0.075}{0.004 - \varepsilon_o}\right)\left(\frac{\varepsilon_o^2}{\varepsilon_c}\right) \quad \text{for } \varepsilon_o \leq \varepsilon_c \leq \varepsilon_u \tag{5.42g}$$

$$\alpha = \frac{\varepsilon_c}{\varepsilon_o}\left(1 - \frac{\varepsilon_c}{3\varepsilon_o}\right) \quad \text{for } 0 \leq \varepsilon_c \leq \varepsilon_o \tag{5.42h}$$

where, $\varepsilon_o = \frac{2f_c'}{E_c}$ is concrete strain corresponding to the maximum concrete stress, $\varepsilon_u = \varepsilon_{ac} = 0.003$.

5. Compute the nominal ultimate moment capacity using Eq. (5.42i), (also see Fig. 5.9b).

$$M_n = A_s f_s (d - \gamma_c) + \psi_f A_f \sigma_{fueff}(d_f - \gamma_c) \tag{5.42i}$$

where,

$$\gamma_c = \beta_1 c \tag{5.42j}$$

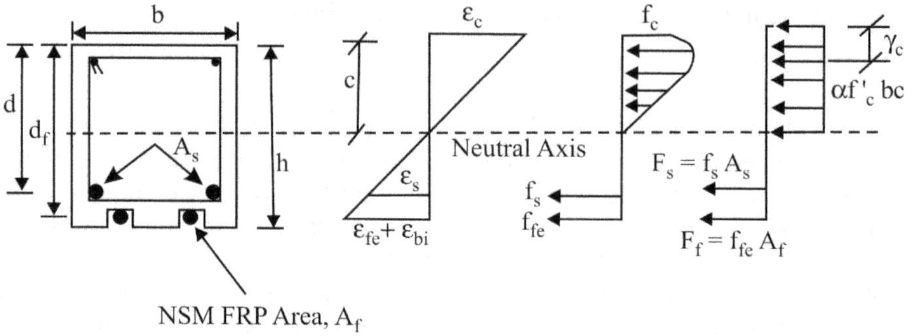

Figure 5.9b. Stress, strain and force distribution across the section at ultimate condition.

$$\beta_1 = \frac{\frac{1}{3} - \frac{\varepsilon_c}{12\varepsilon_o}}{1 - \frac{\varepsilon_c}{3\varepsilon_o}} \quad \text{if } 0 \le \varepsilon_c \le \varepsilon_o \tag{5.42k}$$

$$\beta_1 = 1 - \frac{\varepsilon_c^3 - 5.1\varepsilon_o\,\varepsilon_c^2 - 0.004\varepsilon_o^2 + 0.024\varepsilon_c^2}{\varepsilon_c\left(3.925\varepsilon_o^2 - 10.2\varepsilon_o\,\varepsilon_c - 0.9\varepsilon_c^2 - 0.016\varepsilon_o + 0.048\varepsilon_c\right)} \quad \text{if } \varepsilon_o \le \varepsilon_c \le \varepsilon_u \tag{5.42l}$$

6. Compute the design Ultimate Moment of the cross-section by multiplying the nominal ultimate moment by a reduction factor ϕ_f (Eq. (5.43a)). Also to be taken into account is the novelty of this strengthening technique, an additional reduction factor, ψ_f, should be applied to the strength contribution of FRP. Hence, design moment capacity of the cross-section can be expressed by Eq. (5.43b).

$$\phi_f = \begin{cases} 0.90 & \text{for } \varepsilon_s \ge 0.005 \\ 0.70 + \dfrac{0.20(\varepsilon_s - \varepsilon_y)}{0.005 - \varepsilon_y} & \text{for } \varepsilon_y \le \varepsilon_s \le 0.005 \\ 0.70 & \text{for } \varepsilon_s \le \varepsilon_y \end{cases} \tag{5.43a}$$

$$\phi M_n = \phi_f\left[M_s + \psi_f M_f\right] \tag{5.43b}$$

where, M_s and M_f are the contributions to the moment capacity given by steel and FRP, respectively. The value of ψ_f could be taken as 0.85 as per recommendation of ACI 440.2R-02.

7. Check serviceability: This step presents no difference with respect to the case of externally bonded laminates.

5.14.3. Shear Strengthening

A design approach for computing the shear capacity of RC beams strengthened in shear with NSM bars includes two equations that may be used to obtain the FRP contribution to the shear capacity and taking the lower of the two results. The proposed design equations are briefly summarized as follows:

1. Compute d_{net}: A reduced value is used for the height of the cross-section containing shear reinforcement in the form of NSM rods.

$$d_{net} = d_r - 2c_c \tag{5.44}$$

where, d_r is the height of the shear strengthened part of cross-section and c_c is the concrete cover of the longitudinal reinforcement. In the case of vertical NSM rods, d_r coincides with the length of FRP rods. It may be noted that this reduction approximately accounts for the height of the Mörsch truss being lower than the total height of the beam. It can be assumed that the axis of the upper strut is situated on the resultant of the compressive stresses and the axis of the lower tie coincides with that of the longitudinal steel.

2. Compute V_{1F}: V_{1F} is the FRP shear strength contribution related to bond controlled shear failure in most unfavorable crack position. It is computed using the following assumptions.

   i. Inclination angle of the shear cracks is constant and equal to 45°.
   ii. Distribution of bond stresses along the FRP rods at ultimate is uniform.
   iii. The ultimate bond stress is reached in all the rods intersected by the cracks at ultimate.

$$V_{1F} = 2\pi d_b u_f L_{\text{tot min}} \tag{5.45}$$

The value of $L_{\text{tot min}}$ depends on d_{net}, on the spacing, s, of rods and on inclination.

$$L_{\text{tot min}} = d_{net} - s \text{ if } \frac{d_{net}}{3} < s < d_{net} \tag{5.46a}$$

$$L_{\text{tot min}} = 2d_{net} - 4s \text{ if } \frac{d_{net}}{4} < s < \frac{d_{net}}{3} \tag{5.46b}$$

For 45° inclined rods:

$$L_{\text{tot min}} = (2d_{net} - s)\frac{\sqrt{2}}{2} \text{ if } \frac{2d_{net}}{3} < s < 2d_{net} \tag{5.47a}$$

$$L_{\text{tot min}} = 2\sqrt{2}(d_{net} - s) \text{ if } \frac{d_{net}}{2} < s < \frac{2d_{net}}{3} \tag{5.47b}$$

3. Check if calculation of V_{2F} is necessary

$$\text{If } d_{net} < 0.002\frac{d_b E_b}{u_f} \quad \text{for vertical rods} \tag{5.48a}$$

$$\text{If } d_{net} < \sqrt{2}\left(0.001\frac{d_b E_b}{u_f}\right) \quad \text{for 45° inclined rods} \tag{5.48b}$$

Then calculation of V_{2F} is not necessary. If Eq. (5.48) is not satisfied, V_{2F} has to be computed. In Eq. (5.48), E_b is the elastic modulus of the NSM FRP bars.

4. Compute V_{2F} (if necessary): V_{2F} is the FRP strength contribution corresponding to a maximum FRP strain of 4000 micron (i.e. 0.004). This limit is suggested to maintain the shear integrity of the concrete. V_{2F} has to be computed in the most unfavorable crack position, that it is the position in which it is minimum. It can be shown that, for vertical rods, the minimum value is given by Eq. (5.49).

$$V_{2F} = 2\pi\, d_b\, u_f\, L_i \quad \text{if } \frac{d_{\text{net}}}{2} < s < d_{\text{net}} \tag{5.49a}$$

$$V_{2F} = 2\pi\, d_b\, u_f L_i \frac{3d_{\text{net}} - 4s}{d_{\text{net}}} \quad \text{if } \frac{d_{\text{net}}}{4} < s < \frac{d_{\text{net}}}{2} \tag{5.49b}$$

In the case of 45° rebars, it is:

$$V_{2F} = 2\pi\, d_b\, u_f L_i \quad \text{if } d_{\text{net}} < s < 2d_{\text{net}} \tag{5.50a}$$

$$V_{2F} = 2\pi\, d_b\, u_f L_i \frac{3d_{\text{net}} - 2s}{d_{\text{net}}} \quad \text{if } \frac{d_{\text{net}}}{2} < s < d_{\text{net}} \tag{5.50b}$$

In the above equations, L_i is the effective length of an FRP rod caused by the crack corresponding to a strain of 0.004 and is given by Eq. (5.51).

$$L_i = 0.001 \frac{d_b E_b}{u_f} \tag{5.51}$$

5. Compute $V_{\text{FRP}} = \text{Min}(V_{1F}, V_{2F})$
6. Check that limits on V_{FRP} are satisfied. Limits on the value of V_{FRP} and of the sum $(V_s + V_{\text{FRP}})$ indicated by ACI440-2R-02 should be extended to the case of NSM strengthening also as their rationale is of general validity.
7. Compute the shear capacity of the beam: The nominal shear strength (Eq. (5.52)) of an RC beam strengthened with an FRP system can be computed as the sum of the shear strength of the concrete, the shear strength provided by the steel shear reinforcement and the contribution of the FRP reinforcement.

$$V_n = V_c + V_s + V_{\text{FRP}} \tag{5.52}$$

8. Compute design shear strength: The design shear strength (Eq. (5.53)) is obtained by applying a strength reduction factor, ϕ_s, to the nominal shear strength. The reduction factor, $\phi = 0.85$ should be maintained for concrete and steel contributions and an additional reduction factor, ψ_f, should be applied to the FRP contribution, to account for the novelty of this strengthening technique.

$$\phi V_n = \phi_s [V_c + V_s + \psi_f V_{\text{FRP}}] \tag{5.53}$$

The value of factor ψ_f should not exceed 0.85.

5.14.4. Anchorage Length Requirement

Based on the limit state philosophy, a design approach for the anchorage length of NSM FRP bars in concrete is suggested as follows:

1. Check that $P_s \le P_1$: At service load level, it should be required that the free-end slip is zero and that the bar is anchored using only the ascending portion of the bond slip relationship. This poses a limit to the service load that can be applied to the bar, P_s, which must be less than or equal to P_1. The value of P_1 being function of the calibrated local bond slip relationship.

2. Find the anchorage length at Ultimate Limit States (ULS), $L_{an,u}$, i.e., the embedment length needed to anchor the bar under the design load at the ULS, P_u. To account for uncertainties in the bond behavior, the curves of the bond failure load as a function of the embedment length should be scaled homotetically by an appropriate reduction factor. Entering the reduced curve with factored load, P_u, the corresponding anchorage length at the ULS can be found.
3. Find the anchorage length at service, $L_{an,s}$
   For $P_s \le P_1$, the embedment length needed is given by Eq. (5.54).

$$L_{an,s} = l_m \left(\frac{P_s}{P_1} \right)^{\frac{1-\alpha}{1+\alpha}} \tag{5.54}$$

   An additional check to be performed at service load level is that the loaded end slip is not larger than a limiting value compatible with aesthetic and/or durability requirements. This limit has been proposed as 0.4 mm based on the results available in literature. As the value of maximum end slip (s_m) found for NSM reinforcement is always less than 0.4 mm, this condition is automatically satisfied as soon as the service load is less than or equal to P_1.
4. Compute the anchorage length.

Finally, the anchorage length of the bar should be maximum of $L_{an,u}$ and $L_{an,s}$.

5.15. ACI 440.2R-02 Design Approach for NSM FRP Strengthening

Design of strengthening system using NSM FRP bars as per ACI 440.2R-02 standard is similar to that used for external FRP plate or fabric systems described earlier in conjunction with the concrete tooth model. Details of ACI 440.2R-02 design procedure are presented below for the sake of completeness.

5.15.1. Flexural Design Approach

This section provides guidance on the procedure for calculating the flexural strengthening effect of adding longitudinal NSM FRP reinforcement to the tension face of a reinforced concrete member. The concepts given in this section apply to the strengthening process of rectangular sections reinforced in tension zone with non-prestressed steel bars. The concepts described here can be used in non-rectangular sections also such as T-sections and I-sections, and to members with compression reinforcement. In the case of prestressed members, strain compatibility with respect to the state of strain in the stressed member, should be used to evaluate the FRP contribution. In addition, failure modes controlled by rupture of prestressing tendons should also be considered.

Assumptions: The following assumptions are made in calculating the flexural resistance of a section strengthened with an externally applied FRP system:

1. Design calculations are based on the actual dimensions, internal reinforcing steel arrangement and material properties of the existing member being strengthened.
2. The strains in the reinforcement and concrete are directly proportional to the distance from the neutral axis, that is, a plane section before loading remains plane after loading.
3. The maximum usable compressive strain in the concrete is 0.003.

4. The tensile strength of concrete is neglected.
5. The FRP reinforcement has a linear elastic stress strain relationship to failure.
6. Perfect bond exists between the concrete and external FRP reinforcement.

Section shear strength: When NSM reinforcement is used to increase the flexural strength of a member, it should be ensured that the member has sufficient shear strength to resist the shear forces associated with the increased flexural capacity. The potential for shear failure of the section to the required shear strength should be considered by comparing the design shear strength of section to the required shear strength. If additional shear capacity is required, the member should be upgraded to the meet the new shear demands.

Existing substrate strain: In case all the loads on a member including its self-weight and any prestressing forces are not removed before installation of FRP reinforcement, the strain developed at the substrate where FRP will be installed should be considered in the design as initial strains and should be excluded from the strain in the FRP bars. The initial strain level on the bonded substrate can be determined from an elastic analysis of the existing member, considering all loads that will be acting on the members during the installation of the FRP system.

5.15.2. Nominal Flexural Strength

The nominal flexural capacity of concrete member strengthened with NSM FRP bars can be determined based on strain compatibility, internal force equilibrium and the controlling mode of failure. The internal strain and stress distribution for a rectangular section under flexure at an ultimate stage is shown in Fig. 5.10.

The nominal flexural capacity of a reinforced concrete section with NSM FRP reinforcement can be computed using Eq. (5.55). Moreover, an additional reduction factor $\psi_f = 0.85$ is applied to the flexural strength contribution of the NSM FRP reinforcement. The value of depth of neutral axis is calculated as for FRP externally strengthened section using equilibrium equation (Eq. (5.15)).

$$M_n = A_s f_s \left(d - \frac{\beta_1 c}{2} \right) + \psi_f A_f f_{fe} \left(d_f - \frac{\beta_1 c}{2} \right) \tag{5.55}$$

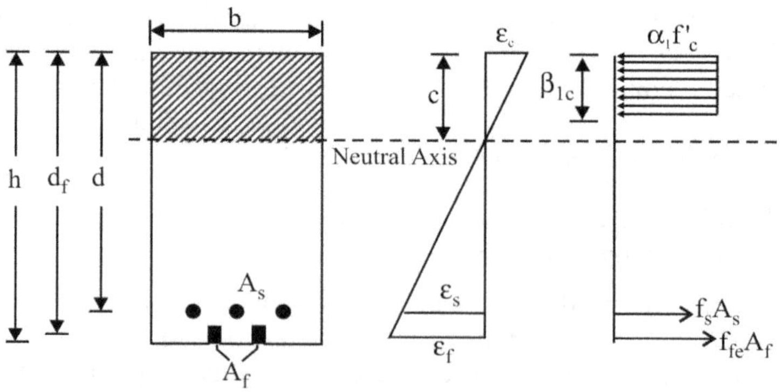

Figure 5.10. Strains and stresses at ultimate failure level of NSM FRP reinforced rectangular section.

5.15.2.1. Failure modes

The flexural strength of a section depends on the controlling flexural failure mode. For NSM FRP strengthened sections, the following failure modes should be investigated.

i. Crushing of the concrete in compression before yielding of the reinforcing steel.
ii. Yielding of the steel in tension followed by rupture of the FRP bars.
iii. Yielding of the steel in tension followed by concrete crushing.
iv. Shear/Tension de-lamination of the concrete cover (i.e., cover delamination).

It may be noted that concrete crushing is assumed to occur if the compressive strain in the concrete reaches its maximum usable strain ($\varepsilon_c = \varepsilon_{cu} = 0.003$) before steel yield. Similarly, FRP rupture is assumed to occur if the strain in the FRP reaches its design rupture strain ($\varepsilon_f = \varepsilon_{fu}$). Furthermore, cover delamination or FRP debonding can occur if the force in the FRP bars cannot be sustained by the concrete substrate.

The calculations of strains using strain compatibility, stresses using constitutive relations, force resultants and checks for ductility as well as serviceability are similar to that used for FRP externally strengthened members discussed earlier in this chapter.

5.16. Design for Shear Strength

The approach used for estimating the shear strength of NSM FRP strengthened members is the same as that used in the case of externally bonded FRP laminates and/or fabric sheets discussed earlier in this chapter. However, several parameters influence the contribution of NSM FRP bars to the shear capacity (V_f) such as quality of bond and FRP rebar type, groove dimensions and conditions of support. While computing V_f, two aspects should be taken into account. The first aspect refers to the FRP shear strength related to bond controlled shear failure while the second aspect refers to the shear resisted by NSM FRP bars when the maximum strain in the bar reaches 0.004. This strain limit is recommended to maintain the shear integrity of the concrete to avoid large shear cracks that could compromise the aggregate interlock mechanism.

Assumptions: To evaluate the nominal shear strength of NSM FRP reinforced concrete sections, the following assumptions are made:

i. The slope of the shear crack is assumed to be at 45° from the axis of the beam.
ii. The bond stresses are uniformly distributed along the effective length of the FRP bar at the ultimate.
iii. The ultimate bond stress is reached in all the bars intersected by the crack.

5.16.1. Ultimate Shear Strength

The nominal and design shear strengths of NSM FRP reinforced beams are given by Eqs. (5.52) and (5.53), respectively. The shear strength (V_f) provided by the NSM FRP reinforcement can be determined from the tensile stress in the FRP across the assumed crack. The shear contribution of the FRP shear reinforcement is then given by Eq. (5.56).

$$V_f = 2\pi d_b \tau_b L_{\text{tot}} \quad \text{for circular bars} \tag{5.56a}$$

$$V_f = 4[a+b]\tau_b L_{tot} \quad \text{for rectangular bars} \tag{5.56b}$$

where, d_b is the nominal bar diameter, a and b represent the cross-sectional dimensions for rectangular bars and τ_b represents the average bond strength of the bars crossed by a shear crack expressed as a function of d_b, bonded length and ultimate tensile load. It may be noted that the experimental data available on No. 3 carbon FRP deformed bars demonstrate that τ_b can be taken as 6.9 MPa when used with epoxy based resin in a groove size at least 1.5 times the bar diameter. Based on ACI 440 committee's consensus, when τ_b is not known, a value of 6.9 MPa could be assumed for τ_b.

L_{tot} can be expressed by Eq. (5.57) as follows:

$$L_{tot} = \sum_i L_i \tag{5.57}$$

where L_i as shown in Fig. 5.11 represents the length of each single NSM FRP bar crossed by a 45° shear crack and expressed by Eq. (5.58a).

$$L_i = \begin{cases} \text{Min}\left(l_{0.004}, s.i\right) & i = 1 \ldots \dfrac{n}{2} \\ \text{Min}\left(l_{0.004}, d_{net} - s.i\right) & i = \dfrac{n}{2} + 1 \ldots n \end{cases} \tag{5.58a}$$

In the case of shear of NSM bars inclined at angle, α, from the longitudinal axis of the member, Eq. (5.58b) can be used to determine the value of L_i.

$$L_i = \begin{cases} \dfrac{s}{\cos\alpha + \sin\alpha} i \le l_{0.004} & i = 1 \ldots \dfrac{n}{2} \\ d_{net} - \dfrac{s}{\cos\alpha + \sin\alpha} \le l_{0.004} & i = \dfrac{n}{2} + 1 \ldots n \end{cases} \tag{5.58b}$$

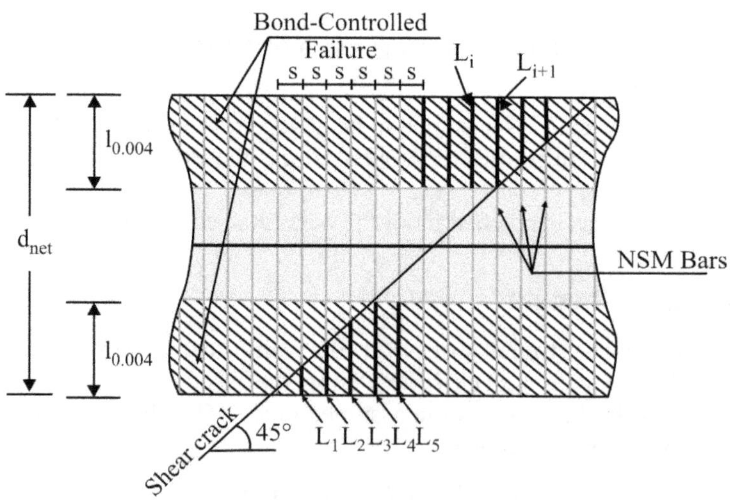

Figure 5.11. Generic shear crack used in definition of L_i.

There are two limitations of Eq. (5.58): (1) The first limitation takes into account for the shear integrity of the concrete by limiting at 0.004 the maximum strain in the FRP reinforcement. From the force equilibrium condition for circular bars,

$A_b(0.004E_b) = \pi d_b l_{0.004}\tau_b$, we have

$$l_{0.004} = 0.001\frac{d_b E_b}{\tau_b} \quad \text{for circular bars} \tag{5.59a}$$

Similarly, from the force equilibrium condition of rectangular bars, we have

$$l_{0.004} = 0.002\frac{ab}{a+b}\frac{E_b}{\tau_b} \quad \text{for rectangular bars} \tag{5.59b}$$

where A_b and E_b represent area and elastic modulus of one FRP bar. (2) The second limitation of Eq. (5.58) takes into account for bond as the shear controlling failure mechanism and represents the minimum effective length of an FRP bar crossed by a shear crack. It is expressed by $s.i$ or $d_{net} - s.i$ (for vertical bars) and $s.i/(\cos\alpha + \sin\alpha)$ or $d_{net} - s.i/(\cos\alpha + \sin\alpha)$ (for inclined bars) depending on the value assumed by the term, n as given by Eq. (5.60).

$$n = \frac{d_{net}}{s} \quad \text{for vertical bars} \tag{5.60a}$$

$$n = \frac{d_{eff}(1+\cot\alpha)}{s} \quad \text{for inclined bars} \tag{5.60b}$$

where, d_{eff} represents the vertical length of d_{net} and is given by Eq. (5.60c).

$$d_{eff} = d_r \sin\alpha - 2c_c \tag{5.60c}$$

In Eq. (5.60), n is taken as the smallest integer of the number in question. For example, $n = 32/3 = 10.7$ then $n = 10$. The term s is the center to center distance between vertical bars. The value of spacing (i.e., s) of FRP shear reinforcement placed perpendicular to the axis of member shall not exceed $d_{net}/2$ or 600 mm. From Fig. 5.11, it is observed that six rods (L_1, L_2, L_3 and three symmetric on the other side) are controlled by bond, while the five remaining rods are controlled by the strain limitation of 0.004.

The value of d_{net} (Eq. (5.61)) is a reduced value for the effective length of bar which takes into account the formation of vertical flexural cracks in shear region that could compromise the bond between the FRP bar and surrounding concrete.

$$d_{net} = d_r - 2c_c \quad \text{for vertical bars} \tag{5.61a}$$

$$d_{net} = d_r - \frac{2c_c}{\sin\alpha} \quad \text{for inclined bars} \tag{5.61b}$$

where d_r is the actual length of the bar and c_c is the clear concrete cover of the internal longitudinal reinforcement as shown in Fig. 5.12. The reduced length also takes into account of Mörsch's shear theory in which the tension force is located at the level of longitudinal reinforcement and the compression force is located close to the top of the cross-section. As a result, the effective section height is less than total section height. Moreover, d_{net}, also allows for the formation of a development length for the top portion

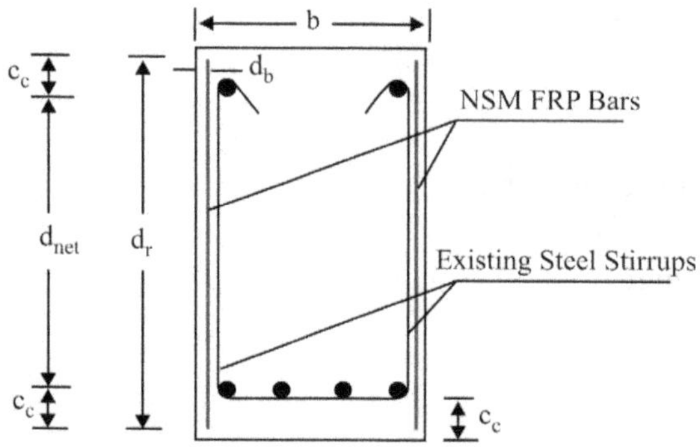

Figure 5.12(a). NSM FRP bars used as shear reinforcement in rectangular section.

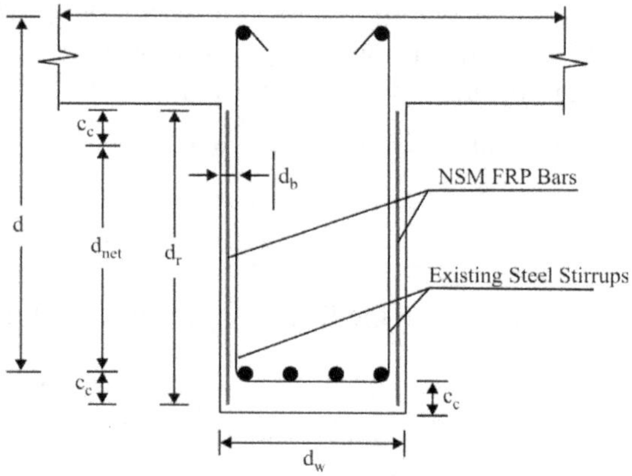

Figure 5.12(b). NSM FRP bars used as shear reinforcement in T-sections.

of the bar where hooks are not possible. It may be noted that the total reinforcement contribution taken as sum of both steel and FRP contributions should satisfy the condition expressed in Eq. (5.30), i.e., $V_s + V_f \leq 0.66\sqrt{f_c'}b_w d$.

5.17. Serviceability

As it is known that the excessive deflection may interfere with the use of the structure causing cracking of partitions and/or malfunctioning of other non-structural elements. The immediate deflection (δ) can be calculated as per ACI 440.2R-02 standard. For cracked cross-sections, an equivalent moment of Inertia defined by Branson and later simplified by ACI 318 can be used as given in Eq. (5.62).

$$I_e = \left(\frac{M_{cr}}{M_a}\right)^3 I_g + \left[1 - \left(\frac{M_{cr}}{M_a}\right)^3\right] I_{cr} \tag{5.62}$$

For beams having two continuous ends, the average equivalent moment of inertia as given in ACI 440.2R-02 standard is expressed by Eq. (5.63).

$$I_{e,\text{average}} = 0.70 I_{em} + 0.15\left(I_{e1} + I_{e2}\right) \tag{5.63}$$

The long-term deflection (Δ) due to sustained load and creep of the concrete can be calculated using Eq. (5.64).

$$\Delta = (1 + \lambda)\delta \tag{5.64}$$

where,

$$\lambda = \frac{\xi}{1 + 50\rho'} \tag{5.65}$$

In Eq. (5.65), ρ' is compression reinforcement ratio. The crack width can be determined using Gergely–Lutz equation that has been described in Chapter 4.

5.18. Detailing

The minimum dimensions of the grooves to be made for installing the NSM FRP bars of circular and/or rectangular bars are presented in Fig. 5.13. As shown in Fig. 5.13, the minimum dimension of the grooves should be taken as at least 1.5 times the diameter of the FRP bar; when rectangular bar having dimensions of $a \times b$ is used, the minimum groove size should be taken as $3.0a \times 1.5b$ where a is smallest dimension of rectangular bar. It may be noted that in many situations, the minimum groove dimension could be the result of installation requirements rather than engineering. For example, a 5 mm. groove may be the smallest possible because: (i) no saw blade can be less than this size, and (ii) minimum amount of adhesive needed.

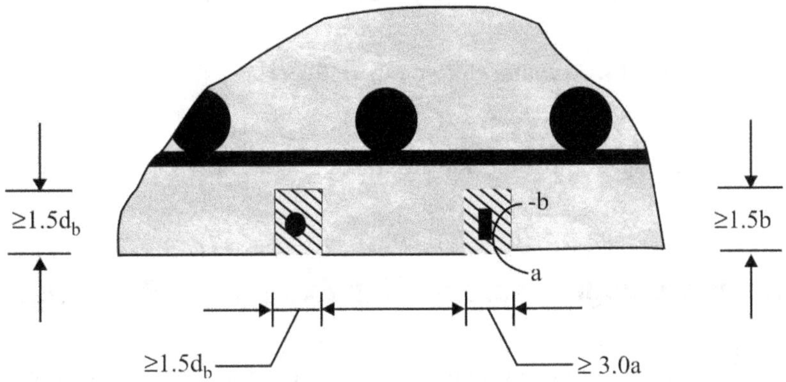

Figure 5.13. Minimum dimensions of the grooves.

5.19. Development Length of NSM FRP Bars

It has been shown by research that the development length of FRP bars is highly dependent on strip dimensions, groove size, concrete and adhesive properties, internal steel reinforcement ratio, reinforcement configuration and type of loading. In general, the bond properties between FRP reinforcement and concrete are similar to that of steel reinforcement and depend on FRP type, elastic modulus, surface deformation and shape of FRP bar. It has been also found that the development length increases by increasing the internal steel reinforcement ratio and decreases with the increase of either the concrete compressive strength and/or the groove size. The equilibrium condition of an FRP bar with a length equal to its development length, l_d, is shown in Fig. 5.14. The axial force in the bar is resisted by the shear stresses τ_b acting on the surface of the bar. Assuming a triangular stress distribution, the average bond stress can be express as $\tau_b = 0.5\tau_{max}$ and the expression for predicting the development length is given by Eq. (5.66).

$$l_d = \frac{d_b}{4(0.5\,\tau_{max})} f_{fu} \quad \text{for circular bars} \tag{5.66a}$$

$$l_d = \frac{ab}{2(a+b)(0.5\tau_{max})} f_{fu} \quad \text{for rectangular bars} \tag{5.66b}$$

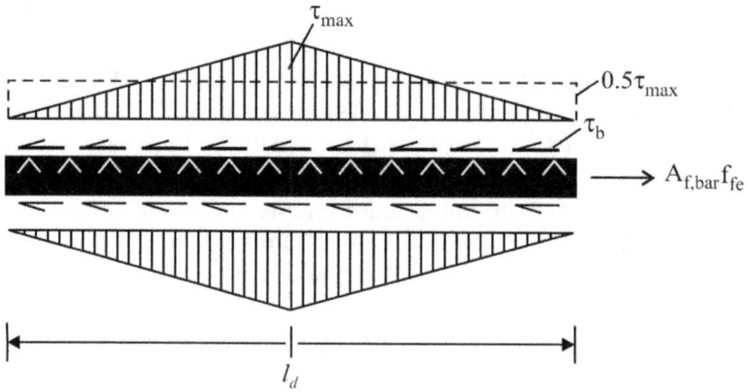

Figure 5.14. Transfer of force in an FRP bar.

It may be noted that when the controlling failure mode is not known, the value of τ_{max} can be taken as 3.45 MPa.

5.20. ISIS Canada Design Approach for External FRP Strengthening

In this section, fundamentals of external strengthening requirements and design procedure as per ISIS Canada design guidelines have been presented. The first and foremost thing for external strengthening using FRP is to determine the causes of deficiency in structures. Deficiencies in the existing structures can be due to many factors. The followings are the main factors leading to the deterioration of structures worldwide.

1. **Environmental factors:** The chloride induced corrosion of conventional reinforcing steel in concrete, freeze thaw cycling and wet dry cycling can contribute to the development of cracking and hence, deterioration of structures.
2. **Increase of design loads:** Increase of design loads may make the structure unsuitable for upgraded version of design loads and hence require strengthening.
3. **Upgrade of design guidelines:** Due to change in design guidelines for example, change from the working stress to limit states design methods may make the structure unsafe as per the current design procedure and hence, the structure may require external strengthening.
4. **Increase of traffic volume and loads:** Due to increase in traffic volume and loads on highways and on roads, structures may need upgrade in strength and stiffness and hence, require external strengthening.

To develop an appropriate FRP strengthening an existing structure should first be assessed to determine the condition of concrete, to determine the causes of deficiencies, to establish the existing load carrying capacity of the structures and to evaluate the feasibility of using externally bonded FRP systems for repair. The evaluation of existing structures should be carried out with extreme care and should look for the following details.

- As built drawings including all past modifications;
- Actual size of the concrete elements;

- Actual properties of the existing materials including surface tensile strength of concrete;
- Location, size and causes of cracks and spalls;
- Location and extent of any corrosion of reinforcing steel;
- Quantity and location of the existing reinforcing steel; and
- An appropriate evaluation of the applied loads.

It must be noted that the condition of the concrete surface is one of the most critical aspects to be considered when strengthening a structure. The concrete must be able to transfer the load from the existing structure to the FRP system through shear stresses that develop in the adhesive matrix. In the following section, flexural strengthening mechanism for beams and one way slab is described including assumptions made for developing design equations.

5.20.1. Flexural Strengthening of Beam and One-Way Slab

As FRP plates or sheets are bonded to the tensile surfaces of reinforced concrete beams. The flexural strength of beams and/or slabs strengthened using FRP can be evaluated based on the following assumptions.

1. FRPs are perfectly linear elastic materials.
2. Plane sections remain plane.
3. Perfect bond exists between concrete and steel reinforcement and between FRP reinforcement and concrete.
4. Adequate anchorage and development length is ensured for the FRP reinforcement
5. Concrete compressive stesss–strain curve is parabolic and concrete has no strength in tension.
6. Initial strains in the section at the time of strengthening can in most cases can be ignored.

Material resistance factors: As per recommendations of ISIS Canada Design Manual No. 4, the materials resistance factors for various materials are as follows:

Steel: $\phi_s = 0.85$ for building

$= 0.90$ for bridges

Concrete: $\phi_c = 0.60$ for building

$= 0.75$ for bridges

FRP: The material resistance factor for FRP depends on the type of FRP materials used, variability of material characteristics, intended use, effect of sustained load and various durability considerations. Typical values of material resistance factor for CFRP and GFRP are given as follows:

$$\phi_F = 0.7 - 0.78 \quad \text{for CFRP}$$

$$= 0.6 - 0.76 \quad \text{for CFRP}$$

However, in ISIS Canada Design Manual No. 4, the recommended values of resistance factor for CFRP and GFRP are 0.75 and 0.5, respectively.

Failure modes: The following four failure modes are considered for developing the design equations for flexural strength of FRP strengthened reinforced concrete members.

- Concrete crushing before yielding of the reinforcing steel;
- Steel yielding followed by concrete crushing;
- Steel yielding followed by FRP rupture; and
- Debonding of FRP reinforcement at the FRP or concrete interface.

As per ISIS design manual No. 4, it is assumed that the fourth failure mode (i.e., debonding of FRP reinforcement) will not occur because in practice this failure mode may be precluded by the use of specialized anchorage techniques such as mechanical anchorage systems.

5.20.2. Flexural Design Approach

The flexural design of rectangular beams and one way slabs can be conducted using the concept of strain compatibility and equilibrium. The strain and stress distribution over the cross-section is given in Figs. 5.15 and 5.16. As shown in Fig. 5.15, the strain distribution in an FRP strengthened concrete beam at failure is more complicated than that in a conventionally reinforced beam because the concrete and steel will be subjected to some initial strains, ε_{ci} and ε_{si}, due to the self-weight of the beam and any live load applied to the structure during strengthening. ε_{cl} and ε_{sl} are the additional strains that are due to additional loading after strengthening with FRP and ε_{frp} is the strain in the FRP at failure. As per ISIS Canada design approach, the initial strains in concrete and steel are assumed to be negligible and so the stress and strain distribution can be approximated as shown in Fig. 5.16. However, in practice, it is important to consider the effect of initial strains in the concrete member prior to strengthening as explained earlier in the case of ACI 440.2R-02 design standards.

From the equilibrium of internal forces (Fig. 5.16), we have

$$C_c = T_s + T_{\text{frp}} \tag{5.67}$$

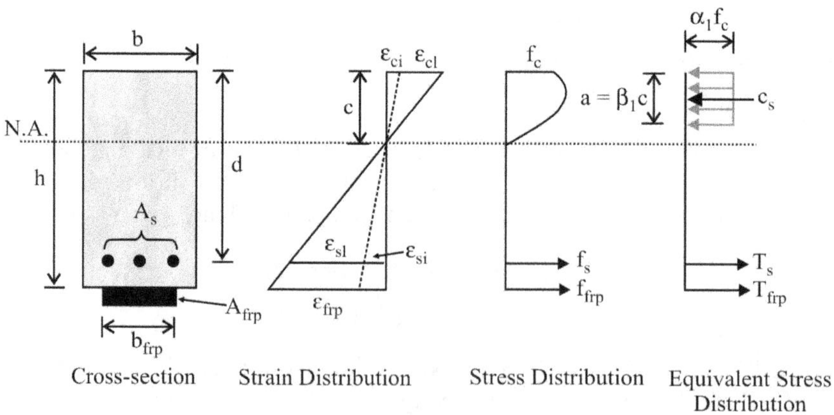

Figure 5.15. Strain and stress distribution for a singly reinforced concrete beam strengthened in flexure with externally bonded FRP laminates (with initial strains before strengthening considered).

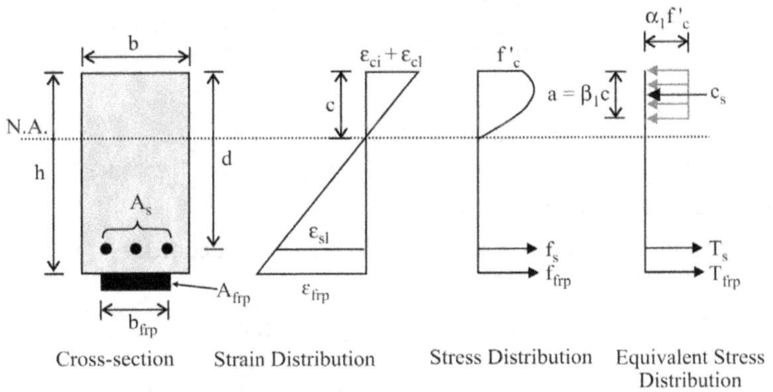

Figure 5.16. Assumed strain and stress profiles for a singly reinforced concrete beam strengthened in flexure with externally bonded FRP materials (with initial strains at the time of strengthening neglected).

where stress resultants, C_c, T_s, and T_{frp} are given as follows:

$$C_c = \phi \alpha_1 f'_c \beta_1 bc \tag{5.68}$$

$$T_s = \phi_s A_s f_s \quad \text{with } f_s \leq f_y \tag{5.69}$$

$$T_{\text{frp}} = \phi_{\text{frp}} A_{\text{frp}} E_{\text{frp}} \varepsilon_{\text{frp}} \quad \text{with } \varepsilon_{\text{frp}} \leq \varepsilon_{\text{frp}u} \tag{5.70}$$

Using Eqs. (5.67) to (5.70), the depth of neutral axis can be determined for assumed strain in the extreme compression fiber. As per the ISIS Canada design guidelines, the assumed strain distribution at failure is given in Fig. 5.17. The procedure to compute the moment of resistance of FRP reinforced concrete section is given here in the following section.

5.20.2.1. Analysis procedure to compute moment of resistance (ISIS Canada)

The procedure for analysis of an externally strengthened reinforced concrete beam (with tension steel only) is performed using the above principles as follows:

1. Assume the section fails by compression of the concrete at the extreme compression fibre, after yielding of the longitudinal reinforcing steel. This means that the strain in the compression fibre of the cross-section is ε_{cu} = 0.0035 (Fig. 5.17), the assumed failure strain for concrete in compression. Thus, from strain compatibility, we have:

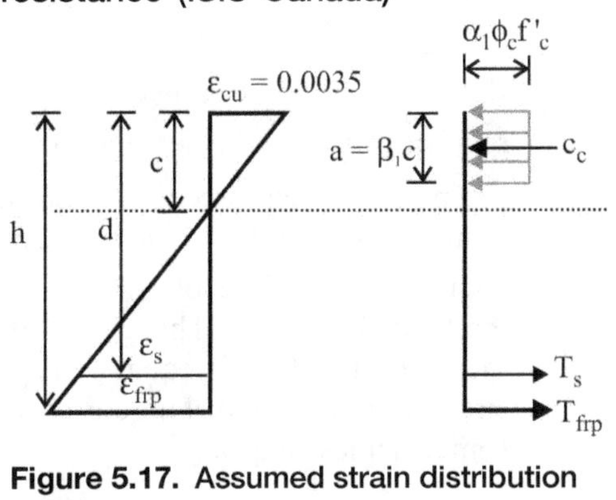

Figure 5.17. Assumed strain distribution at failure.

$$\text{Strain in FRP, } \varepsilon_{frp} = \varepsilon_{cu}\frac{h-c}{c} \tag{5.71}$$

$$\text{Strain in steel, } \varepsilon_s = \varepsilon_{cu}\frac{d-c}{c} \tag{5.72}$$

Assume that the steel has yielded so that $f_s = f_y$.

2. Determine the concrete compressive stress block factors in accordance with ISIS Canada Design Manual No. 4 or CSA A23.3 or CHBDC:

$$\alpha_1 = 0.85 - 0.0015 f'_c \geq 0.67 \tag{5.73}$$

$$\beta_1 = 0.97 - 0.0025 f'_c \geq 0.67 \tag{5.74}$$

3. Determine the depth of the neutral axis, c, by using Eqs. (5.67) to (5.70):
4. After depth of neutral axis, c, is known, check to see if the strain in the FRP has exceeded its tensile failure strain:

$$\text{If } \varepsilon_{frp} = \varepsilon_{cu}\frac{h-c}{c} > \varepsilon_{frpu} \quad \text{then go to Step 6} \tag{5.75a}$$

$$\text{If } \varepsilon_{frp} = \varepsilon_{cu}\frac{h-c}{c} \leq \varepsilon_{frpu} \quad \text{then go to Step 5} \tag{5.75b}$$

5. The factored moment resistance can be obtained from the following equation:

$$M_r = \phi_s f_y A_s\left(d - \frac{a}{2}\right) + \phi_{frp} E_{frp} A_{frp} \varepsilon_{frp}\left(h - \frac{a}{2}\right) \tag{5.76}$$

where, $$a = \beta_1 c \tag{5.77}$$

To avoid sudden and brittle failure of the externally strengthened member, we ensure that the internal steel reinforcement has yielded.

$$\text{If } \varepsilon_s = \varepsilon_{cu}\frac{d-c}{c} > \varepsilon_y \quad \text{then O.K.} \tag{5.78}$$

If $\varepsilon_s = \varepsilon_{cu}\frac{d-c}{c} < \varepsilon_y$, then the amount of the FRP reinforcement could be reduced. Recalculate the strain in steel reinforcement ensuring that the steel reinforcement yields in the final design.

6. Assume that failure occurs by tensile failure of the FRP (Fig. 5.18). This means that the strain in the FRP at failure is equal to $\varepsilon_{frp} = \varepsilon_{frpu}$ and the strain in the extreme concrete compression fibre is somewhat less than ε_{cu}.

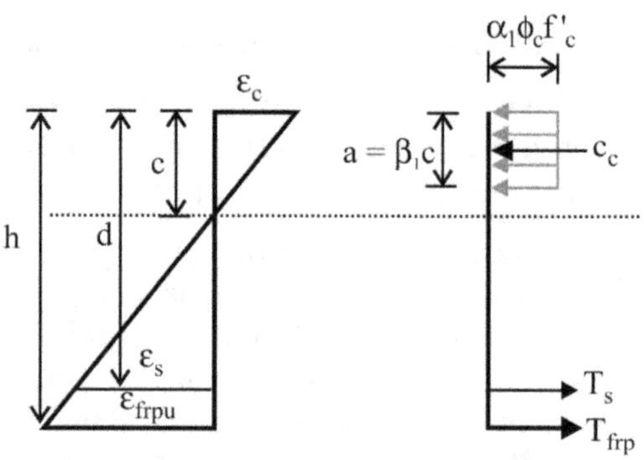

Figure 5.18. Assumed strain distribution at failure corresponding to FRP rupture.

7. Determine the depth of the neutral axis, c, using:

$$C_c = Ts + T_{\text{frp}} \rightarrow \phi_s \alpha_{\text{frp}} f_y A_s + \phi_{\text{frp}} E_{\text{frp}} A_{\text{frp}} \varepsilon_{\text{frp}} \tag{5.79}$$

8. Verify that the strain at the extreme compression fibre is less than ε_{cu}:

$$\varepsilon_c = \varepsilon_{\text{frpu}} \frac{c}{h-c} < \varepsilon_{cu} \tag{5.80}$$

9. Calculate the factored moment resistance of the cross-section using:

$$M_r = \phi_s f_y A_s \left(d - \frac{a}{2} \right) + \phi_{\text{frp}} E_{\text{frp}} A_{\text{frp}} \varepsilon_{\text{frp}} \left(h - \frac{a}{2} \right) \tag{5.81}$$

5.20.2.2. Compression reinforcement and T-beams

For members reinforced in flexure with steel in both tension and compression, a similar analysis procedure can be used as that presented in previous section. The only change required in the analysis is the addition of a compressive stress resultant to account for the presence of the compression reinforcing steel. This has been shown in Fig. 5.19, from which design equations can easily be derived in a similar way as in the previous section.

For T-beams, as is the case in the design and analysis of conventionally reinforced concrete beams, there are two possible cases to consider: the neutral axis is in the flange and the neutral axis is in the web. If the neutral axis is in the flange, then the beam can be treated as an equivalent rectangular beam using the procedures presented earlier. If, however, the neutral axis is in the web and the beam behaves as a T-beam, then the compression zone can be subdivided as shown in Fig. 5.20. Strain compatibility can then be used to derive design equations in a similar fashion as shown previously.

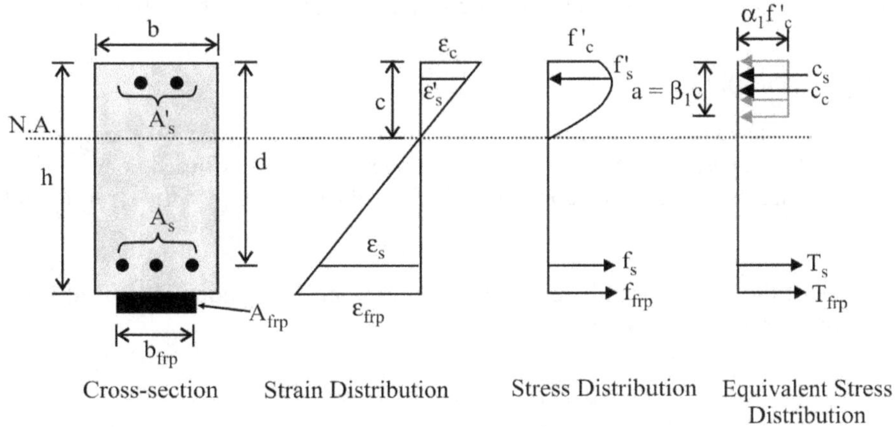

Figure 5.19. Strain and stress profiles for a doubly reinforced rectangular concrete beam strengthened in flexure with externally bonded FRP materials.

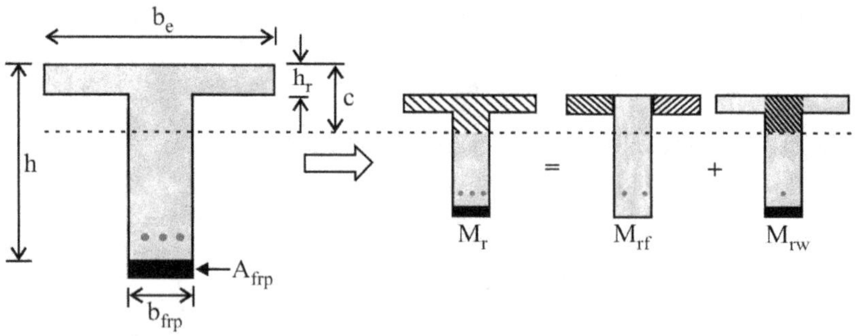

Figure 5.20. Subdividing T-beam cross-section's moment of resistance, M_r, into the flange moment resistance, M_{rf}, and the web moment resistance, M_{rw}.

5.21. ISIS Canada Design Guidelines for Shear Strengthening

As shown in Fig. 5.21, the shear strength of beams can be enhanced by applying FRP materials to its side faces externally or in the form of U-wrap. The U-wraps have an additional advantage of improving the anchorage of flexural external FRP reinforcements when placed over the flexural sheets or strips. Moreover, FRP shear reinforcements can be applied as continuous sheets or in strips of finite width. It must be noted that the external FRP shear reinforcements act in a manner similar to internal steel stirrups by bridging shear cracks to increase the shear capacity of the concrete. To avoid the possible failure of the FRP sheets due to stress concentrations at the corners of the beam, the corners should be rounded to a minimum radius of 15 mm. The material resistance factors could be taken as that prescribed for flexural strengthening with externally bonded FRP materials.

It is known that the shear failure occurs in a very brittle mode. Hence, to avoid the sudden collapse of structure by a shear failure, the deformation of the FRP shear reinforcement should be controlled which in turn will severely limit the potential for sudden shear failure. The following section entitled 'Design Principles' focuses on design of externally bonded shear reinforcements for building applications.

5.21.1. Design Principles

The factored shear resistance of a reinforced concrete member is given by the following expression, which is similar to the equation commonly used for conventionally reinforced concrete beams. The shear resistance, V_r, is given as the sum of the contributions from the concrete, V_c, the steel, V_s and the FRP, V_{frp}:

$$V_r = V_c + V_s + V_{\text{frp}} \tag{5.82}$$

The contributions of the concrete and steel can be determined using equations used for conventionally reinforced concrete structures as follows:

$$V_c = 0.2\lambda\phi_c\sqrt{f_c'}\,b_w d \quad \text{for } d \le 300 \text{ mm} \tag{5.83a}$$

$$V_c = \left(\frac{260}{1000+d}\right)\lambda\phi_c\sqrt{f_c'}\,b_w d \quad \text{for } d \ge 300 \text{ mm} \tag{5.83b}$$

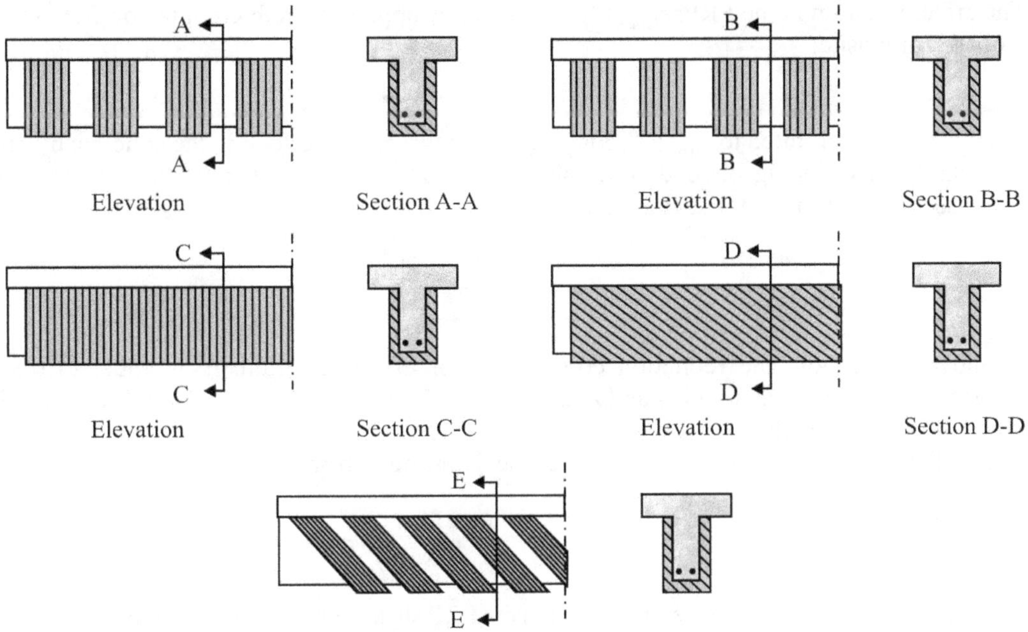

Figure 5.21. Various schemes for strengthening for shear by externally bonded FRP reinforcements.

But

$$V_c > 0.1\lambda\phi_c\sqrt{f_c'}b_w d \tag{5.83c}$$

$$V_s = \frac{\phi_s f_y A_v d}{s} \tag{5.84}$$

In the above expressions, A_v and s are the area and spacing of the transverse steel shear reinforcement, respectively. The shear contribution of the externally bonded FRP, V_{frp}, can be determined using the following expression, which is similar to the equation used for the contribution of the internal reinforcing steel:

$$V_{\text{frp}} = \frac{\phi_{\text{frp}} A_{\text{frp}} E_{\text{frp}} \varepsilon_{\text{frpe}} d_{\text{frp}} \left(\sin\beta + \cos\beta\right)}{s_{\text{frp}}} \tag{5.85}$$

where,

$$A_{\text{frp}} = 2t_{\text{frp}} w_{\text{frp}} \tag{5.86}$$

In the above expressions, s_{frp}, w_{frp}, and β are the spacing, width and angle of the shear reinforcement to the longitudinal axis of the beam, respectively. For full surface FRP shear reinforcement, $w_{\text{frp}} = s_{\text{frp}}$.

The effective depth of the FRP stirrups, d_{frp}, is taken as the distance from the free end of the FRP shear reinforcement underneath the slab to the bottom of the internal steel stirrups. For the rare case of a completely wrapped member, d_{frp} is taken as the total height of the section.

The effective strain in the FRP, ε_{frpe}, is determined by applying a reduction factor, R, to the ultimate strain of the composite:

$$\varepsilon_{frpe} = R\varepsilon_{frpu} \leq 0.004 \tag{5.87}$$

The effective strain is limited to $\varepsilon_{frpe} \leq 0.004$ to ensure aggregate interlock in the concrete by preventing shear cracks from widening beyond acceptable limits. The reduction factor, R, is determined by an equation based on experimental data as follows:

$$R = \alpha\lambda_1 \left[\frac{f_c'^{\,2/3}}{\rho_{frp} E_{frp}} \right]^{\lambda_2} \tag{5.88}$$

In the above expression, the reduction co-efficient for effective strain, α, is equal to 0.8 and the experimentally derived parameters λ_1 and λ_2 are: $\lambda_1 = 1.35$ and $\lambda_2 = 0.30$ for carbon FRPs; and $\lambda_1 = 1.23$ and $\lambda_2 = 0.47$ for glass FRPs.

The FRP shear reinforcement ratio, ρ_{frp}, can be determined from:

$$\rho_{frp} = \left(\frac{2t_{frp}}{b_w} \right) \left(\frac{w_{frp}}{s_{frp}} \right) \tag{5.89}$$

A second limit is imposed on the effective strain in the FRP shear reinforcement to avoid failure by sudden debonding of the FRP reinforcement. Obviously, this limit does not apply to fully wrapped specimens. The limiting strain in the FRP shear reinforcement to prevent debonding failure is described by:

$$\varepsilon_{frpe} = \frac{\alpha k_1 k_2 L_e}{9525} \tag{5.90}$$

where, $\alpha = 0.8$ and the parameters k_1 and k_2 are given by:

$$k_1 = \left[\frac{f_c'}{27.65} \right]^{2/3} \tag{5.91}$$

$$k_2 = \frac{d_{frp} - n_e L_e}{d_{frp}} \tag{5.92}$$

The parameter n_e in the above expression is the number of free ends of the FRP stirrup on the side of the beam (i.e., 1 for a U-wrap and 2 for side plates). The effective anchorage length, L_e, can be determined using the following equation:

$$L_e = \frac{25350}{(t_{frp} E_{frp})^{0.58}} \tag{5.93}$$

If the FRP is applied in strips, as opposed to a continuous FRP sheet, the maximum band spacing is defined by:

$$s_{frp} \leq w_{frp} + \frac{d}{4} \tag{5.94}$$

where, d is the depth of the internal steel reinforcement. Finally, for buildings, the maximum allowable shear strengthening is described by the following expression:

$$V_r \leq V_c + 0.8\lambda\phi_c\sqrt{f_c'}b_w d \tag{5.95}$$

where, λ is 1.0 for normal density concrete, 0.85 for semi-lightweight concrete and 0.75 for lightweight concrete.

5.22. External Strengthening of Columns

As shown in Fig. 5.22, the FRP wraps could be used to externally strengthen the existing concrete columns under pure compressive loads by circumferential confinement. When the column is subjected to axial load, it shortens longitudinally but dilates (expands) laterally. This dilation causes tensile stress to develop in the FRP wrap and this tensile stress *confines* the concrete and places it in a state of triaxial (3-D) stress. The result of this stress condition is that both the load capacity and deformation capability of the concrete in the column are significantly improved, leading to stronger and more ductile structural members. As per the ISIS Canada design guidelines, the design of concrete columns strengthened with externally bonded FRP wraps is performed using empirical equations derived primarily from test data. The applicability of the procedures presented herein is currently limited to the following types of members and load conditions:

1. Strengthening of relatively undamaged concrete columns.
2. Strengthening of short columns subjected to concentric axial loading.
3. Fibers oriented perpendicular to the column axis (circumferentially).

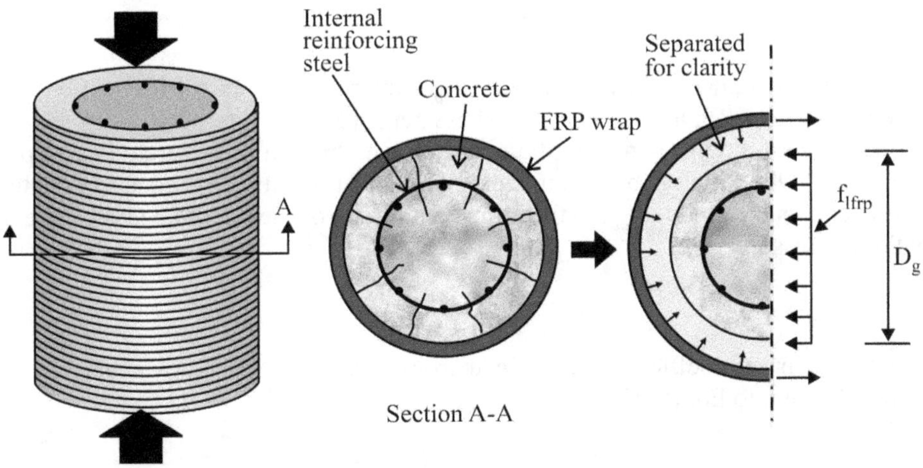

Figure 5.22. Schematic showing confinement mechanism for axial strengthening of circular reinforced concrete columns using externally bonded FRP wraps.

5.22.1. Slenderness Limits of Circular Columns

The procedures presented herein are valid only for columns where slenderness effects (which can contribute to non-linear behaviour and to buckling) can be ignored. For conventionally reinforced circular concrete columns, the limiting slenderness for short columns in pure compression is defined in the ISIS Canada Design Manual:

$$\frac{l_u}{D_g} \leq \frac{6.25}{\sqrt{P_f / f_c' A_g}} \tag{5.96}$$

where, A_g = gross cross-sectional area of the column
f_c' = specified concrete strength
P_f = factored axial load
l_u = unsupported length of the member
D_g = column diameter

When a column is strengthened in compression with externally bonded FRP wraps, its axial load capacity is increased by the confining effect of the wrap. However, the buckling strength of column which depends on the modulus of elasticity of the concrete is not significantly improved by wrapping.

Under the increased axial load it is possible that the same column may become slender. Hence, it is important to ensure that any FRP-strengthened column remains short under the updated and increased value of P_f for the equations that follow to apply.

5.22.2. Confinement

The lateral confining pressure exerted by an FRP wrap, $f_{l\text{frp}}$, on a circular concrete column at ultimate can be calculated using the following expression, which is derived on the basis of equilibrium of the confined concrete core (see Fig. 5.22):

$$f_{\ell\text{frp}} = \frac{2N_b \phi_{\text{frp}} f_{\text{frp}u} t_{\text{frp}}}{D_g} \tag{5.97}$$

In Eq. (5.97), N_b is the number of layers of FRP; ϕ_{frp} is the resistance factor for the FRP wrap; $f_{\text{frp}u}$ is the ultimate strength of the FRP; and t_{frp} is its thickness per layer.

The effect of confining pressure on the compressive strength of concrete in compression can be described by Eq. (5.98) which is based on experimental test results. This equation relates the ultimate strength of the confined concrete, f_{cc}', to the ultimate strength of the unconfined concrete, f_c', for a given lateral confining pressure:

$$f_{cc}' = f_c' + k_1 f_{\ell\text{frp}} \tag{5.98}$$

where k_1 is an empirical coefficient determined from tests. ISIS Design Manual No. 4 suggests the following modification to Eq. (5.98):

$$f_{cc}' = f_c' + k_1 f_{\ell\text{frp}} = f_c'(1 + \alpha_{pc} \omega_w) \tag{5.99}$$

where α_{pc} is a performance coefficient that depends on a number of factors such as the FRP type, concrete strength and member size. Currently, it is recommended that the value of α_{pc} be taken as 1.0. ω_w is referred to as the volumetric confinement ratio and is determined using:

$$\omega_w = \frac{\rho_{\text{frp}}\phi_{\text{frp}} f_{\text{frpu}}}{\phi_c f_c'} = \frac{2 f_{l\text{frp}}}{\phi_c f_c'} \tag{5.100}$$

In the preceding expression:

$$\rho_{\text{frp}} = \frac{4N_b t_{\text{frp}}}{D_g} \tag{5.101}$$

Thus, Eqs. (5.97), (5.99) and (5.100) can be used to determine the increased ultimate compressive strength of FRP-confined concrete.

5.22.2.1. Confinement limits

The limits on the imposed lateral pressure are as follows:

Minimum confinement pressure: The effectiveness of the confinement pressure depends on the degree to which ductility can be developed by the member. It is required that the sufficient FRP be provided to develop a minimum confinement pressure to ensure adequate ductility of:

$$f_{l\text{frp}} \geq 4 \text{ MPa} \tag{5.102}$$

Maximum confinement pressure: In reinforced concrete design, spirals in excess of requirement are of no practical interest because of the accompanying large deformations and extensive cracking and spalling which occur. FRP confinement eliminates spalling and limits cracking, but large deformations must still be prevented. Hence, the factored strength of an FRP confined column should not exceed the nominal strength of the unconfined concrete. Thus,

$$k_e \alpha_1 \phi_c f_{cc}' \leq \alpha_1 f_c' \tag{5.103}$$

where k_e is a strength reduction factor, taken as 0.85 for ductile columns, to account for unexpected eccentricities and α_1 is the equivalent rectangular stress block factor for concrete. The above equation can be combined with Eqs. (5.99) and (5.100) and rearranged to give the maximum allowable confinement pressure:

$$f_{\ell\text{frp}} \leq \frac{f_c'}{2\alpha_{pc}} \left(\frac{1}{k_e} - \phi_c \right) \tag{5.104}$$

Axial load resistance: The factored axial load resistance for an FRP-confined reinforced concrete column, $P_{r\max}$, is given by an equation similar to that suggested for conventionally reinforced concrete columns, with the exception that the concrete compressive strength, f_c', is replaced by the confined concrete compressive strength, f'_{cc}:

$$P_{r\max} = k_e [\alpha_1 \phi_c f_{cc}' (A_g - A_s) + \phi_s f_y A_s] \tag{5.105}$$

It may be noted that the externally bonded FRP fabric sheets can be used to strengthen short rectangular reinforced concrete columns in pure compression. However, FRP wrapping is less effective as a strengthening mechanism for non-circular columns especially rectangular ones. In the absence of design guidelines for strengthening the rectangular columns, ACI 440.2R-02 guidelines as described earlier in this chapter could be used for strengthening design of rectangular columns.

5.23. Fundamentals of Seismic Retrofit of Columns

The energy absorption rather than load carrying capacity is the main concern for a column subjected to seismic loading. The energy absorption capacity may be increased by RC jacketing or steel jacketing; however, jacketing may lead to additional earthquake forces in the column since they increase column stiffness significantly. In this section, two retrofit methods for RC bridge columns using FRP Composites: (1) Strength Oriented Retrofit and (2) Ductility Oriented Retrofits are discussed below.

Strength-oriented retrofit: In this case, FRP plates are longitudinally bonded to increase the flexural strength of columns. In this case, fibers are in the longitudinal direction. A typical diagram showing load deflection characteristics of strength as well as ductility oriented retrofit is given in Fig. 5.23. It may be noted that the longitudinal bonding of FRP has occasionally been considered when the amount of longitudinal steel reinforcement is insufficient.

Ductility-oriented retrofit: In this case, FRPs with fibers in the circumferential direction are wrapped around the column (lateral bonding of the FRP) to enhance its ductility.

5.23.1. Potential Failure Modes

For effective design of FRP retrofit, it is essential to have the thorough understanding of the potential failure modes of columns under seismic loading situation. The main failure modes of the columns subjected to seismic loading situations are given in the following:

Shear failure: The most undesirable mode of failure is shear failure. Typically, this failure mode is characterized by diagonal cracks in the concrete followed by the rupture or opening up of transverse steel reinforcement and then buckling of longitudinal steel reinforcement. This failure mode is brittle or explosive in nature and hence catastrophic. In seismic retrofit, existing columns with insufficient or poorly detailed steel reinforcement should be shear strengthened so that shear failure is no longer the critical failure mode.

Flexural plastic hinge failure: This failure mode is also common in earthquakes and usually occurs at column ends and is limited to a small region. This failure mode is characterized by the spalling of concrete cover, failure of transverse steel reinforcement and buckling of longitudinal reinforcement. Furthermore, this failure mode is accomplished by large inelastic flexural deformation and is thus, more ductile and desirable than the brittle shear failure mode. In seismic retrofit of RC columns using FRP composites, the FRP provides lateral confinement to prevent the spalling of cover concrete and to restrain the longitudinal steel reinforcement from buckling so that more ductile responses can be developed and larger inelastic deformation can be sustained.

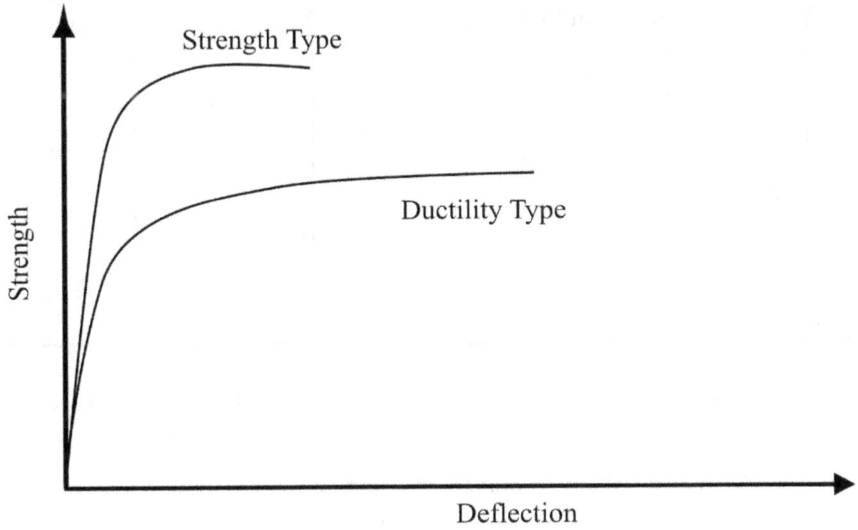

Figure 5.23. Strength-oriented and ductility-oriented seismic retrofits.

Lap splice failure: This type of failure occurs in columns in which the longitudinal steel reinforcement is lap spliced in the maximum moment regions near the column ends. Under seismic loads, the lap splice may break down leading to loss of structural integrity for sustaining large inelastic deformation to achieve required energy absorption capacity. In seimic retrofit, FRP composites can be used to provide a clamping force for the lap spliced longitudinal steel reinforcement so that lap splice failure is prevented and large inelastic deformation can be sustained.

Flexural shear failure of columns with longitudinal reinforcement cut-offs: A mode of failure, that is likely for columns with some of the longitudinal reinforcing bars cut-offs outside the potential plastic hinge regions at column ends, has also been identified. Flexural failure may occur at a cut-off section rather than at the ends of the column.

Footing failure: This failure is less frequently encountered during earthquakes and is caused due to: (a) inadequate pile capacity, (b) inadequate footing flexural strength, and (c) inadequate column footing joint shear resistance. It may be noted that footing strengthening using FRP composites does not appear to have been explored so far.

5.23.2. Flexural Ductility of Retrofitted Columns

The purpose of seismic retrofit of RC columns is to achieve a sufficient level of deformation ductility to dissipate seismic energy before one of the failure modes becomes critical. In order to quantify the ductility, the ductility factors are defined. In particular, two types of ductility factors are considered in structural design, (i) Displacement ductility factor and (ii) Curvature ductility factor.

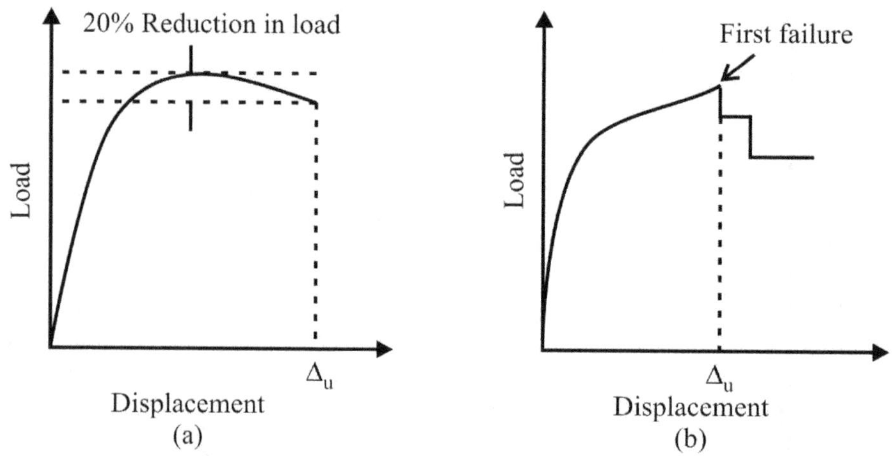

Figure 5.24. Load deflection curves showing ultimate deflections.

Displacement ductility factor: It is defined as the ratio of the displacement at ultimate condition to displacement at first yield. Usually, displacement at ultimate condition could be taken as the one corresponding to the load 80% of the peak load located on the load deflection response after the peak load (see Fig. 5.24). However, in the case of progressive failure, the ultimate displacement refers to the load at which first failure occurs. Mathematically, displacement ductility is expressed by Eq. (5.106).

$$\text{Displacement ductility factor, } \mu_\Delta = \frac{\Delta_u}{\Delta_y} \tag{5.106}$$

where, Δ_u and Δ_y are displacements at ultimate and at the first yield, respectively.

Curvature ductility factor: It is defined as the ratio of curvature at ultimate condition to the curvature at first yield. Here curvature at ultimate refers to the load 80% of the peak load which is located on the load deflection curve after the peak load (see Fig. 5.24). However, in the case of progressive failure, the ultimate curvature refers to the load at which first failure occurs. Mathematically, curvature ductility is defined by Eq. (5.107).

$$\text{Curvature ductility, } \mu_\phi = \frac{\phi_u}{\phi_y} \tag{5.107}$$

where, ϕ_u and ϕ_y are curvature at ultimate and curvature at first yield, respectively.

The difference in ultimate deflection and deflection at first yield is called plastic displacement while the difference in ultimate curvature and curvature at yield is called plastic curvature. The plastic displacement and plastic curvature are expressed by Eqs. (5.108) and (5.109), respectively. It may be noted that the plastic curvature, ϕ_p, is assumed to be constant within a plastic region.

$$\Delta_p = \Delta_u - \Delta_y \tag{5.108}$$

$$\phi_p = \phi_u - \phi_y \tag{5.109}$$

The relationship between displacement ductility and curvature ductility factor is given by the following factor.

$$\mu_\Delta = 1 + 3\left(\mu_\phi - 1\right)\frac{L_P}{L}\left(1 - 0.5\frac{L_P}{L}\right) \tag{5.110}$$

where, L is the distance from the critical section of the plastic hinge region to the point of contraflexure (i.e., shear span)

L_p is length of plastic hinge region and is expressed by Eq. (5.111).

$$L_P = 0.08L + 0.022 f_{yl} d_{bl} \geq 0.044 f_{yl} d_{bl} \tag{5.111}$$

where f_{yl} and d_{bl} are the yield strength and diameter of longitudinal steel reinforcing bars. The plastic rotation is defined by Eq. (5.112).

$$\text{Plastic rotation, } \theta_P = L_P \phi_P \tag{5.112}$$

5.23.3. Shear Strength Contributions

The shear strength contribution of concrete is given by Eq. (5.113), which relates the effective shear area and a constant, k (Table 5.4), depending on the required curvature ductility factor.

$$V_c = k\sqrt{f_c'}A_e \tag{5.113}$$

where A_e = effective shear area = $0.8A_g$, and $k = 0.25$ for the region outside the plastic hinge area.

Table 5.4. Concrete shear strength contribution factor.

Curvature ductility	Concrete shear strength contribution factor, *k* (inside plastic hinge region)	
	Uni-axial ductility	Bi-axial ductility
0–2	0.25	0.25
4	0.25	0.14
5	0.19	0.083
7	0.083	0.071
12	0.050	0.042
13	0.048	0.042
14	0.042	0.042

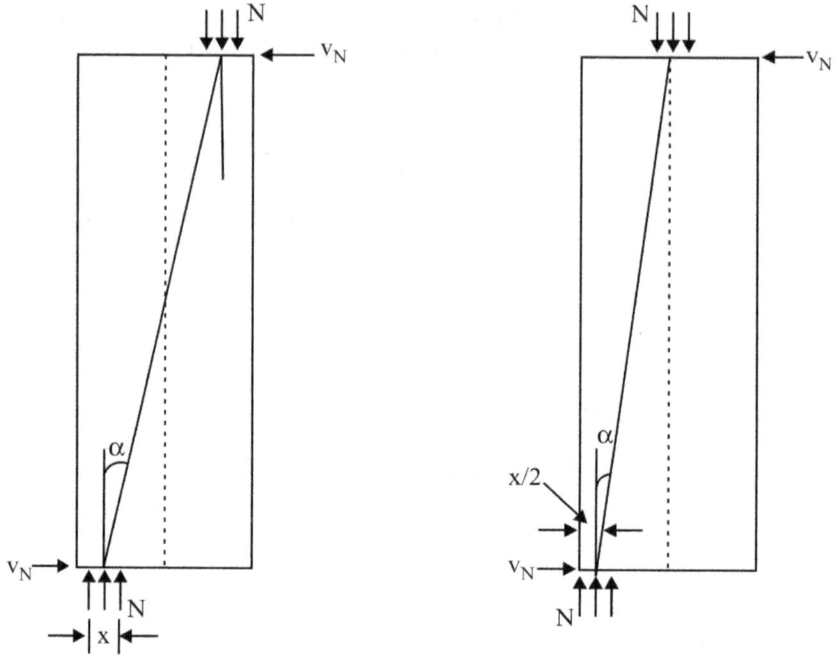

Figure 5.25. Schematics of single and double bending.

The shear strength contribution of axial load component is given by Eq. (5.114).

$$V_n = N \tan\alpha = \frac{d-x}{2L} \text{ N} \tag{5.114}$$

where d is the diameter of circular column or section depth in the lateral load direction; and L is the shear span. For a cantilever, L is column height whereas for column in double bending, L is half of column height (Fig. 5.25).

The shear Strength contributions (V_s) of internal steel ties or transverse reinforcement for a circular and rectangular column sections are given by Eqs. (5.115) and (5.116), respectively.

Circular column section:

$$V_s = \frac{\pi}{2}\frac{A_{sh} f_{yh} D'}{s}\cot\theta \tag{5.115}$$

where, A_{sh} = cross-sectional area of helical ties transverse steel reinforcement;

f_{yh} = yield strength of transverse reinforcement;

D' = diameter of core of the column between centers of peripheral hoop;

s = the spacing of transverse steel reinforcement;

θ = inclination of critical shear crack.

Rectangular section:

$$V_s = \frac{A_{sh} f_{yh} D'}{s} \cot\theta \tag{5.116}$$

The shear contribution of FRP jacket for circular and rectangular sections is given by Eqs. (5.117) and (5.118), respectively.

Circular columns:

$$V_{\text{frp}} = \frac{\pi}{2} f_{\text{frp},e}\, t_{\text{frp}}\, d \cot\theta \tag{5.117}$$

Rectangular columns:

$$V_{\text{frp}} = 2 f_{\text{frp},e}\, t_{\text{frp}}\, d \cot\theta \tag{5.118}$$

where, $f_{\text{frp},e}$ is effective stress in FRP jacket whereas, t_{frp}, is thickness of FRP.

The required thickness of FRP jacket for circular column section is given by Eq. (5.119) whereas for rectangular section, the jacket thickness is given by Eq. (5.120).

$$t_{\text{frp}} = \frac{\dfrac{V_o}{\phi_V} - [V_c + V_N + V_s]}{\dfrac{\pi}{2} f_{\text{frp},e} d \cot\theta} \tag{5.119}$$

$$t_{\text{frp}} = \frac{\dfrac{V_o}{\phi_V} - [V_c + V_N + V_s]}{2 f_{\text{frp},e} d \cot\theta} \tag{5.120}$$

where, V_o is column shear demand based on the full flexural overstrength in the potential plastic hinge region; ϕ_V is shear strength reduction factor and is taken as equal to 0.85. The expression for shear demand is given by Eq. (5.121).

$$V_o = 1.5 \frac{M_i}{L} \tag{5.121}$$

where, M_i is ideal flexural strength of column section based on measured material strength.

Effective stress in FRP jacket,

$$f_{\text{frp},e} = 0.001 E_{\text{frp}} \leq \varepsilon_{\text{frp,rup}} \tag{5.122}$$

In Eq. (5.122), E_{frp} and $\varepsilon_{\text{frp,rup}}$ are modulus of elasticity and rupture strain of FRP jacket.

5.23.4. Flexural Plastic Hinge Confinement

For a given displacement ductility, μ_Δ the curvature ductility, μ_ϕ, can be found and the required ultimate strain of confined concrete in the extreme compression fiber, ε_{cc}, can be determined using Eq. (5.123).

$$\varepsilon_{cc} = \phi_u x = \mu_\phi \phi_y x \tag{5.123}$$

where ϕ_y is curvature corresponding to the first yield; and x is the depth to the neutral axis at ultimate. The required thickness of FRP corresponding to the plastic hinge condition is given by Eq. (5.124).

$$t_{\text{frp}} = 0.09 \frac{d(\varepsilon_{cc} - 0.004) f'_{cc}}{\phi_F f_{\text{frp},e} \varepsilon_{\text{frp,rup}}} \tag{5.124}$$

where, $f'_c = f'_{cc}$ is confined compressive strength of concrete; and ϕ_F is flexural strength reduction factor = 0.9.

5.23.5. Lap Splice Clamping

To prevent the splice failure, sufficient clamping pressure is required. This required lateral confining pressure (f_l) is given by Eq. (5.125).

$$f_l = \frac{A_{bl} f_{yl}}{\left[\frac{p}{2n} + 2(d_{bl} + c)\right], L_s} \tag{5.125}$$

where, p is the length of internal cracking line surrounded by the n longitudinal bars; A_{bl} is area of a single longitudinal bar; f_{yl}, d_{bl}, yield strength and diameter of longitudinal bars; c is thickness of concrete cover to the main longitudinal bar; Ls is length of lap splice; $\frac{p}{2n} + 2(d_{bl} + c)$ is perimeter of cracking around a longitudinal bar when debonding of lap splice bars occurs.

For circular columns:

$$f_l = \frac{0.002 E_{\text{frp}} t_{\text{frp}}}{d} \tag{5.126}$$

$$\frac{p}{2n} + 2(d_{bl} + c) \tag{5.127}$$

where f_h is the confining pressure provided by the transverse steel reinforcement. For rectangular columns, section should be modified into an ellipse and the term d should be replaced by the average of minor and major axis.

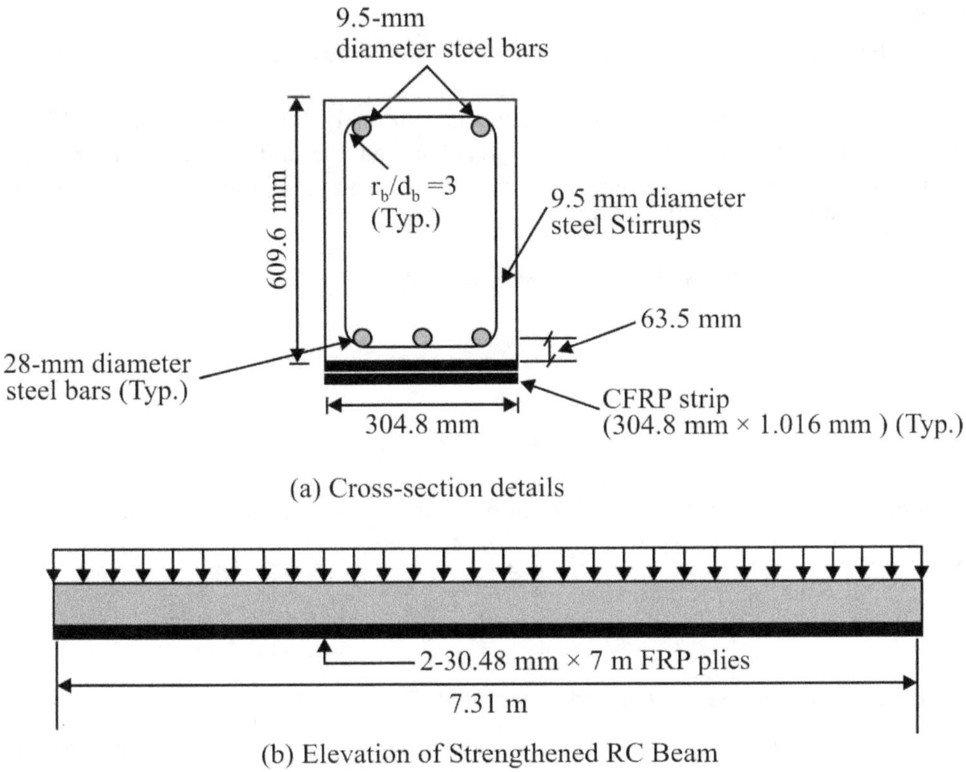

Figure E5.1. Simply supported FRP-strengthened RC beams.

E5.1. Design Example 1

A simply supported concrete beam reinforced with three #9 bars (28-mm diameter) (see Fig. E5.1) is located in unoccupied warehouse and is subjected to a 50% increase in its live load carrying requirements. An analysis of the existing beam indicates that the beam still has sufficient shear strength to resist the new required shear strength and meets the deflection and crack control serviceability requirement. Its flexural strength, however, is inadequate to carry the increased live load. Given: Concrete strength is f'_c = 34.48 MPa and yield strength of steel is 414 MPa. It is proposed to strengthen the existing reinforced concrete beam with the FRP system described in Table E5.1. Design flexural capacity of existing RC beam is 353.3 kN-m.

Table E5.1. Manufacturer reported FRP system properties.

Properties	In SI units
Thickness per ply	1.016 mm
Ultimate tensile strength, f^*_{fu}	0.62 kN/mm^2
Rupture strain, ε^*_{fu}	0.017 mm/mm
Modulus of elasticity, E_f	37 kN/mm^2

Solution

In Table E5.2, the existing loadings, new loadings and associated midspan moments for the beam are summarized.

Although the existing design moment capacity of the beam is given in the problem, design moment capacity of the existing beam is recalculated for understanding the procedure for design moment calculation of existing beam. The stress and strain distribution across the depth of existing beam is given in Fig. E5.2.

Calculation of moment capacity of existing beam (Fig. E5.2):

$$A_s = 3 \times \frac{\pi}{4} \times 28^2 = 1847.26 \text{ mm}^2$$

$$\beta_1 = 0.85 - (0.145 \times 34.48 - 4) \times 0.05 = 0.80$$

Assuming steel yields at ultimate failure of beam from equilibrium condition,

$$f_y A_s = 0.85 f_c' b \beta_1 c \Rightarrow c = \frac{f_y A_s}{0.85 f_c' b \beta_1} = \frac{414 \times 1847.26}{0.85 \times 34.48 \times 304.8 \times 0.80} = 107 \text{ mm}$$

Table E5.2. Loadings and corresponding moments.

Loading/Moment	Existing loads	Anticipated loads
Dead load, w_{DL}	14 N/mm	14 N/mm
Live load, w_{LL}	17 N/mm	1.5 × 1.7 = 25.5 N/mm
Unfactored loads, $w_{DL} + w_{LL}$	31 N/mm	39.5 N/mm
Unstrengthened load limits, $1.2w_{DL} + 0.85w_{LL}$	Not applicable	1.2 × 14 + 0.85 × 25.5 = 38.5 N/mm
M_{DL}	$\frac{w_{DL}l^2}{8} = \frac{14 \times (7.31 \times 1000)^2}{8} = 93.5$ kN-m	93.5 kN-m
M_{LL}	$\frac{w_{LL}l^2}{8} = \frac{17 \times (7.31 \times 1000)^2}{8} = 113.6$ kN-m	170.4 kN-m
Service load moment, M_s	207.1 kN-m	263.9 kN-m
Unstrengthened moment limit, $1.2M_{DL} + 0.85M_{LL}$	Not Applicable	257 kN-m
Factored moment, $M_u = 1.2 M_{DL} + 1.6M_{LL}$	293.96 kN-m	384.84 kN-m

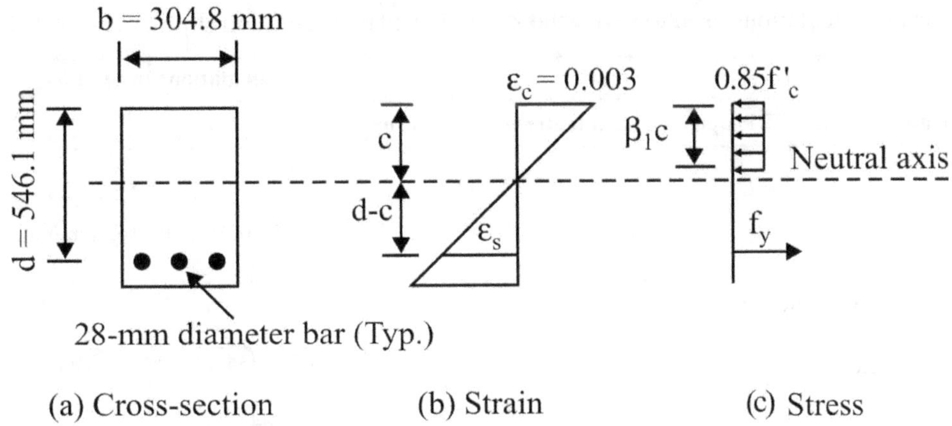

Figure E5.2. Strain and stress distribution for existing beam at ultimate condition.

Balanced depth, $c_b = \dfrac{\varepsilon_u d}{\varepsilon_u + \varepsilon_y} = 0.375d = 0.375 \times 546.1 = 204.8 \text{ mm} > 107 \text{ mm}$

Thus, the tension governs the failure of the existing beam and section is under reinforced. From strain compatibility, the strain in reinforcing steel bars is given by:

$$\varepsilon_s = 0.003 \times \left(\frac{d-c}{c}\right) = \frac{0.003 \times (546.1 - 107)}{107} = 0.01 > 0.005$$

Based on, ε_s, value greater than 0.005, the value of strength reduction factor for the existing beam is given by $\phi = 0.90$ (see Eq. (5.10) and Fig. 5.3).

The nominal moment capacity (M_n) of existing beam is given by:

$$(M_n)_{\text{existing}} = A_s f_y \left(d - \frac{\beta_1 c}{2}\right) = 1847.26 \times 414 \times \left(546.1 - \frac{0.80 \times 107}{2}\right) = 384.91 \text{ kN-m}$$

$$\phi (M_n)_{\text{existing}} = 0.9 \times 384.91 = 346.1 \text{kN-m} \cong 353.3 \text{kN-m} \quad \text{(given value)}$$

$$\phi (M_n)_{\text{existing}} = 0.9 \times 384.91 = 346.1 \text{ kN} > 257 \text{ kN-mm} \quad (1.2M_{DL} + 0.85M_{LL})$$

(unstrengthened moment limit) O.K.

The design calculations used to verify the strengthened configuration (Fig. E5.1) are as follows:

Procedure	Calculations in SI units
Step 1: Calculate the CFRP/Epoxy system material design properties	
$f_{fu} = C_E f_{fu}^*$ $\varepsilon_{fu} = C_E \varepsilon_{fu}^*$	$f_{fu} = 0.95 \times 620 = 589 \text{ N/mm}^2$ $\varepsilon_{fu} = 0.95 \times 0.017 = 0.0162 \text{ mm/mm}$
Step 2: Preliminary calculations	
Modulus of concrete, $E_c = 4733\sqrt{f_c'}$	$E_c = 4733\sqrt{34.48} = 27792 \text{ MPa}$
Reinforcement ratio, $\rho_s = \dfrac{A_s}{bd}$	$\rho_s = \dfrac{1847.26}{304.8 \times 546.1} = 0.0111$
Properties of externally bonded CFRP reinforcements	
CFRP cross-sectional area, $A_f = n t_f w_f$	$A_f = 2 \times 1.016 \times 304.8 = 619.35 \text{ mm}^2$
CFRP reinforcement ratio, $\rho_f = \dfrac{A_f}{bd}$	$\rho_f = \dfrac{619.35}{304.8 \times 546.1} = 0.00372$
Step 3: Determine the existing state of strain on the soffit of beam.	
Assuming the beam is cracked and the only load acting on the beam at the time of the FRP installation are the dead loads. The cracked section analysis of the beam gives the following:	
Modular ratio of steel and concrete is $m_s = \dfrac{E_s}{E_c}$	$m_s = \dfrac{2 \times 10^5}{27792} = 7.20$
Equate the first moments of transformed area of steel to that of concrete about the neutral axis to get neutral axis depth co-efficient, c_o= 180 mm.	
Moment of inertia of the cracked section, $I_{cr} = \dfrac{bc_o^3}{3} + m_s A_s (d - c_o)^2$	$I_{cr} = \dfrac{304.8 \times 180^3}{3} + 7.20 \times 1847.26(546.1 - 180)^2$ $= 2375 \times 10^6 \text{ mm}^4$
Substrate strain, $\varepsilon_{bi} = \dfrac{M_{DL}(h - c_o)}{I_{cr} E_c}$	$\varepsilon_{bi} = \dfrac{93.5 \times 10^6 (609.6 - 180.0)}{2375 \times 10^6 \times 27792} = 0.00060$

Step 4: Determine the bond dependent coefficient of the FRP system	
The dimensionless bond-dependent coefficient for flexure, k_m, is calculated using Eq. (5.7). Compare nE_ft_f to 180 000 Therefore, $k_m = \frac{1}{60\varepsilon_{fu}}\left(1-\frac{nE_ft_f}{360000}\right) \le 0.90 \quad \text{for } nE_ft_f \le 180000$	$nE_ft_f = 2\times 37000\times 1.016 = 75184 \rightleftarrows \text{N/mm} < 180000$ $k_m = \frac{1}{60\times 0.0162}\left(1-\frac{75184}{360000}\right) = 0.82 < 0.90, \quad \text{O.K.}$
Step 5: Estimate, *c*, the depth to the neutral axis at ultimate	
A reasonable estimate of *c* is 0.20*d*. The value of *c* is adjusted after checking equilibrium. $c = 0.20d$	$c = 0.20 \times 546.1 = 109.2 \text{ mm}$
Step 6: Determine the effective level of strain in the CFRP reinforcement	
The effective strain level in the FRP may be found from Eq. (5.8) $\varepsilon_{fe} = \varepsilon_{cu}\left(\frac{h-c}{c}\right) - \varepsilon_{bi} \le k_m\varepsilon_{fu}$ Note that for the neutral axis depth selected, concrete crushing would be the failure mode because the first expression in this equation governs. If the second (limiting) expression governs, then FRP would be in the delamination failure mode.	$\varepsilon_{fe} = 0.003\left(\frac{609.6-109.2}{109.2}\right) - 0.00060$ $= 0.013 \le 0.82\times 0.0162$ $\varepsilon_{fe} = 0.0131 \le 0.0133$
Step 7: Calculate the strain in the existing reinforcing steel	
The strain in the reinforcing steel can be calculated using similar triangles according to Eq. (5.13) $\varepsilon_s = \left(\varepsilon_{fe} + \varepsilon_{bi}\right)\left(\frac{d-c}{h-c}\right)$	$\varepsilon_s = (0.0131 + 0.00060)\left(\frac{546.1-109.2}{609.6-109.2}\right)$ $= 0.012$
Step 8: Calculate the stress level in the FRP and reinforcing steel	
The stresses are calculated using Eqs. (5.9) and (5.14), respectively. $f_{fe} = E_f\varepsilon_{fe}$ $f_s = E_s\varepsilon_s \le f_y$	$f_{fe} = 37000 \times 0.0131 = 484.7 \text{ MPa} < 620 \text{ MPa}, \quad \text{O.K.}$ $f_s = 200 \times 10^3 \times 0.012 = 2400 \text{ MPa} > f_y = 414 \text{ MPa}$ Hence, $f_s = f_y = 414$ MPa

Step 9: Calculate the internal force resultants and check equilibrium	
The force equilibrium is verified by checking the initial estimate of c with Eq. (5.15). Since concrete crushing controls failure, γ can be taken as 0.85. $$c=\frac{A_s f_s + A_f f_{fe}}{\gamma f_c' \beta_1 b}$$	$c=\frac{1847.26\times 414+619.35\times 484.7}{0.85\times 34.48\times 0.80\times 304.8}$ $=149 \text{ mm} > 109.2 \text{ mm, n.g.}$ O.K. $\therefore$ Revise estimate of c and repeat steps 6 through 9 until equilibrium is achieved.
Step 10: Adjust c until force equilibrium is satisfied	
Steps 6 through 9 were repeated several times with different values of c until equilibrium was achieved. The results of the final iteration are: c = 140 mm; $\varepsilon_s = 0.0087$; $f_s = f_y = 414$ MPa; $\varepsilon_{fe} = 0.00949$; $f_{fe} = 351.13$ MPa	$c=\frac{1847.26\times 414+619.35\times 351.13}{0.85\times 34.48\times 0.80\times 304.8}$ $=137.44 \text{ mm} \cong 140 \text{ mm}$ So, take c = 140 mm
Step 11: Calculate design flexural strength of the section	
The design flexural strength is calculated using Eq. (5.16) and strength reduction factor, ϕ. An additional reduction factor, $\psi_f = 0.85$, is applied to the contribution of FRP system. Since $\varepsilon_s = 0.0087 > 0.005$, strength reduction factor, $\phi = 0.90$ is appropriate as per Eq. (5.10). $$\phi M_n = \phi\left[A_s f_s\left(d-\frac{\beta_1 c}{2}\right)+\psi_f A_f f_{fe}\left(h-\frac{\beta_1 c}{2}\right)\right]$$	$\phi M_n = 0.90\left[\begin{array}{l}1847.26\times 414\times\left(546.1-\frac{0.80\times 140}{2}\right)\\ +0.85\times 619.35\times 351.13\times\left(610.62-\frac{0.80\times 140}{2}\right)\end{array}\right]$ $=429.1 \text{ kN-m} > M_u = 384.84 \text{kN-m}$ O.K. Thus, the strengthened section is capable of sustaining the new required moment strength.
Step 12: Check service stresses in the reinforcing steel and FRP	
Calculate the elastic depth to the neutral axis by equating the first moment of areas of the transformed section about the neutral axis. This can be simplified for a rectangular beam without compression reinforcement as follows: $$k=\sqrt{\left(\rho_s\frac{E_s}{E_c}+\rho_f\frac{E_f}{E_c}\right)^2+2\left(\rho_s\frac{E_s}{E_c}+\rho_f\frac{E_f}{E_c}\left(\frac{h}{d}\right)\right)}-\left(\rho_s\frac{E_s}{E_c}+\rho_f\frac{E_f}{E_c}\right)$$	$k=\sqrt{(0.0111\times 7.2+0.00372\times 1.33)^2+2\left(0.0111\times 7.2+0.00372\times 1.33\left(\frac{610.62}{546.1}\right)\right)}-(0.0111\times 7.2+0.00372\times 1.33)$ $=0.337$ $kd = 0.337\times 546.1 = 184 \text{ mm}$
Calculate the stress level in the reinforcing steel using Eq. (5.17) and verify that it is less than recommended limit as per Eq. (5.11). $$f_{s,s}=\frac{\left[M_s+\varepsilon_{bi}A_f E_f\left(h-\frac{kd}{3}\right)\right](d-kd)E_s}{A_s E_s\left(d-\frac{kd}{3}\right)(d-kd)+A_f E_f\left(h-\frac{kd}{3}\right)(h-kd)}$$ $$f_{s,s} \text{ " } 0.80 f_y$$	$f_{s,s}=\frac{\left[\begin{array}{l}263.9\times 10^6+0.0006\times 619.35\\ \times 37000\left(610.62-\frac{184}{3}\right)\end{array}\right](546.1-184)\times 2\times 10^5}{1847.26\times 2\times 10^5\left(546.1-\frac{184}{3}\right)(546.1-184)+619.35\times 37000\left(610.62-\frac{184}{3}\right)(610.62-184)}$ $=279.56 < 0.8\times 414 = 331.2 \text{MPa}$ O.K.

<table>
<tr><td colspan="2">Step 13: Check for creep-rupture stress limit</td></tr>
<tr><td>Assuming that the dead and live loads are the sustained loads. Use Eq. (5.18) to calculate the stress in the FRP and verify that it is less than creep-rupture stress limit (see Table 5.2)

$$f_{f,s} = f_{s,s}\left(\frac{E_f}{E_s}\right)\frac{(h-kd)}{(d-kd)} - \varepsilon_{bi}E_f$$</td>
<td>$$f_{f,s} = 279.56\times\left(\frac{37000}{200000}\right)\frac{(610.62-184)}{(546.1-184)} - 0.0006\times 37000$$
$= 39.84\text{ MPa} < 0.55\times 589 = 323.95\text{ MPa}$ O.K.</td></tr>
<tr><td colspan="2">Step 14: Check for crack width limit</td></tr>
<tr><td>For 414 MPa steel bars, the crack width is given by Eq. (4.20) and is given as follows:

$$w = 1.1\times 10^{-5}\times \beta f_{s,s}\sqrt{d_c A_e}$$
where, $d_c = h - d$

$$A_e = \frac{2d_c b}{\text{No. of bars}}$$
$$\beta = \frac{h-kd}{d(1-k)}$$</td>
<td>$$d_c = 610.62 - 546.1 = 63.5\text{ mm}$$
$$A_e = \frac{2\times 63.5\times 304.8}{3} = 12903.2\text{ mm}^2$$
$$f_{s,s} = 279.56\text{ MPa}$$
$$\beta = \frac{609.6-184}{(546.1-184)} = 1.18$$
$$w = 1.1\times 10^{-5}\times 1.18\times 279.56\times\sqrt[3]{63.5\times 12903.2}$$
$= 0.33\text{ mm} < 0.4\text{ mm}$ O.K.</td></tr>
<tr><td colspan="2">Step 15: Check for long-term deflection</td></tr>
<tr><td>The long-term deflection can be calculated in the same way as of the FRP reinforced RC beams (see Chapter 4) per Eq. (4.29).

$$I_e = \left(\frac{M_{cr}}{M_a}\right)^3 I_g + \left[1-\left(\frac{M_{cr}}{M_a}\right)^3\right] I_{cr} \le I_g$$
where,
$$I_g = \frac{bh^3}{12}$$
$$f_r = 0.62\sqrt{f_c'}$$
$$M_{cr} = \frac{2f_r I_g}{h}$$
$$I_{cr} = \frac{bd^3k^3}{3} + m_f A_f (d_f - kd)^2 + m_s A_s (d-kd)^2$$</td>
<td>$$I_g = \frac{304.8\times 609.6^3}{12} = 575.4\times 10^6\text{ mm}^4$$
$$f_r = 0.62\sqrt{34.48} = 3.64\text{ MPa}$$
$$M_{cr} = \frac{2\times 3.64\times 5754\times 10^6}{609.6\times 10^6} = 68.72\text{ kN-m}$$
$$I_{cr} = \frac{304.8\times 184^3}{3} + 1.33\times 619.35\times(610.62-184)^2$$
$$+ 7.20\times 1847.26\times(546.1-184)^2$$
$$= 2527\times 10^6\text{ mm}^4$$
$$I_e = \left(\frac{68.72}{263.9}\right)^3\times 5754\times 10^6 + \left[1-\left(\frac{68.72}{263.9}\right)^3\right]\times 2527\times 10^6$$
$= 2584\times 10^6\text{ mm}^4 < I_g$ O.K.</td></tr>
</table>

<table>
<tr>
<td>Assuming that the total applied load is sustained load, the instantaneous deflection is given by:
$$(\Delta_i)_{DL+LL} = \frac{5(w_{DL}+w_{LL})^4_{new}}{384E_cI_e}$$
The long-term deflection is given by the following expression and is verified by deflection limits as presented in Chapter 4.
$$\Delta_{LT} = 1.2(\Delta_i)_s < \frac{L}{240}$$</td>
<td>Assuming that total applied load is sustained load,
$$(\Delta_i)_{DL+LL} = \frac{5\times 39.5\times(7.31\times 1000)^4}{384\times 27.792\times 2584\times 10^6} = 20.4 \text{ mm}$$
$$\Delta_{LT} = 1.2\times 20.4 = 24.48 \text{ mm} < \frac{L}{240}$$
where, $\frac{L}{240} = \frac{7310}{240} = 30.46$ mm, and hence O.K.</td>
</tr>
</table>

In addition to the aforementioned design calculations, the following checks are also required for the economy of the material utilization and safety against cover delamination.

Check for cover delamination

To locate the section where the value of the bending moment becomes equal to the cracking moment, the bending moment at any section located at distance x from supported end should be equated to the cracking moment.

So,

$$M_x = \frac{wl}{2}x - \frac{wx^2}{2} = 68.72 \Rightarrow \frac{39.5\times 7.31\times x}{2} - \frac{39.5\times x^2}{2} = 68.72$$

$$x^2 - 7.31x + 3.48 = 0 \Rightarrow x = \frac{7.31\pm\sqrt{7.31^2 - 4\times 1\times 3.48}}{2}$$

$$= 0.52 \text{ m and } 6.80 \text{ m from left end of the beam.}$$

Thus, CFRP strips should be terminated at a distance of about 546.1 mm (i.e., effective depth of the beam, d) from the section of cracking (see ACI 440.2R-02). This means that in this particular problem, the CFRP plate should be extended to both supports and the overall length of the plate is equal to the full span of the beam.

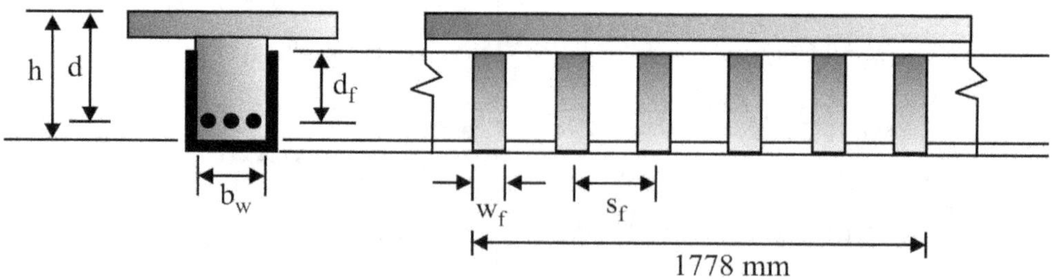

Figure E5.3. Configuration of the CFRP shear strengthening scheme.

Check for shear

$$V_u = 1.2\frac{w_{DL}\,l}{2} + 1.6\frac{w_{LL}\,l}{2}$$

Factored shear force at termination point,

$$V_u = 1.2\times\frac{14\times 7.31}{2} + 1.6\frac{25.5\times 7.31}{2}$$
$$= 210.53\,\text{kN}$$

The concrete shear strength,

$$V_c = \frac{\sqrt{f'_c}}{6}bd = \frac{\sqrt{34.48}}{6}\times\frac{304.8\times 546.1}{1000}$$
$$= 162.9\,\text{kN}$$

Since $\frac{2}{3}V_c = \frac{2}{3}\times 162.9 = 108.6\,\text{kN} < V_u$, provide CFRP U-wraps to reinforce against cover delamination.

E5.2. Design Example 2

A reinforced concrete T-beam is located inside of an office building and is subjected to an increase in its live load carrying capacity. Using detailed analysis, it was shown that an existing building is satisfactory from the point of view of flexural strength. However, the shear strength of the beam is inadequate to carry the increased live load. The nominal shear strength contribution of concrete is 162 kN while the shear strength contributed by steel stirrups is 87.2 kN. The material properties of concrete and CFRP reinforcements are given below:

Concrete

Compressive strength, $f'_c = 20.7\,\text{N/mm}^2$

CFRP system properties:

Thickness of each ply, $t_f = 0.1651$ mm

Ultimate tensile strength, $f^*_{fu} = 3792\,\text{N/mm}^2$

Rupture strain, $\varepsilon^*_{fu} = 0.017$ mm/mm

Modulus of elasticity, $E_f = 227\,535\,\text{N/mm}^2$

Adopt the CFRP external shear strengthening scheme as shown in Fig. E5.3. In the above strengthening scheme, parameters, d, d_f, w_f, and s_f are 558.8 mm, 406.4 mm, 254 mm and 304.8 mm, respectively. Using the design approach for external FRP shear strengthening, show the above three-sided wrap shear strengthening is adequate for the T-beam subjected to the factored shear force, $V_u = 267$ kN. The factored shear force has been computed at distance, d (effective depth of beam) from the support.

Solution

Nominal shear strength of existing beam, $(V_n)_{\text{existing}} = 162.0 + 87.2 = 249.2$ kN

Design shear strength of the existing beam, $\phi(V_n)_{\text{existing}} = 0.85 \times 249.2 = 211.82$ kN

Check for spacing of the CFRP fabric systems,

$$s_f \leq \frac{d}{4} + w_f \quad \Rightarrow \quad 304.8 \text{ mm} \leq \frac{558.4}{4} + 254 = 393.7 \text{ mm}$$

Thus, the provided CFRP spacing is O.K.

Check for limit state of shear strength

To check the limit state of shear strength against the factored shear load, the following steps are followed as per ACI 440.2R-02.

Step 1: Compute the design material properties

Since the beam is located in an enclosed and conditioned space, hence, an environmental factor of 0.95 is used as the material is CFRP/Epoxy system (See Table 5.1 or Table 8.1 of ACI 440.2R-02).

$$f_{fu} = C_E f_{fu}^* = 0.95 \times 3792 = 3602 \text{ N/mm}^2$$

$$\varepsilon_{fu} = C_E \varepsilon_{fu}^* = 0.95 \times 0.017 = 0.0162$$

Step 2: Calculate the effective strain level in the FRP shear reinforcement

Based on onset of delamination for two- or three-sided wraps (see Eq. (5.25)),

$$\varepsilon_{fe} = k_v \varepsilon_{fu} \leq 0.004$$

where,

$$k_v = \frac{k_1 k_2 L_e}{11900 \varepsilon_{fu}} \leq 0.75$$

where, L_e = active bond length parameter and is calculated using Eq. (5.27b).

$$L_e = \frac{23300}{(n t_f E_f)^{0.58}} = \frac{23300}{(1 \times 0.1651 \times 227535)^{0.58}}$$
$$= 51.8 \text{ mm}$$

Modification factor, k_1, is calculated using Eq. (5.28b) for SI units,

$$k_1 = \left(\frac{f_c'}{27}\right)^{2/3} = \left(\frac{20.7}{27}\right)^{2/3}$$
$$= 0.84$$

The modification factor, k_2, is calculated using Eq. (5.29) for U-wrap (three-sided wrap),

$$k_2 = \frac{d_f - L_e}{d_f} = \frac{406.4 - 51.8}{406.4} = 0.873$$

$$k_v = \frac{k_1 k_2 L_e}{11900\,\varepsilon_{fu}} = \frac{0.84 \times 0.873 \times 51.8}{11900 \times 0.0162} = 0.197 \leq 0.75 \qquad \text{O.K.}$$

Thus, effective strain, $\varepsilon_{fe} = k_v \varepsilon_{fu} = 0.197 \times 0.0162 = 0.0032 \leq 0.004$ O.K.

Step 3: Calculate the contribution of CFRP reinforcement to shear strength

The area of shear reinforcement,

$$A_{fv} = 2nt_f w_f = 2 \times 1 \times 0.1651 \times 254 = 83.87 \text{ mm}^2$$

The effective stress in the FRP can be computed using linear stress strain relationship,

$$f_{fe} = \varepsilon_{fe} E_f = 0.0032 \times 227\,335 = 728.1 \text{ N/mm}^2$$

The shear contribution of the CFRP can be computed using Eq. (5.21),

$$V_f = \frac{A_{fv} f_{fe} (\sin\alpha + \cos\alpha) d_f}{s_f} = \frac{83.87 \times 728.1 \times (1+0) \times 406.4}{304.8 \times 1000} = 81.42\,\text{kN}$$

Step 4: Calculate the shear strength of the section

The design shear strength can be computed using Eqs. (5.19) and (5.20),

$$\phi V_n = \phi(V_c + V_s + \psi_f V_f) = 0.85 \times (162 + 87.2 + 0.85 \times 81.42) = 270.6\,\text{kN} > 267 \text{ kN} (V_u)$$

Thus, the strengthened section is capable of sustaining the required shear strength.

Step 5: Check the total shear reinforcement limit

$$V_s + V_f = 87.2 + 81.42 = 168.62\,\text{kN}$$

From the known value of $V_c = 162 \text{ kN} = \dfrac{\sqrt{f_c'}}{6}\left(\dfrac{b_w d}{1000}\right)$

Then $0.66\sqrt{f_c'} b_w d = \dfrac{0.66 \times 162 \times 6 \times 1000}{1000} = 641.52 \text{ kN}$

Thus, $V_s + V_f \leq 0.66\sqrt{f_c'} b_w d$ O.K.

E5.3. Design Example 3: Shear Strengthening Using CFRP Laminates—A Case Study

Problem Statement

During construction of a reinforced concrete beam, half of the steel reinforcement was accidentally omitted. The beam was originally designed with #3 stirrups (9.5 mm diameter) spaced at 152 mm center-to-center. The yield strength of the steel reinforcement is 413.7 MPa and the concrete strength is 27.58 MPa. Calculate the amount of externally bonded CFRP fabric sheet required to restore the beam to its original shear design capacity. Properties of the CFRP sheet are given in Table E5.3. The method of installation will be a U-wrap using laminates made of CFRP sheet. Possible fiber orientations are 0°, 90°, and 45°. The effective width of the beam is 228 mm, depth 381 mm. The effective depth (d) of the beam is 343 mm.

Table E5.3. Typical properties of CFRP fabric (fiber reinforced systems™).

Property	Value
Width, mm	304.8
Thickness, mm	0.178
Ultimate tensile strength, MPa	2758
Average modulus of elasticity, GPa	228
Average ultimate strain (%)	1.8

Solution

Design procedure

The general design methodology for shear strengthening of RC beam using CFRP laminates is presented along with comparison with experimental results in the following text. This problem forms a case study on shear strengthening of typical RC beams.

Shear strength of concrete and steel reinforcements

$$V_c = \frac{\sqrt{f_c'}b_w d}{6} = \frac{\sqrt{27.58} \times 228 \times 343}{6} = 68.45 \text{ kN}$$

$$V_s, \text{ Design} = \frac{A_v f_y d}{s} = \frac{2 \times 70.88 \times 413.7 \times 343}{152} = 132.3 \text{ kN}$$

$$V_s, \text{ As-built} = \frac{A_v f_y d}{s} = \frac{2 \times 70.88 \times 413.7 \times 343}{304} = 66.2 \text{ kN}$$

$$\begin{aligned} \phi V_n, \text{ Design} &= 0.85(V_c + V_s, \text{ Design}) \\ &= 0.85(68.45 + 132.3) \\ &= 170.6 \text{ kN} \end{aligned}$$

$$\phi V_n, \text{ As-built} = 0.85(V_c + V_s, \text{ As-built}) = 114.45 \text{ kN}$$

Case I: Assume 45° lay-up of strengthening (continuous U-wraps)

Contribution of CFRP reinforcement:

$$V_{\text{frp}} = \frac{A_{fv} f_{fe}(\sin\beta + \cos\beta) d_{\text{frp}}}{S_{\text{frp}}}$$

Cross-sectional area of CFRP fabric sheets = A_{fv}

$$A_{fv} = 2nt_{\text{frp}}w_{\text{frp}} = 2 \times 1 \times 0.178 \times 305 = 108.6 \text{ mm}^2$$

$$\beta = 45°$$

$$d_{\text{frp}} = d = 343 \text{ mm}$$

$S_{\text{frp}} = w_{\text{frp}} = 305$ mm (since strengthening pattern is continuous)

$$f_{fe} = Rf_{fu}$$

Factor R based on effective stress:

$$R = 0.5622(\rho_f E_f)^2 - 1.2188(\rho_f E_f) + 0.778$$

where,

$$p_f = \frac{2t_{\text{frp}}}{b_w} \times \frac{w_{\text{frp}}}{S_{\text{frp}}} = \frac{2 \times 0.178}{228} \times \frac{305}{305} = 1.56 \times 10^{-3}$$

$$\rho_f E_f = 1.56 \times 10^{-3} \times 228 = 0.36 \text{ GPa} < 1.1 \text{ GPa} \qquad \text{O.K.}$$

$$R = 0.5622(0.36)^2 - 1.2188(0.36) + 0.778$$

$$= 0.412$$

Factor R based on bond failure mechanism:

Effective bond length, L_e

$$L_e = \frac{461.3}{(E_f t_{\text{frp}})^{0.58}} = \frac{461.3}{[228 \times 0.178]^{0.58}} = 53.8 \text{ mm}$$

$$w_{fe} = d_{\text{frp}} - L_e = 343 - 53.8 = 289.2 \text{ mm}$$

$$R = \frac{0.0042(f_c')^{2/3} w_{fe}}{(E_f t_{\text{frp}})^{0.58} \varepsilon_{fu} d_{\text{frp}}} = \frac{0.0042(27.58)^{2/3} \times 289.2}{(228 \times 0.178)^{0.58} \times 0.018 \times 343} = 0.210$$

Factor R based on limiting strain:

$$R = \frac{0.004}{\varepsilon_{fu}} = \frac{0.004}{0.018} = 0.222$$

The lowest value of factor R is based on bond failure mechanism. Hence, $R = 0.210$ controls design.

Effective stress, $f_{fe} = (R \times f_{fu}) = (0.210 \times 2758) = 579.2$ MPa

$$V_{\text{frp}} = \frac{108.6 \times 579.2 \times (\sin 45^\circ + \cos 45^\circ) \times 343}{305} = 100.0 \text{ kN}$$

Check maximum shear reinforcement:

$$V_s + V_{\text{frp}} = \frac{2\sqrt{f_c'}}{3} b_w d$$

$$V_{\text{frp}} = 100.0 \text{ kN} < \frac{2\sqrt{27.58}}{3 \times 1000} \times 228 \times 343 - 66.2 = 207.6 \text{ kN} \qquad \text{O.K.}$$

Determine shear capacity of CFRP reinforced beam:

$$\phi V_n = 0.85(V_c + V_s) + 0.70 V_{\text{frp}}$$

$$= 0.85(68.45 + 66.2) + 0.70 \times 100.0$$

$$= 184.5 \text{ kN} > \phi V_n, \text{ design} = 170.6 \text{ kN} \qquad \text{O.K.}$$

Evaluate the difference between experimental and theoretical shear strength values:

Average experimental shear strength = 209.5 kN

Theoretical shear strength = 184.5 kN

$$\% \text{ difference} = \frac{209.5 - 184.5}{209.5} \times 100 = 11.9\%$$

Thus, one layer of 45° (continuous U-wrap) provides adequate shear strengthening. However, the design equations underestimate the actual shear strength of Beam-45° by about 11.9%.

Case II: Assume 0°/90° lay-up of strengthening (continuous U-wraps)

Cross-sectional area (A_{fv}) of each of fabric sheet with fiber orientation of 0° and 90°

$$A_{fv} = 2nt_{\text{frp}}w_{\text{frp}} = 2 \times 1 \times 0.178 \times 305 = 108.6 \text{ mm}^2$$

Factor R based on effective stress:

$$R = 0.5622(\rho_f E_f)^2 - 1.2188(\rho_f E_f) + 0.778$$

where,

$$\rho_f = \frac{2t_{\text{frp}}}{b_w} \times \frac{w_{\text{frp}}}{S_{\text{frp}}} = \frac{2 \times 2 \times 0.178}{228} \times \frac{305}{305} = 3.1 \times 10^{-3}$$

$$\rho_f E_f = 3.1 \times 10^{-3} \times 228 = 0.709 \text{ GPa} < 1.1 \text{ GPa} \qquad \text{O.K.}$$

$$R = 0.5622(0.707)^2 - 1.2188(0.707) + 0.778 = 0.197$$

Factor R based on bond failure mechanism:

Effective bond length, L_e

$$L_e = \frac{461.3}{(E_f t_{\text{frp}})^{0.58}} = \frac{461.3}{[228 \times 2 \times 0.178]^{0.58}} = 36 \text{ mm}$$

$$w_{fe} = d_{\text{frp}} - L_e = 343 - 36 = 307 \text{ mm}$$

$$R = \frac{0.0042\,(f_c')^{2/3}\, w_{fe}}{(E_f t_{\text{frp}})^{0.58}\, \varepsilon_{fu} d_{\text{frp}}} = \frac{0.0042\,(27.58)^{2/3} \times 307}{(228 \times 2 \times 0.178)^{0.58} \times 0.018 \times 343} = 0.149$$

Factor R based on limiting strain:

$$R = \frac{0.004}{\varepsilon_{fu}} = \frac{0.004}{0.018} = 0.222$$

The minimum value of factor R is based on bond failure mechanism.

Hence, $R = 0.149$ controls design.

Effective stress, $f_{fe} = R \times f_{fu} = (0.149 \times 2758) = 410.9$ MPa

Total shear contribution of 0°/90° lay-up:

The shear contribution of CFRP sheet with 0° fiber orientation, V_{f0}

$$V_{f0} = \frac{108.6 \times 410.9 \times (\sin 0^\circ + \cos 0^\circ) \times 343}{305} = 50.2 \text{ kN}$$

The shear contribution of CFRP sheet with 90° fiber orientation, V_{f90}

$$V_{f90} = \frac{108.6 \times 410.9 \times (\sin 90^\circ + \cos 90^\circ) \times 343}{305} = 50.2 \text{ kN}$$

Thus, total shear contribution of 0°/90° lay-up

$$V_{\text{frp}} = V_{f0} + V_{f90} = 50.2 + 50.2 = 100.4 \text{ kN}$$

Check maximum shear reinforcement:

$$V_s + V_{\text{frp}} < \frac{2\sqrt{f_c'}}{3} b_w d$$

$$V_{\text{frp}} = 100.4 \text{ kN} < \frac{2\sqrt{27.58}}{3 \times 1000} \times 228 \times 343 - 66.2 = 207.6 \text{ kN} \qquad \text{O.K.}$$

Determine shear capacity of CFRP reinforced beam:

$$\phi V_n = 0.85(V_c + V_s) + 0.70 V_{\text{frp}} = 0.85(68.45 + 66.2) + 0.70 \times 100.4$$

$$= 184.7 \text{ kN} > \phi V_n, \quad \text{Design} = 170.6 \text{ kN} \qquad \text{O.K.}$$

Evaluate the difference between experimental and theoretical shear strength values:

Average experimental shear strength = 200 kN

Theoretical shear strength =184.7 kN

$$\%\,\text{difference} = \frac{200.0 - 184.7}{200.0} \times 100 = 7.7\%$$

Thus, 0°/90° lay-up provides adequate shear strengthening. Note that the shear resistance contributions of a single 45° lay-up and 0°/90° lay-up are almost the same, which is consistent with experimental results (see Table E5.4). However, the design equations underestimate the actual strength of Beam-0°/90° by about 7.7%.

Table E5.4. Ultimate shear capacities of test beams.

Beam designation	Shear Capacity, kN				Increase in shear capacity due to initial strengthening (%)	Increase in shear capacity due to supplemental strengthening (%)	
	Initial strengthening	Supplemental strengthening					
	Test #1*	Test #1	Test #2	Test #3	Test #1	Test #2	Test #3
Unstrengthened beam	182.5	—	175.8	146.4†	—	—	—
Beam-45°	212.7	—	229.2	178.0‡	16.6	30.4	21.6
Beam-0°/90°	—	209.2	218.0	158.9§	14.6	24.0	8.5
Beam-0°/90°/45°	—	242.5	259.4	202.5¶	32.9	47.6	38.3

* Test #1 for Beam-45° was conducted after initial strengthening, while for other strengthened beams Test #1, Test #2 and Test #3 were conducted after supplemental strengthening.

† Control beam failed by diagonal shear cracking.

‡ Beam-45° failed by rupture of CFRP sheet along with pulling of concrete cover on sides of beam.

§ Beam-0°/90° failed by rupture of CFRP sheets at seams between CFRP sheets on sides of beam.

¶ Beam-0°/90°/45° failed by rupture of CFRP sheets along with side cover delamination.

Case III: Assume three layers of fabric sheet having 0°/90°/45° lay-up

Cross-sectional area (A_{fv}) of each fabric sheet with fiber orientation of 0°, 90°, and 45°

$$A_{fv} = 2nt_{\text{frp}}w_{\text{frp}} = 2 \times 1 \times 0.178 \times 305 = 108.6 \text{ mm}^2$$

Factor R based on effective stress:

$$R = 0.5622(\rho_f E_f)^2 - 1.2188(\rho_f E_f) + 0.778$$

where,

$$\rho_f = \frac{2t_{\text{frp}}}{b_w} \times \frac{w_{\text{frp}}}{S_{\text{frp}}} = \frac{2 \times 3 \times 0.178}{228} \times \frac{305}{305} = 4.68 \times 10^{-3}$$

$$\rho_f E_f = 4.68 \times 10^{-3} \times 228 = 1.067 \text{ GPa} < 1.1 \text{ GPa} \qquad \text{O.K.}$$

$$R = 0.5622(1.067)^2 - 1.2188(1.067) + 0.778 = 0.118$$

Factor R based on bond failure mechanism:

Effective bond length, L_e

$$L_e = 28.5 \text{ mm} \frac{461.3}{(E_f t_{\text{frp}})^{0.58}} = \frac{461.3}{[228 \times 3 \times 0.178]^{0.58}} = 28.5 \text{ mm}$$

$$w_{fe} = d_{\text{frp}} - L_e = 343 - 28.5 = 314.5 \text{ mm}$$

$$R = \frac{0.0042 (f_c')^{2/3} w_{fe}}{(E_f t_{\text{frp}})^{0.58} \varepsilon_{fu} d_{\text{frp}}} = \frac{0.0042 (27.58)^{2/3} \times 314.5}{(228 \times 3 \times 0.178)^{0.58} \times 0.018 \times 343} = 0.121$$

Factor R based on limiting strain:

$$R = \frac{0.004}{\varepsilon_{fu}} = \frac{0.004}{0.018} = 0.222$$

The minimum value of factor R is based on effective stress. Hence, $R = 0.118$ controls design.

Effective stress, $f_{fe} = R \times f_{fu} = (0.118 \times 2758) = 325.4$ MPa

Total shear contribution of 0°/90°/45° *lay-up:*

The shear contribution of CFRP sheet with 0° fiber orientation, V_{f0}

$$V_{f0} = \frac{108.6 \times 325.4 \times (\sin 0° + \cos 0°) \times 343}{305} = 39.7 \text{ kN}$$

The shear contribution of CFRP sheet with 90° fiber orientation, V_{f90}

$$V_{f90} = \frac{108.6 \times 325.4 \times (\sin 90° + \cos 90°) \times 343}{305} = 39.7 \text{ kN}$$

The shear contribution of CFRP sheet with 45° fiber orientation, V_{f45}

$$V_{f45} = \frac{108.6 \times 325.4 \times (\sin 45° + \cos 45°) \times 343}{305} = 56.2 \text{ kN}$$

Thus, total shear contribution of 0°/90°/45° lay-up

$$V_{\text{frp}} = V_{f0} + V_{f90} + V_{f45} = 39.7 + 39.7 + 56.2 = 135.6 \text{ kN}$$

Check maximum shear reinforcement:

$$V_s + V_{\text{frp}} < \frac{2\sqrt{f_c'}}{3} b_w d$$

$$V_{\text{frp}} = 135.6 \text{ kN} < \frac{2\sqrt{27.58}}{3 \times 1000} \times 228 \times 343 - 66.2 = 207.6 \text{ kN} \qquad \text{O.K.}$$

Determine shear capacity of CFRP strengthened beam:

$$\phi V_n = 0.85(V_c + V_s) + 0.70V_{\text{frp}} = 0.85(68.45 + 66.2) + 0.70 \times 135.6$$

$$= 209.4 \text{ kN} > \phi V_n, \text{ Design} = 170.6 \text{ kN} \qquad \text{O.K.}$$

Evaluate difference in experimental and theoretical shear strength values:

Average experimental shear strength = 234.8 kN

Theoretical shear strength = 209.4 kN

$$\% \text{ difference} = \frac{234.8 - 209.4}{234.8} \times 100 = 10.8\%$$

Thus, it is seen that the shear capacity of Beam-0°/90°/45° is the maximum and provides adequate shear strengthening. However, the design equations underestimate the strength of Beam-0°/90°/45° by about 10.8%.

The followings are the notations used in this example:

A_{fv} = cross-sectional area of each CFRP sheet, mm^2

A_v = cross-sectional area of steel stirrups, mm^2

d = effective depth of beam, mm

d_{frp} = effective depth of CFRP sheet, mm

E_f = modulus of elasticity of CFRP sheet, GPa

f_c' = specified compressive strength of concrete, MPa

f_{fe} = effective stress in CFRP sheet, MPa

f_{fu} = ultimate tensile strength of CFRP sheet, MPa

f_y = yield strength of steel, MPa

L_e = effective bond length, mm

n = number of CFRP sheets of a specific fiber orientation

R = ratio of effective stress to ultimate tensile strength of CFRP sheet

s = center-to-center distance between adjacent steel stirrups, mm

S_{frp} = spacing between adjacent CFRP sheets, mm

t_{frp} = thickness of CFRP sheet, mm

V_c = shear resistance provided by concrete with steel flexural reinforcement, kN

V_{frp} = shear resistance provided by CFRP sheet, kN

V_{f0} = shear contribution of CFRP sheet with fibers oriented at 0°, kN

V_{f45} = shear contribution of CFRP sheet with fibers oriented at 45°, kN

V_{f90} = shear contribution of CFRP sheet with fibers oriented at 90°, kN

V_n = nominal shear strength of a reinforced concrete section, kN

V_s = shear resistance provided by steel stirrups, kN

w_{frp} = width of CFRP sheet, mm

w_{fe} = effective width of CFRP sheet, mm

ε_{fu} = design rupture strain of CFRP sheet

β = orientation of CFRP fibers with respect to longitudinal axis of beam, degrees

ϕ = shear strength reduction factor

E5.4. Design Example 4: A Case Study Problem

Problem statement

A simply supported beam (shown in Fig. E5.4) is to be strengthened using CFRP strips to increase its ultimate load carrying capacity from 142.4 kN to 200.3 kN. This beam is provided with adequate stirrups to prevent premature shear failure in the beam. Properties of strengthening materials (CFRP CarboDur strips) are given in Table E5.5. Slump and compressive strengths of the supplied concrete mix are given in Table E5.6. The beam is located in an enclosed, air conditioned space. A strengthening system will be designed and experimental results will be compared to theoretical values for the increase in load carrying capacity of the beam due to the addition of CFRP strips.

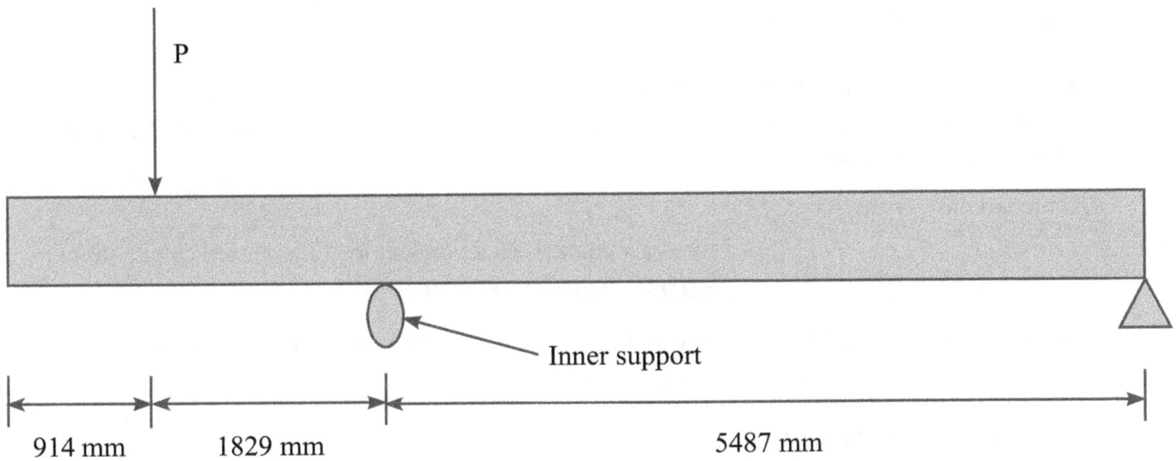

Figure E5.4. Experimental set-up of test beams.

Table E5.5. Mechanical properties of strengthening materials.

Property \ Type	CFRP strips	CFRP fabric	Epoxy (for strips)	Epoxy (for fabric)
	Sika CarboDur 50	SikaWrap Hex 103° C	SikaDur 30	SikaDur Hex 300/306
Tensile strength, GPa	2400	960	24.8	72.3
Modulus of elasticity, GPa	149	73.1	4.46	3.16
Failure strain (%)	1.4	1.3	1.0	4.8
Shear strength, MPa	—	—	24.8	123.4
Dimensions, mm	50 wide by 1.2 thick	1.0	—	—

Table E5.6. Slump and compressive strengths of the supplied concrete mix.

Sampling time during pour	Slump, mm	7 days strength, MPa
Beginning	215	32.2
Middle	178	30.5
End	165	31.0

Design procedure

Step 1: Estimate the flexural demand for CFRP strips

Existing ultimate moment capacity of the critical section = 1.829 × 142.4 = 260.5 kN-m

Ultimate flexural demand = 1.829 × 200.3 = 366.3 kN-m

Additional moment capacity required = (366.3 – 260.5) kN-m = 105.8 kN-m

From trial and error, it is found that 3 strips are not sufficient to provide above required moment capacity. Hence, three strips at the top and two strips on the sides of the beam will be provided as shown in Fig. E5.5 for Beam3/2.

Step 2: Compute the design material properties

Since the beam is located in an enclosed air conditioned space, an environmental reduction factor of 0.95 is suggested.

Design strength of the strips, $f^*_{fu} = C_E f^*_{fu}$

$= \text{Environmental reduction factor} \times \text{Guaranteed tensile strength}$

$= 0.95 \times 2400 = 2280 \text{ MPa}$

Design rupture strain, $\varepsilon_{fu} = C_E\, \varepsilon^*_{fu} = \text{Environmental reduction factor} \times \text{rupture strain}$

$= 0.95 \times 0.014 = 0.013$

Step 3: Compute concrete substrate strain (ε_{bi})

Since there is no other load except the dead load of the beam prior to bonding of the strips to the top of the beam, only the dead load is considered when computing the substrate strain.

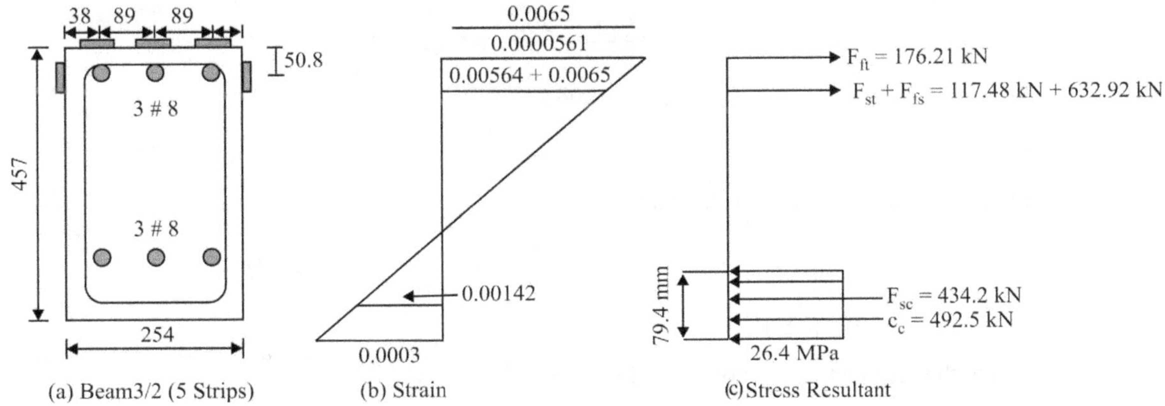

Figure E5.5. Strain and stress resultants across the cross-section.

Assuming the density of concrete is 25 kN/m^3, the self-weight of the beam can be found as,

$$W_d = 0.254 \times 0.457 \times 25 = 2.90 \text{ kN/m}$$

The factored moment at the inner support section due to the self-weight of the beam

$$= 1.2 \times W_d \times (2.743^2)/2$$
$$= 1.2 \times 2.90 \times (2.743^2)/2$$
$$= 13.1 \text{ kN-m}$$

Gross moment of inertia of the section, $I_g = bh^3/12 = 254 \times 457^3/12 \text{ mm}^4 = 2020 \times 10^6 \text{ mm}^4$

Modulus of elasticity of concrete, $E_c = 26.4$ GPa

Initial substrate strain, $\varepsilon_{bi} = \text{Moment} \times y_{\max} / (I_g \times E_c)$

$$= 13.1 \times 10^6 \times 228.5/(2020 \times 10^6 \times 26.4 \times 10^3)$$
$$= 5.61 \times 10^{-5}$$

Step 4: Calculate the effective strains in CFRP strips, tension and compression steel at the ultimate limit state

Effective strain in top-strips, $\varepsilon_{fe} = 0.003\dfrac{(h-c)}{c} - \varepsilon_{bi} < k_m \varepsilon_{fu}$

Since the primary mode of failure in this beam is the onset of delamination, a value of 0.5 is suggested for the effective design strain factor (k_m) of the strips, based on experimental strain results for category-II beams. This will prevent the onset of delamination prior to flexural failure of the concrete.

Taking $k_m = 0.5$, $k_m \varepsilon_{fu} = 0.5 \times 0.013 = 0.0065$

Assume $c = 96.5$ mm

$$\varepsilon_{fe} = 0.003 \times \frac{(457 - 96.5)}{96.5} - 5.61 \times 10^{-5}$$
$$= 0.0011 > 0.0065$$

So, take $\varepsilon_{fe} = 0.0065$.

The effective strain at top of the side-strips,

$$= \frac{0.003}{96.5}(457 - 96.5 - 25.4) - \frac{5.61 \times 10^{-5}}{(457 - 96.5)}(457 - 121.92)$$
$$= 0.010 > 0.0065$$

So, take the strain at the top of the side-strips = 0.0065.

Strain at the bottom of the side-strips

$$= \frac{0.003}{96.5}(457 - 96.5 - 76.2) - \frac{5.61 \times 10^{-5}}{(457 - 96.5)}(457 - 172.72)$$
$$= 8.794 \times 10^{-3} > 6.5 \times 10^{-3}$$

So, take the strain at the bottom of the side-strips = 6.5×10^{-3}

Strain in tension steel, $\varepsilon_{st} = (0.0065 + 5.61 \times 10^{-5}) \times \frac{(406.4 - 96.5)}{(457 - 96.5)} = 5.64 \times 10^{-3}$

Strain in compression steel, $\varepsilon_{sc} = \frac{0.003}{96.5} \times (96.5 - 50.8) = 1.42 \times 10^{-3}$

Step 5: Calculate the effective stresses in the CFRP strips, tension and compression steels at the ultimate limit state

Effective stress in CFRP strips,

$$f_{fe} = E_f \times \varepsilon_{fe} = 149\,000 \times 0.0065 = 968.5 \text{ MPa} < 2280 \text{ MPa} \qquad \text{O.K.}$$

Effective stress in tension steel,

$$f_{st} = E_s \times \varepsilon_{st} = 2 \times 10^5 \times 5.64 \times 10^{-3} = 1128 \text{ MPa} > 414 \text{ MPa}$$

Take $f_{st} = f_y = 414$ MPa (Yield stress of steel)

Effective stress in compression steel,

$$f_{sc} = E_s \times \varepsilon_{sc} = 2 \times 10^5 \times 1.42 \times 10^{-3} = 284 \text{ MPa}$$

Step 6: Check for the value of the depth to the neutral axis (*c*) from the equilibrium of internal forces

From the internal force (see Fig. E5.5) equilibrium, we have

$$C_c + F_{sc} = F_{ft} + F_{st} + F_{fs}$$
$$\gamma f'_c \beta_1 bc + n_c A_{sc} f_{sc} - \gamma f'_c A_{sc} = n_t A_{st} f_{st} + A_f f_{fe}$$

Take $\gamma = 0.85$

$$\beta_1 = 0.85 - (0.145 f_c' - 4) \times 0.05$$
$$= 0.85 - (0.145 \times 31 - 4) \times 0.05 = 0.825 \text{ for } f_c' = 31 \text{ MPa}$$

width of the beam, $b = 254$ mm.

Substituting the values of the parameters into the above equation, we have

$$0.85 \times 31 \times 0.825 \times 254 \times c = 3 \times 509.6 \times 414 + 5 \times 60.65 \times 968.5 + 0.85 \times 31 \times 3 \times 509.6 - 3 \times 509.6 \times 284$$
$$c = 96.5 \text{ mm} = \text{assumed value of } c$$

Take $c = 96.5$ mm.

Step 7: Calculate the additional nominal moment capacity provided by strips

Additional moment capacity provided by strips

$$= \psi A_f f_{fe}\left(h - \frac{\beta_1 c}{2}\right) + \psi A_f f_{fe}\left(h - 50.8 - \frac{\beta_1 c}{2}\right)$$
$$= 0.85 \times 3 \times 60.65 \times 968.5 \times \left(457 - \frac{0.825 \times 96.5}{2}\right)$$
$$+ 0.85 \times 2 \times 60.65 \times 968.5 \times \left(457 - 50.8 - \frac{0.825 \times 96.5}{2}\right)$$
$$= 99.08 \times 10^6 \text{ N-mm} = 99.08 \text{ kN-m}$$

The percentage difference in the theoretical and experimental values for the increase in moment capacity due to CFRP strips $= \dfrac{(99.08 - 105.9)}{105.9} \times 100 = -6.4\%$

Step 8: Check for ductility of the strengthening system

Maximum tensile strain developed, $\varepsilon_{st} = 0.00564 > 0.005$.

Hence, the strengthened system is ductile.

In the above example, the following notations have been used.

A_f = area of FRP reinforcement, mm^2

A_{st} = total area of tension steel bars, mm^2

A_{sc} = total area of compression steel bars, mm^2

b = width of a rectangular cross-section, mm

c = distance from extreme compression fiber to the neutral axis, mm

C_E = environmental reduction factor

C_c = resultant compression force in concrete, kN

d = distance from extreme compression fiber to centroid of tension reinforcement, mm.

E_c = modulus of elasticity of concrete, MPa

E_f = modulus of elasticity of CFRP, MPa

Es = modulus of elasticity of steel, MPa

f'_c = specified compressive strength of concrete, MPa

f_{fe} = effective stress in the CFRP strip, MPa

f^*_{fu} = guaranteed tensile strength of an CFRP strip as reported by the manufacturer, MPa

f_{fu} = ultimate design tensile strength of CFRP strip, MPa

f_{sc} = stress in longitudinal compression steel, MPa

f_{st} = stress in longitudinal tension steel, MPa

f_y = specified yield stress of non-prestressed steel reinforcement, MPa

F_{fs} = resultant force in side-strips, MPa

F_{ft} = resultant force in top-strips, MPa

F_{sc} = resultant compression force in compression reinforcement, MPa

F_{st} = resultant tensile force in tension reinforcement, MPa

h = overall depth of the beam, mm

I_g = gross moment of inertia of cross-section of the beam, mm^4

k_m = bond-dependent coefficient for flexure

n_c = number of compression reinforcing bars

n_t = number of tension reinforcing bars

y_{max} = distance of extreme fiber of the beam from the neutral axis, mm

β_1 = ratio of the depth of the equivalent stress block to the depth of the neutral axis

ε_{bi} = strain level in the concrete susbstrate at the time of FRP installation

ε_{fe} = effective design strain in CFRP reinforcement

ε^*_{fu} = guaranteed rupture strain of CFRP reinforcement

ε_{fu} = design rupture strain of CFRP reinforcement

ε_{sc} = strain in longitudinal compression steel

ε_{st} = strain in longitudinal tension steel

γ =Whitney's stress parameter

ψ = partial strength reduction factor

E5.5. Design Example 5

Problem statement

The cross-sectional details of an exterior RC beam of an office building is shown in Fig. E5.6. The beam is 4.4 m long and resting on column supports at its both ends with an effective span of 4.2 m. The beam is reinforced with three 20-mm diameter main longitudinal steel bars and 9.5 mm diameter steel shear stirrups at c/c spacing of 100 mm. Two 10-mm diameter steel hanger bars are provided at the top of section to support the stirrups. Note these two hanger bars are not provided to increase the strength and stiffness of the RC beam and hence their contribution to structural strength and stiffness are neglected. In addition, the beam is reinforced on its soffit by two layers of CFRP strips as shown in Figs. E5.6 and E5.7. The cross-section dimensions of RC beam are 450 mm height and 350 mm width. The effective concrete concrete cover to the main longitudinal steel bars is 50 mm. The material properties of various constituent materials are as follows:

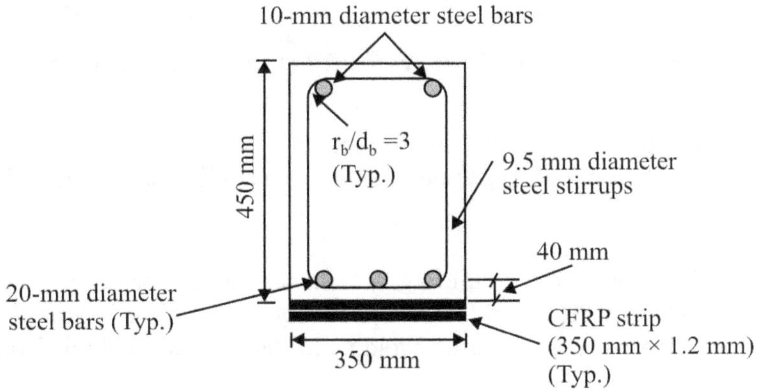

Figure E5.6. Cross-section details of rectangular simply supported beam.

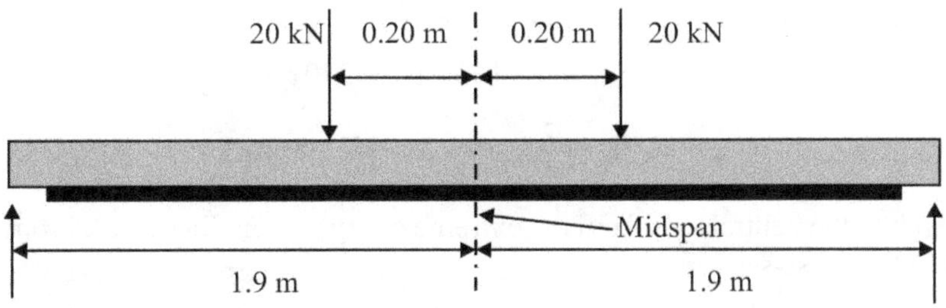

Figure E5.7. Elevation and loading details of CFRP reinforced RC beam.

Concrete

28-day Characteristics Strength, $f_c' = 31\,\text{MPa}$

Modulus of Elasticity, $E_c = 26.4$ GPa

Crushing Strain, $\varepsilon_u = 0.0035$ mm/mm

CFRP Strips

Properties specified by Manufacturer's:

Thickness of each strip =1.2 mm

Tensile strength, $f_{fu}^* = 2,400\,\text{MPa}$

Rupture Strain, $\varepsilon_{fu}^* = 0.015$ mm/mm

Modulus of Elasticity, $E_f = 150$ GPa

Steel main bars and Stirrups:

Yield strength, $f_y = 414$ MPa

Young's modulus, $E_s = 200$ GPa

Assume perfect bond between two layers of CFRP strips and evaluate the followings as per ISIS design guidelines:

1. Design moment capacity of strengthened beam.
2. Check whether the beam is strong enough to take the applied loads (Fig. E5.7).

Solution

Assume cast in place concrete for building. Resistance factors for concrete, CFRP, and steel are $\phi_c = 0.6$; $\phi_F = 0.8$ for CFRP; $\phi_s = 0.85$, respectively. Assume the failure of the beam due to crushing of concrete in compression after yielding of the internal steel reinforcement:

$$\alpha_1 = 0.85 - 0.0015f'_c = 0.85 - 0.0015 \times 31 = 0.80 > 0.67 \qquad \text{O.K.}$$

So, $\alpha_1 = 0.80$

$$\beta_1 = 0.97 - 0.0025f'_c = 0.97 - 0.0025 \times 31 = 0.089 > 0.67 \qquad \text{O.K.}$$

So, $\beta_1 = 0.89$

$$A_f = 2 \times 350 \times 1.2 = 840\,\text{mm}^2$$

$$A_s = \frac{\pi}{4} \times 20^2 \times 3 = 942.48\,\text{mm}^2$$

Let the depth to the neutral axis from the extreme compression fiber is 'c'. From the equilibrium of forces in the cross-section,

$$\phi_c \alpha_1 f_c' \beta_1 b c = \phi_s A_s f_y + \phi_F E_f A_f \varepsilon_{\text{frp}}$$

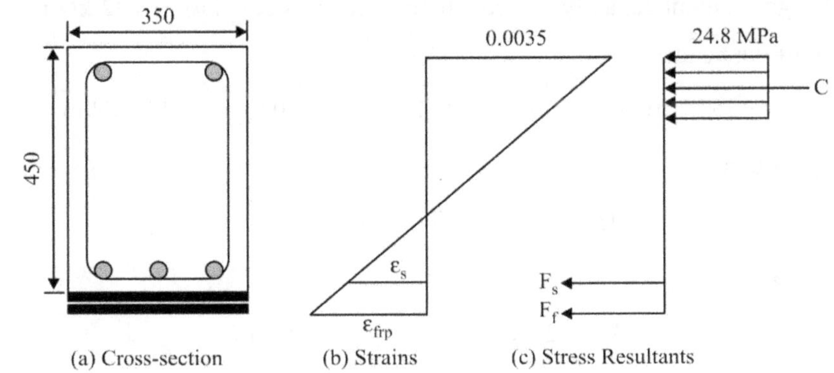

Figure E5.8. Strains and Stress resultants across the depth of cross-section.

$$0.6\times 0.80\times 31\times 0.89\times 350\times c = 0.85\times 942.48\times 414 + 0.8\times 150\times 10^3\times 840\times\left[\frac{0.0035(450-c)}{c}\right]\Rightarrow$$

$$4635.12c = 221098.77 + 352800\left[\frac{(450-c)}{c}\right]\Rightarrow 4635.12c = 221098.77 + 352800\left[\frac{(450-c)}{c}\right]$$

$$\Rightarrow c^2 - 47.70c + 76.11c - 34251.54 = 0$$

$$\Rightarrow c = \frac{-28.41+\sqrt{(28.41)^2 + 4\times 1\times 34251.54}}{2}$$

$$= 171.41 \text{ mm}$$

$$\varepsilon_{\text{frp}} = 0.0035\times\frac{(450-171.41)}{171.41} = 0.0056 < 0.015$$

So the assumption of failure of concrete crushing is correct.

$$\varepsilon_s = 0.0035\times\frac{(400-171.41)}{171.41} = 0.0047 > \varepsilon_y = 0.002$$

Thus the steel has yielded and $f_s = f_y = 414$ MPa. This also indicates that the amount of CFRP reinforcement is appropriate.

Factored moment of resistance of strengthened beam is given by:

$$M_r = \phi_s A_s f_y\left(d-\frac{a}{2}\right) + \phi_F E_f A_f \varepsilon_{\text{frp}}\left(h-\frac{a}{2}\right)$$

$$M_r = 0.85\times 414\times 942.48\times\left(400-\frac{0.89\times 171.47}{2}\right) + 0.8\times 150\times 10^3\times 0.0056\left(450-\frac{0.89\times 171.47}{2}\right)$$

$$= 318.32 \text{ kN-m}$$

So, the design moment capacity of strengthened beam is equal to 318.32 kN-m.

Dead load of the beam,

$$W_d = (25 \text{ kN/m}^3) \times (0.30 \text{ m}) \times (0.450 \text{ m}) \times (1.0 \text{ m}) = 3.9375 \text{ kN/m}$$

Dead load moment,

$$M_D = \frac{3.9375 \times 3.8^2}{8} = 7.10 \text{ kN-m}$$

Live load moment,

$$M_L = 20 \times (1.9 - 0.20) = 34 \text{ kN-m}$$

Factored Moment,

$$M_u = 1.5\, M_D + 1.5\, M_L = 1.5(7.10 + 34) = 61.65 \text{ kN-m} < 318.32 \text{ kN-m} \qquad \text{O.K.}$$

Hence the beam is strong enough to carry the applied loads.

E5.6. Design Example 6

A deficient cantilever RC beam of 5 m span is to be strengthened to carry updated live load of 6 kN/m in addition to its self-weight. The beam is located in an aggressive environmental condition. The cross-section of beam is 200 mm (wide) by 300 mm (deep) and is reinforced with 2 Nos. of 16-mm diameter steel bars of Grade Fe-415 in tension zone and 2 Nos. of 10-mm diameter bars in compression zone. Design suitable NSM CFRP strengthening systems for flexure using 10-mm diameter CFRP rebars and Epoxy bonding materials. Material properties of CFRP/Epoxy and concrete are given below. Assume clear concrete cover to the main longitudinal reinforcement bars equal to 42 mm.

Material properies

CFRP/EPOXY system

Specified tensile strength = 2500 MPa

Specified rupture strain = 0.015 mm/mm

Concrete

Cylinder strength, $f_c' = 31\,\text{MPa}$;

Cube Strength, $f_{cu} = 39\,\text{MPa}$;

Bond Strength, $\tau_{bs} = 7\,\text{MPa}$; $\tau_{bf} = 6\,\text{MPa}$

Solution

Environmental condition is aggressive. So environmental reduction factor, $C_E = 0.85$ for CFRP bars.

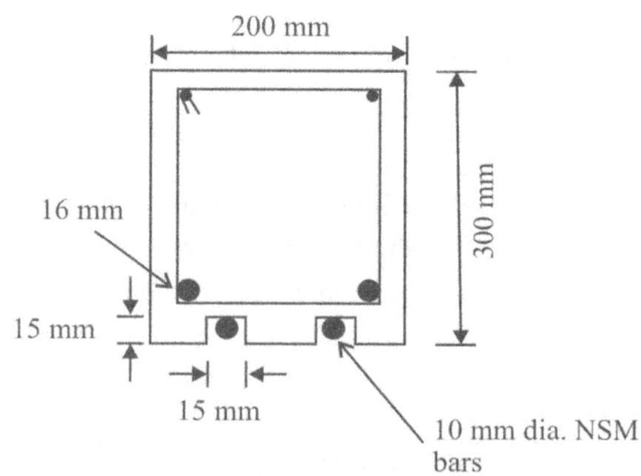

Figure E5.9 Cross-section details of NSM FRP reinforced beams.

Self weight of beam = 25 kN/m^3 × 0.2 m × 0.3 m × 1 m = 1.5 kN/m

$$M_u = 1.2M_D + 1.6M_L = 1.2\left(\frac{1.5\times 5^2}{2}\right) + 1.6\left(\frac{6\times 5^2}{2}\right) = 142.5\,\text{kN-m}$$

Design Steps: Design steps are provided as per ACI 440.2R-02 Standards.

Step I: Compute Design material properties

$$f_{fu} = 0.85\times 2500 = 2125\,\text{MPa}$$
$$\varepsilon_{fu} = 0.85\times 0.015 = 0.013\,\text{mm/mm}$$

Step II: Compute dead load and existing substrate strain

Dead load = 1.5 kN/m

Assuming dead load being the only load sustained before strengthening.

$$M_d = \frac{W_d l^2}{2} = \frac{1.5\times 5^2}{2} = 18.75\,\text{kN-m}$$

$$E_c = 4733\sqrt{31} = 26352\,\text{MPa}$$

Modular ratio of steel and concrete, $m_s = \dfrac{E_s}{E_c} = \dfrac{200\,\text{GPa}}{26.4\,\text{GPa}} = 7.6$

$$A_s = 2\times\frac{\pi}{4}\times 16^2 = 402\text{mm}^2$$

$$d = 300 - 42 - 16/2 = 250 \text{ mm}$$

Let the neutral axis depth of unstrengthened beam is c_o. Then we have,

$$\frac{bc_o^2}{2} = m_s A_s(d - c_o) \Rightarrow \frac{200\times c_o^2}{2} = 7.6\times 402(250 - c_o)$$

$$c_o^2 + 30.552c_o - 7638 = 0 \Rightarrow c_o = \frac{-30.552 + \sqrt{30.552^2 + 4\times 1\times 7638}}{2} = 73.4 \text{ mm}$$

Transformed moment of Inertia of unstrengthened beam section,

$$I_{tro} = \frac{bc_o^3}{3} + m_s A_s (d - c_o)^2 = \frac{200\times 73.4^3}{3} + 7.6\times 402\times(250 - 73.4)^2 = 121.65\times 10^6 \text{ mm}^4$$

$$d_f = 300 - \frac{15}{2} = 292.5 \text{ mm}$$

Existing substrate strain, $\varepsilon_{bi} = \dfrac{M_o\left(d_f - c_o\right)}{E_c I_{tro}} = \dfrac{18.75\times 10^6\left(292.5 - 73.4\right)}{121.65\times 10^6\times 26352} = 1.2815\times 10^{-3}$

Step III: Compute effective stress in NSM FRP bars

Effective area of tension reinforcement,

$$A_e = \frac{2d_c b}{\text{No. of bars}} = \frac{2\times 50\times 200}{2} = 10000\,\text{mm}^2$$

$$f_{ct} = 0.36\sqrt{f_{cu}} = 0.36\sqrt{39} = 2.25\,\text{MPa}$$

Total perimeter of steel bars,

$$\sum O_s = 2\times \pi \times d_{bs} = 2\times \pi \times 16 = 100.5\,\text{mm}$$

Total perimeter of NSM FRP bars,

$$\sum O_f = 2\times \pi \times d_{bf} = 2\times \pi \times 10 = 62.8\,\text{mm}$$

Minimum crack spacing,

$$l_{\min} = \frac{A_e f_{ct}}{u_s \sum O_s + u_f \sum O_f} = \frac{10000\times 2.25}{(7\times 100.5 + 6\times 62.8)} = 20.8\,\text{mm}$$

$$l_{\max} = 2l_{\min} = 2\times 20.8 = 41.6\,\text{mm}$$

Since $l_{\min} < 50$ mm,

$$L_{P2} = 1.86 l_{\min}^2 - 127 l_{\min} + 2436 = 1.86(20.8)^2 - 127\times 20.8 + 2436$$
$$= 599.1\text{ mm, say } 599\text{ mm}$$
$$L_P = \text{Min}(L_{P1}, L_{P2}) = 0.599\text{ m}$$

$$\sigma_{f\text{del},\max} = \frac{2bL_P l_{\max}}{3n\pi d_b^2 h'} f_{ct}$$

where, $h' = 50 - \frac{16}{2} - \frac{15}{2} = 34.5\,\text{mm}$

$$\sigma_{f\text{del},\max} = \frac{2\times 200\times 599\times 41.6}{3\times 2\times \pi\times 10^2 \times 34.5} = 153.3\,\text{MPa}$$

Let the depth to the neutral axis for the strengthened beam is 'c' and equal to 80 mm. The effective strain in NSM bar based on crushing of concrete mode of failure is given by,

$$\varepsilon_{fe} = \frac{0.003(d_f - c)}{c} - \varepsilon_{bi}$$
$$= \frac{0.003(292.5 - 80)}{80} - 1.2815\times 10^{-3} = 6.6873\times 10^{-3}$$

$$E_f = \frac{2500}{0.015} = 166.7\,\text{GPa}$$

Effective stress in FRP bars = 166.7 × 10^3 × 6.6873 × 10^{-3} = 1 114.55 MPa > 153.3 MPa (i.e., the stress corresponding to delamination). Hence, failure of the beam will be based on delamination of the bars.

$$\text{So, } \varepsilon_{fe} = \frac{153.3}{166.7\times10^3} = 9.20\times10^{-4}$$

Strain in concrete,

$$\varepsilon_c = \frac{(9.20\times10^{-4} + 1.2815\times10^{-3})}{(292.5-80)}\times 80 = 8.28\times10^{-4} < 0.003 \quad (\text{i.e., } \varepsilon_u)$$

$$\varepsilon_0 = \frac{2f_c'}{E_c} = \frac{2\times31}{26352} = 2.353\times10^{-3} > \varepsilon_c$$

Since $\varepsilon_c < \varepsilon_o < \varepsilon_u$, hence

$$\alpha = \left(\frac{\varepsilon_c}{\varepsilon_o}\right) - \frac{1}{3}\left(\frac{\varepsilon_c}{\varepsilon_o}\right)^2 = \frac{8.28\times10^{-4}}{2.353\times10^{-3}} - \frac{1}{3}\left(\frac{8.28\times10^{-4}}{2.353\times10^{-3}}\right)^2 = 0.31$$

$$\varepsilon_s = \frac{(\varepsilon_{fe} + \varepsilon_{bi})\,(d-c)}{(d_f - c)} = \frac{(2.2015\times10^{-3})(250-80)}{(292.5-80)} = 1.7612\times10^{-3}$$

$$f_s = 200\times10^3\times1.7612\times10^{-3} = 352.24\,\text{MPa} < 414\,\text{MPa}\,(f_y) \qquad \text{O.K.}$$

$$A_f = 2\times\frac{\pi}{4}\times10^2 = 157\,\text{mm}^2$$

From equilibrium condition,

$$c = \frac{A_s f_s + f_{fe}A_f}{\alpha f_c' b} = \frac{(402\times352.24 + 153.3\times157)}{0.31\times31\times200} = 86\,\text{mm} > 80\,\text{mm} \quad \text{(Assumed Value), N.G.}$$

Let c = 83 mm. Strain in concrete,

$$\varepsilon_c = \frac{(9.20\times10^{-4} + 1.2815\times10^{-3})}{(292.5-83)}\times 83 = 8.72\times10^{-4} < 0.003 \quad (\text{i.e., } \varepsilon_u)$$

$$\varepsilon_0 = 2.353\times10^{-3} > \varepsilon_c$$

Since $\varepsilon_c < \varepsilon_o < \varepsilon_u$, hence

$$\alpha = \left(\frac{\varepsilon_c}{\varepsilon_o}\right) - \frac{1}{3}\left(\frac{\varepsilon_c}{\varepsilon_o}\right)^2 = \frac{8.72\times10^{-4}}{2.353\times10^{-3}} - \frac{1}{3}\left(\frac{8.72\times10^{-4}}{2.353\times10^{-3}}\right)^2 = 0.325$$

$$\varepsilon_s = \frac{(\varepsilon_{fe} + \varepsilon_{bi})(d-c)}{(d_f - c)} = \frac{(2.2015\times10^{-3})(250-83)}{(292.5-83)} = 1.75\times10^{-3}$$

$$f_s = 200 \times 10^3 \times 1.75 \times 10^{-3} = 351\,\text{MPa} < 414\,\text{MPa}\,(f_y) \qquad \text{O.K.}$$

$$A_f = 157\,\text{mm}^2$$

From equilibrium condition,

$$c = \frac{A_s f_s + f_{fe} A_f}{\alpha f_c' b} = \frac{(402 \times 351 + 153.3 \times 157)}{0.325 \times 31 \times 200} = 82\,\text{mm} \cong 83\,\text{mm (Assumed Value)} \qquad \text{O.K.}$$

So take $c = 83$ mm.

Step IV: Compute the nominal moment capacity of section

$$M_n = A_s f_s (d - \gamma_c) + \psi_f A_f f_{fe} (d_f - \gamma_c)$$

where, $\gamma_c = \beta_1 c$

$$\beta_1 = \frac{\dfrac{1}{3} - \dfrac{\varepsilon_c}{12\varepsilon_o}}{1 - \dfrac{\varepsilon_c}{3\varepsilon_o}} = \frac{\dfrac{1}{3} - \dfrac{8.72 \times 10^{-4}}{12 \times 2.353 \times 10^{-3}}}{\left(1 - \dfrac{8.72 \times 10^{-4}}{3 \times 2.353 \times 10^{-3}}\right)} = 0.345$$

$$\gamma_c = 0.345 \times 83 = 28.64\,\text{mm}$$

$$M_n = 402 \times 351 \times (250 - 28.64) + 0.85 \times 157 \times 153.3 \times (292.5 - 28.64)$$

=36.63 kN-m, much less than the required moment capacity (i.e., 142.5 kN-m)

$$\phi = 0.70 \quad \text{as } \varepsilon_s < \varepsilon_y$$

Hence, the load on this beam is restricted to $1.2M_D + 1.6M_L = \phi M_n = 0.70 \times 36.63$

$$1.6 \times \frac{w_L l^2}{2} = 0.7 \times 36.63 - 1.2 \times 18.75 \Rightarrow w_L = 0.16\,\text{kN/m}$$

So the live load should be restricted to 0.16 kN/m.

Design for Shear

Total u.d.l. = 1.5 + 0.16 = 1.66 kN/m

$V_u = 1.66 \times 5 = 8.3$ kN (As per revised load)

$V_u = 7.5 \times 5 = 37.5$ kN (As per orginial load)

Shear strength of concrete, $V_c = \dfrac{\sqrt{f_c'}}{6} bd = \dfrac{\sqrt{31}}{6} \times \dfrac{200 \times 250}{1000} = 46.4$ kN

Assuming 2-legged 9.5 mm diameter steel stirrups at 1000 mm c/c.

$$V_s = \frac{A_{sv} f_{sv} d}{s}$$

where,

$$f_{sv} = 0.002\,E_s = 0.002 \times 200 \times 10^3 = 400\,\text{MPa}$$

$$A_{sv} = 2 \times \frac{\pi}{4} \times 9.5^2 = 141.76\,\text{mm}^2$$

Assuming 2-legged 9.5 mm diameter, steel stirrups at 1000 mm c/c.

$$V_s = \frac{A_{sv} f_{sv} d}{s} = \frac{141.76 \times 400 \times 250}{1000} = 11.90\,\text{kN}$$

Contribution of concrete and steel stirrups = 46.4 + 11.90 = 58.3 kN > 37.5 kN
Hence, NSM FRP shear reinforcement is not required.

E5.7. Design Example 7

An existing circular column of 400 mm diameter is reinforced with 12 Nos. of 20 mm diameter longitudinal bars and 9.5 mm helical ties at 400 mm pitch. The column is 4.0 m long and is subjected to single bending. If displacement ductility requirement is 4, using ACI 440.2R-02 guidelines, design suitable strengthening system to avoid failure of column due to shear and plastic hinge formation. Neglect the contribution of axial compression force due to the shear strength of the column. Find the axial load carrying capacity of strengthened column. Material Properties are given as follows:

Steel

Yield Strength, f_y = 414 MPa;
Modulus of Elasticity,
E_s = 200 GPa

Concrete

$f_c' = 36\,\text{MPa},\; E_c = 28\,\text{GPa}$

GFRP/EPOXY system

Thickness of each fabric sheet =1.3 mm

Guaranteed tensile strength,
$f_{fu}^* = 600\,\text{MPa}$

Guaranteed rupture strain,

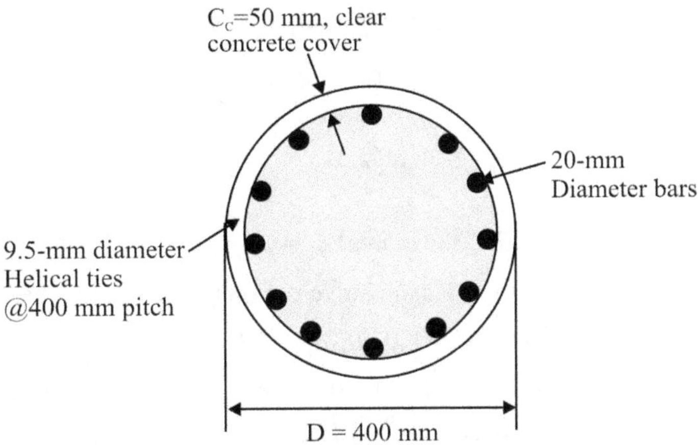

Figure E5.10 Cross-section of circular columns

$\varepsilon_{fu}^* = 0.020$ mm/mm

$E_f = 26\,000$ MPa

Solution

Design Steps

Step 1: Compute the design material properties

$$f_{fu} = C_E f_{fu}^* = 0.75 \times 600 = 450\,\text{MPa}$$

$$\varepsilon_{fu} = C_E \varepsilon_{fu}^* = 0.75 \times 0.020 = 0.015\,\text{mm/mm}$$

Step 2: Calculate the effective strain level in GFRP shear reinforcement

Assuming GFRP is continuously wrapped along the length of column,

The effective strain, $\varepsilon_{fe} = 0.004 \le 0.75 \times \varepsilon_{fu} = 0.75 \times 0.015 = 0.0113$

Hence, take $\varepsilon_{fe} = 0.004$

Step 3: Calculate the ideal moment capacity of column section and thickness of GFRP for shear strengthening

Cross-section area of steel bars,

Cross-section area of steel ties, $A_{sh} = \frac{\pi}{4} \times (9.5)^2 \times 2 = 141.8\,\text{mm}^2$

Center to center distance between the tie bars, $D' = 400 - 50 - 50 + 9.5 = 309.5\,\text{mm}$

Assuming $\theta = 45°$

Shear strength contribution of steel ties,

$$V_s = \frac{\pi}{2} \times \frac{A_{sh} f_{yh} D'}{s} \text{Cot}\ \theta = \frac{\pi}{2} \times \frac{141.8 \times 414 \times 309.5 \times 1.0 \times 10^{-3}}{400} = 71.4\,\text{kN}$$

$$A_g = \frac{\pi}{4} \times (400)^2 = 125\,664\,\text{mm}^2$$

Effective shear area, $A_e = 0.8A_g = 0.8 \times 125\,664 = 100\,531\ \text{mm}^2$

Length of plastic hinge region (L_P) for single bending case, i.e., for $L = 4$ m

$L_P = 0.08L + 0.022 f_{yl} d_{bl} \ge 0.044 f_{yl} d_{bl}$

$= 0.08 \times 4000 + 0.022 \times 414 \times 20 = 502.16 \ge 0.044 \times 414 \times 20 = 364.32$ mm

So, $L_p = 502.16$ mm say 502 mm

$$\mu_\Delta = 1 + 3(\mu_\phi - 1)\frac{L_P}{L}\left(1 - 0.5\frac{L_P}{L}\right) \Rightarrow 4 = 1 + 3(\mu_\phi - 1)\left(\frac{502.0}{4000}\right)\left(1 - 0.5 \times \frac{502}{4000}\right) \Rightarrow \mu_\phi = 9.5$$

Now

$$\mu_\phi = 7.0, \quad k_c = 0.083$$

$$\mu_\phi = 15.0, \quad k_c = 0.042$$

So, for $\mu_\phi = 9.5, k_c = 0.083 - \frac{(0.083-0.042)}{(15-7)}(9.5-7.0) = 0.070$ for inside the plastic hinge region;

$k_c = 0.25$ for outside the plastic hinge region

Shear Strength Contribution of Concrete

i. *Inside the plastic hinge region:*

$$V_c = \frac{0.07\sqrt{36}}{1000} \times 100531 = 42.2 \text{ kN}$$

ii. *Outside the plastic hinge region:*

$$V_c = \frac{0.25\sqrt{36}}{1000} \times 100531 = 151 \text{ kN}$$

Step 4: Calculate the ideal moment capacity, M_i

Assuming 'c' lies above the top compression steel,

$$\beta_1 = 0.85 - \frac{\left(\frac{36}{6.895} \times 1000 - 4000\right)}{1000} \times 0.05 = 0.79$$

From equilibrium of column section, we have

$$A_s f_y = 0.85 f_c' \beta_1 b c$$

$$3770 \times 414 = 0.85 \times 36 \times 0.79 \times 400 \times c \Rightarrow c = 161 \text{mm}$$

Area of steel in tension zone = 7 × 314 mm^2
Taking, $b = 400$ mm

$$7 \times 414 \times 314 = 0.85 \times 36 \times 0.79 \times 400c + 0.85 \times 36 \times 5 \times 314 \times 7.14 \Rightarrow c = 94 \text{mm}$$

So neutral axis lies in the region of 1st and 2nd rebars.

$$\text{Strain in top compression bars} = \frac{0.003}{94}(94 - 50) = 0.0014$$

$$\text{Strain in the middle tensile bars} = \frac{0.003}{94}(200 - 94) = 0.004 > 0.002$$

For maximum moment, assume all tension bars are subjected to yield force

$$M_i = 9 \times 414 \times 314 \left[\left(60 + \frac{(340-60)}{2} \right) - \frac{0.79 \times 94}{2} \right] = 190.6 \text{kN-m}$$

Step 5: Column shear demand based on full flexural overstrength

$$V_o = 1.5\left(\frac{M_i}{L}\right) = 1.5 \times \frac{190.6}{4} = 71.5\,\text{kN}$$

$$\frac{V_o}{\psi_f \phi_V} = \frac{71.5}{0.95 \times 0.85} = 88.55\,\text{kN}$$

Thickness of FRP jacket for circular column

i. *Inside the plastic hinge region:*

$$t_{\text{frp}} = \frac{\left[88.55 - (42.22 + 71.4)\right]}{\frac{\pi}{2} \times (0.004 \times 26000) \times 400 \times \cot 45^\circ} = -4 \times 10^{-4} \cong \text{No jacket required}$$

Similarly, it can be shown that no jacket is required outside the plastic hinge region.

Step 6: Flexural hinge confinement (plastic hinge condition)

$$t_{\text{frp}} = \frac{0.09h(\varepsilon_{cc} - 0.004) f'_{cc}}{\phi_F f_{fu} \varepsilon_{fu}}$$

$$\varepsilon_{cc} = 0.004 + \frac{2.8\rho_v f_{fu} \varepsilon_{fu}}{f'_{cc}}$$

Let the length of plastic hinge region,

$$L_v^i = 1.5 \times 400 = 600\,\text{mm}$$

Provide 10 layers of FRP

$$\rho_{fv} = \frac{A_{fv} L_v^i}{A_g L} = \frac{\frac{\pi}{4}\left[(400 + 1.3 \times 20)^2 - 400^2\right] \times 600}{\frac{\pi}{4} \times 400^2 \times 4000} = 0.020$$

Required ultimate confined compressive strain of concrete,

$$\varepsilon_{cc} = 0.004 + \frac{2.8 \times 0.020 \times 450 \times 0.015}{1.5 \times 36} = 0.011$$

Required thickness,

$$t_{\text{frp}} = \frac{0.09 \times 400 \times (0.011 - 0.004) \times 1.5 \times 36}{0.9 \times 450 \times 0.015} = 2.24 < 10 \times 1.3 = 13\text{ mm};$$ hence, O.K.

Length of primary confinement region

$$\begin{aligned} L_{c1} &\geq L_P = 502\,\text{mm} \\ &\geq 0.5d = 0.5 \times 400 = 200\,\text{mm} \\ &\geq L/8 = 4000/8 = 500\,\text{mm} \end{aligned}$$

So take L_{c1} = 502 mm and Hence, L_{c2} = 502 mm. Provide t_{frp} = 13 mm (10 layers) in primary region and 13/2 = 6.5 mm (5 layers) in secondary region.

Step 7: Calculate the axial load capacity of confined column

$$f'_{cc} = f'_c\left[2.25\sqrt{1+7.9\frac{f_l}{f'_c}} - 2\frac{f_l}{f'_c} - 1.25\right]$$

$$\text{Reinforcement ratio, } \rho_f = \frac{\frac{\pi}{4}\left[(400+20\times 1.3)^2 - 400^2\right]}{\frac{\pi}{4}\times 400^2} = 0.1342$$

Efficiency factor, k_a =1.0

$$\rho_g = \frac{A_{st}}{A_g} = \frac{3770}{125664} = 0.03$$

$$f_l = \frac{k_a\rho_f f_{fe}}{2} = \frac{k_a\rho_f\varepsilon_{fe}E_f}{2} = \frac{1.0\times 0.1342\times 0.004\times 26000}{2} = 6.9784 \quad \text{say } 7.0\ \text{N/mm}^2$$

$$f'_{cc} = 36\left[2.25\sqrt{1+7.9\times\frac{7.0}{36}} - 2\times\frac{7.0}{36} - 1.25\right] = 70\ \text{N/mm}^2$$

$$(\phi P_n)_{\text{confined column}} = 0.8\phi\left[0.85\psi_f f'_{cc}(A_g - A_{st}) + f_y A_{st}\right]$$

$$= 0.8\times 0.7\times\left[0.85\times 0.95\times 70(125664-3770) + 414\times 3770\right]\times 10^{-3} = 4732\,\text{kN}$$

So axial load capacity of the circular column confined with FRP is equal to 4732 kN.

E5.8. Design Example 8

A simply supported precast RC beam of 5 m effective span is resting on 25 cm thick wall on both ends. The beam is having cross-section of 200 mm (width) × 300 mm (depth) and is reinforced with 4 Nos. of 10-mm diameter GFRP bars and 9.5-mm diameter two legged GFRP stirrups at 120 mm center-to-center spacing. Take radius of bend in stirrups equal to 3 times the stirrup bar diameter. Assume beam is located in interior space. Determine the design moment capacity of the beam as per ISIS Canada Design Standards.

Material properties

Concrete

f'_c = 30 MPa, E_c = 27 GPa

GFRP Bars

Guaranteed tensile strength, $f^*_{fu} = 600\,\text{MPa}$,

Guaranteed rupture strain, $\varepsilon^*_{fu} = 0.030\,\text{mm/mm}$, $E_f = 60\,\text{GPa}$

Solution

Beam is located in interior space.

As per ISIS Canada Design Guidelines,

$$\phi_{\text{FRP}} = 0.4(\text{for GFRP});\ \phi_c = 0.65 \quad (\text{Precast Concrete Section})$$

$$A_{\text{frp}} = 4 \times \frac{\pi}{4} \times 10^2 = 314\,\text{mm}^2$$

Required concrete cover, $2.5d_b = 2.5 \times 10 = 25$ mm or 40 mm

So, assume clear concrete cover, C_c to be equal to 40 mm.

Effective depth of beam, $d = h - \text{cover} - \frac{d_b}{2} = 300 - 40 - \frac{10}{2} = 255\,\text{mm}$

GFRP reinforcement ratio

$$\rho_{\text{frp}} = \frac{A_{\text{frp}}}{bd} = \frac{314}{200 \times 255} = 6.16 \times 10^{-3} = 0.00616$$

Balanced reinforcement ratio

$$\rho_{\text{frp}b} = \alpha_1 \beta_1 \frac{\phi_c}{\phi_{\text{FRP}}} \frac{f_c'}{f_{\text{frp}u}} \left[\frac{\varepsilon_{cu}}{\varepsilon_{cu} + \varepsilon_{\text{frp}u}} \right]$$

$$\alpha_1 = 0.85 - 0.0015 f_c' \geq 0.67 \Rightarrow 0.85 - 0.0015 \times 30 = 0.805 \quad \text{say } 0.81$$

$$\beta_1 = 0.97 - 0.0025 f_c' \geq 0.67 \Rightarrow 0.97 - 0.0025 \times 30 = 0.895 \quad \text{say } 0.90$$

$$\rho_{\text{frp}b} = \alpha_1 \beta_1 \frac{\phi_c}{\phi_{\text{FRP}}} \frac{f_c'}{f_{\text{frp}u}} \left[\frac{\varepsilon_{cu}}{\varepsilon_{cu} + \varepsilon_{\text{frp}u}} \right] = 0.81 \times 0.90 \times \frac{0.65}{0.4} \times \frac{30}{600} \times \left[\frac{0.0035}{0.0035 + 0.030} \right] = 0.00618$$

Check for the mode of failure

$$\rho_{\text{frp}} = 0.00616 < \rho_{\text{frp}} = 0.00618$$

Therefore section failure is governed by tension failure mode. Assume the depth to the neutral axis, $c = 70$ mm, then from strain compatibility, we have

$$\frac{\varepsilon_c}{c} = \frac{\varepsilon_{\text{frp}u}}{d - c} \Rightarrow \varepsilon_c = \frac{0.030}{(255 - 70)} \times 70 = 0.0114 = 11.4 \times 10^{-3} > 0.0035 \quad (\text{i.e., concrete crushing})$$

$$\alpha = \alpha_1 = 0.81$$

$$\beta = \beta_1 = 0.90$$

$$T = \phi_{\text{frp}} A_{\text{frp}} f_{\text{frp}u} = 0.4 \times 314 \times 600 = 75360\,\text{N}$$

$$C = \alpha \phi_c f_c' \beta c b = 0.81 \times 0.65 \times 30 \times 0.90 \times 70 \times 200 = 199017\,\text{N} > \text{T}$$

So, let the reduced depth of neutral axis, $c = 26$ mm.

$$\varepsilon_c = \frac{0.030}{(255-26)} \times 26 = 0.0034, \text{ So } \alpha_1 = 0.81 \text{ and } \beta_1 = 0.90$$

$$C = 0.81 \times 0.65 \times 30 \times 0.90 \times 26 \times 200 = 73920.6\,\text{N}$$

So, take $c = 26$ mm

$$M_r = \phi_{\text{frp}} A_{\text{frp}} f_{\text{frp}u}\left(d - \frac{\beta c}{2}\right) = 0.4 \times 314 \times 600\left(255 - \frac{0.9 \times 26}{2}\right) \times 10^{-6} = 18.33\,\text{kN-m}$$

E5.9. Design Example 9

A deficient simply supported RC beam is to be strengthened to carry a nominal ultimate mid-span load of 22 kN using four-point loading system, in addition to its self-weight on a long-term basis. The cross-section of beam is 180 mm wide and 260 mm deep with clear concrete cover to the main reinforcement of 42 mm. The beam is to be used for an interior application. The total span of beam is 3.0 m. The beam is resting on two columns having width of 25 cm along the length of the beam. The effective span of four-point loading spreader system centered on mid-span is 0.81 m. The beam is initially reinforced with two Grade 60 (415 MPa yield strength), 16 mm diameter steel bars longitudinally. The shear reinforcement is provided with 9.5 mm diameter two-legged steel stirrups of the same material properties as longitudinal reinforcement. Design a suitable near surface mounted 10 mm diameter CFRP Leadline/ Epoxy system for NSM strengthening. The specified tensile strength of CFRP bar is 2860 MPa, while specified rupture strain of GFRP bars is 0.019. Modulus of elasticity of Leadline bar is 147 GPa. The cubic and cylindrical strengths of concrete are 39 and 31 MPa. The bond strength of steel and CFRP bars are 8.274 and 6.895 MPa, respectively. Assume any other data suitably. Check for Serviceability is not required. Modulus of elasticity of steel = 200 GPa

Solutions

Let us strengthen the RC beam with two NSM FRP Bars as shown in Fig. E5.11.

Effective concrete cover to the main longitudinal bars = 42 + 16/2 = 50 mm

Effective depth of deficient RC beam = 260 – 50 = 210 mm

Effective span of the beam = 3 – 0.25 = 2.75 m

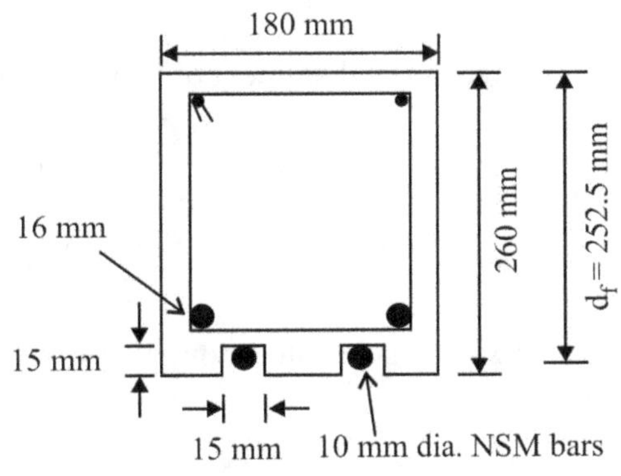

Figure E5.11. Cross-section details of NSM FRP reinforced RC beam.

Design steps

Step 1: Compute the design material properties

Design strength of CFRP Leadline bar, $f_{fu} = C_E f^*_{fu} = 0.95 \times 2860 = 2717$ MPa

Design rupture strain, $\varepsilon_{fu} = C_E \varepsilon^*_{fu} = 0.95 \times 0.019 = 0.018$

Step 2: Compute the dead load and existing substrate strain, ε_{bi}

Dead load of the beam, $w_d = 25 \times 0.26 \times 0.18 = 1.17$ kN/m

Unfactored dead load moment (assuming dead load being the only load sustained before strengthening) $= \dfrac{w_d l^2}{8} = \dfrac{1.17 \times 2.75^2}{8} = 1.11$ kN-m

Modulus of elasticity of concrete, $E_c = 4733\sqrt{f_c'} = 4733\sqrt{31} = 26352\,\text{MPa}$

Modular ratio of steel and concrete, $ms = \dfrac{200}{26.4} = 7.6$

Strengthening pattern:

Take 2 bars of 10-mm diameter leadline bars with groove size equal to $1.5d_b \times 1.5d_b =$ 15 mm × 15 mm.

$$d_f = 260 - \frac{15}{2} = 252.5 \text{ mm}$$

$$A_s = 2 \times \frac{\pi}{4} \times 16^2 = 402 \text{ mm}^2$$

Let the neutral axis depth of unstrenghened beam is c_o,

$$\frac{b c_o^2}{2} = m_s A_s (d - c_o)$$

$$\frac{180 \times c_o^2}{2} = 7.6 \times 402 \times (210 - c_o)$$

$$c_o^2 + 33.95 c_o - 7128.8 = 0 \Rightarrow c_o = \frac{-33.95 + \sqrt{(33.95)^2 + 4 \times 7128.8}}{2} = 69.1 \text{ mm}$$

Transformed moment of inertia of unstrengthened beam section,

$$I_{\text{tro}} = \frac{b c_o^3}{3} + m_s A_s (d - c_o)^2$$

$$= \frac{180 \times 69.1^3}{3} + 7.6 \times 402 \times (210 - 69.1)^2 = 8045 \times 10^4 \text{ mm}^4$$

Existing substrate strain,

$$\varepsilon_{bi} = \frac{M_o (d_f - c_o)}{E_c I_{\text{tro}}} = \frac{1.11 \times 10^6 (252.5 - 69.1)}{26352 \times 8045 \times 10^4} = 9.6 \times 10^{-5}$$

Step 3: Compute effective stress in NSM FRP bars

Effective area of tension reinforcement,

$$A_e = \frac{2 \times d_c \times b}{\text{No. of bars}} = \frac{2 \times 50 \times 180}{2} = 9000\,\text{mm}^2$$

$$f_{ct} = 0.36\sqrt{f_{cu}} = 0.36\sqrt{39} = 2.25\,\text{MPa}$$

$$\sum O_s = 2\pi d_{bs} = 2\times\pi\times16 = 100.5\,\text{mm}$$

$$\sum O_f = 2\pi d_{bf} = 2\times\pi\times10 = 62.8\,\text{mm}$$

$$l_{\min} = \frac{A_e f_{ct}}{u_s\sum O_s + u_f\sum O_f} = \frac{9000\times2.25}{8.274\times100.5+6.895\times62.8} = 16\,\text{mm}$$

$$l_{\max} = 2l_{\min} = 2\times16 = 32\,\text{mm}$$

$$L_{P1} = \frac{2.75}{2} - \frac{0.81}{2} = 0.97\,\text{m}$$

Since $l_{\min} < 50$ mm

$$L_{P2} = 1.86l_{\min}^2 - 127l_{\min} + 2436$$
$$= 1.86\times16^2 - 127\times16 + 2436 = 880.16\,\text{mm}$$

$$L_P = \min\left(L_{P1}, L_{P2}\right) = 0.88\,\text{m}$$

$$\sigma_{f\text{del},\max} = \frac{2bL_P l_{\max}}{3n\pi d_b^2 h'} f_{ct}$$

Here,

$$h' = 50 - \frac{16}{2} - \frac{15}{2} = 34.5$$

$$\sigma_{f\text{del},\max} = \frac{2\times180\times880\times32}{3\times2\times\pi\times10^2\times34.5}\times2.25 = 350.8\,\text{MPa}$$

Let the depth to the neutral axis for the strenghthened beam is "c" and equal to 80 mm. Effective strain in NSM bar based on crushing mode of failure is given by:

$$\varepsilon_{fe} = \frac{0.003(d_f - c)}{c} - \varepsilon_{bi} = \frac{0.003(252.5-80)}{80} - 9.6\times10^{-5} = 6.37\times10^{-3}$$

Effective stress in NSM bars based on crushing of concrete =

$$6.37\times10^{-3}\times147\times10^3 = 936.4\,\text{MPa} > 350.8\,\text{MPa}$$

Hence, failure of the beam will be based on delamination of bars,

$$\varepsilon_{fe} = \frac{350.8}{147\times10^3} = 2.39\times10^{-3}$$

Strain in concrete, $\varepsilon_c = \dfrac{\left(2.39\times10^{-3} + 9.6\times10^{-5}\right)}{(252.5-80)}\times80 = 1.15\times10^{-3} < 0.003$ (i.e., ε_u)

$$\varepsilon_o = \frac{2f'_c}{E_C} = \frac{2\times 31}{26352} = 2.353\times 10^{-3} > \varepsilon_c$$

$$\alpha = \left(\frac{\varepsilon_c}{\varepsilon_o}\right) - \frac{1}{3}\left(\frac{\varepsilon_c}{\varepsilon_o}\right)^2 = \left(\frac{1.15\times 10^{-3}}{2.353\times 10^{-3}}\right) - \frac{1}{3}\left(\frac{1.15\times 10^{-3}}{2.353\times 10^{-3}}\right)^2 = 0.41$$

$$\varepsilon_s = \frac{(\varepsilon_{fe}+\varepsilon_{bi})(d-c)}{(d_f-c)} = \frac{2.4860\times 10^{-3}\times(210-80)}{(252.5-80)} = 1.8735\times 10^{-3}$$

$$f_s = 200\times 10^3\times 1.8735\times 10^{-3} = 374\,\text{MPa} < f_y \qquad \text{O.K.}$$

$$A_f = 2\times\frac{\pi}{4}\times 10^2 = 157\,\text{mm}^2$$

From equilibrium,

$$c = \frac{A_s f_s + f_{fe}A_f}{\alpha f'_c b} = \frac{402\times 374.7 + 350.8\times 157}{0.41\times 31\times 180} = 89.90\,\text{mm} > 80\,\text{mm, n.g.}$$

Assume, $c = 85$ mm. Repeat the above steps,

$$\varepsilon_{fe} = 2.39\times 10^{-3};\ \varepsilon_c = 1.26\times 10^{-3};\ \alpha = 0.44;\ \varepsilon_s = 1.855\times 10^{-3};\ f_s = 371\,\text{MPa}$$

$$c = 83\text{ mm} \approx 85\text{ mm} \quad (\text{assumed value of } c)$$

So take, $c = 85$ mm.

E5.10. Design Example 10

An existing simply supported T-beam supports a uniformly distributed service (unfactored) dead load of 19.0 kN/m including its own weight and a uniformly distributed service live load of 23.4 kN/m. The concrete strength is 27.6 MPa, and yield strength of steel stirrups is 276 MPa. The overall height of beam is 650 mm, flange width is 910 mm, stem width is 300 mm, slab thickness is 150 mm, effective depth, d is 610 mm, and clear concrete cover, c is 40 mm. The original shear design called for 10 mm double-leg stirrups with the following spacing: (i) one at 75 mm from support; (ii) seven at 150 mm; (iii) three at 200 mm; and nine at 300 mm as shown in Fig. E5.12.

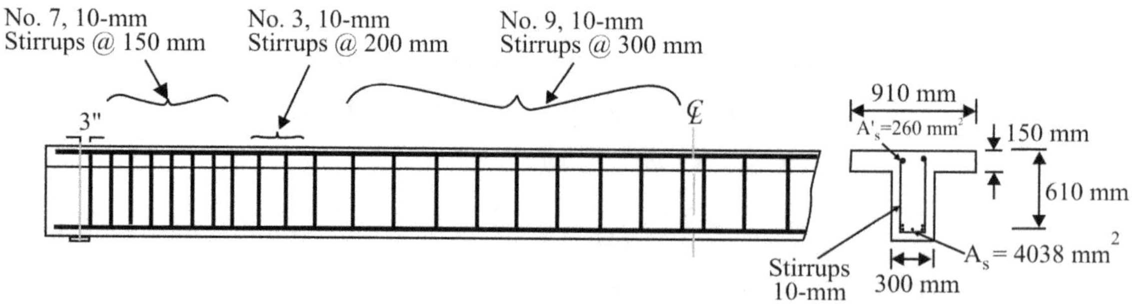

Figure E5.12. Steel stirrups in beam.

The original flexural design called for two layers of longitudinal grade 60 steel bars used as main reinforcement; the bottom layer includes three 35 mm steel bars; the top layer includes two 25 mm steel bars. Two 12 mm steel bars were used as compression reinforcement to hold the stirrups.

The service live load needs to be increased from 23.4 to 32.0 N/mm. Flexural capacity of the beam does not need to be improved, while an upgrading is needed in shear. The new ultimate shear value calculated at d from support is 324.7 kN. Design for shear strengthening using NSM FRP rebars of 6.35 mm diameter CFRP bars of 500 mm length, having ultimate guaranteed tensile strength, $f_{fu}^* = 2068$ MPa; Modulus of elasticity, $E_f = 124$ GPa, and ultimate guaranteed elongation strain, $\varepsilon_{fu}^* = 0.017$.

Solutions

FRP bars will be inserted between the existing steel stirrups leading to center to center spacing of $s = 80$ mm.

The number (n) of FRP bars crossed by a 45° shear crack is given by:

$$n = \frac{d_{\text{eff}}(1+\cot\alpha)}{s} = \frac{(d_r \sin\alpha - 2c_c)(1+\cot\alpha)}{s}$$

For NSM FRP bars to be vertical, $\alpha = 90°$, so

$$n = \frac{(500\,\text{mm}\sin(90) - 2\times 40)\times(1+\cot(90))}{80} = 5.25 \Rightarrow n = 5$$

The length where FRP bar strain governs, i.e., $l_{0.004}$ is given by:

$$l_{0.004} = 0.001\frac{d_b E_f}{\tau_b} = 0.001\frac{(6.35\,\text{mm})(124\,000\,\text{MPa})}{6.9\,\text{MPa}} = 114\,\text{mm}$$

$$\frac{n}{2} = \frac{5}{2} = 2.5 \Rightarrow 2$$

$$L_1 = \min(l_{0.004}, s.1) = \min(114, 80\times 1) = 80\,\text{mm}$$

$$L_2 = \min(l_{0.004}, s.2) = \min(114, 80\times 2) = 114\,\text{mm}$$

$$d_{\text{net}} = d_r - 2c_c = 500 - 2\times 40 = 420\,\text{mm}$$

$$L_3 = \min(l_{0.004}, d_{\text{net}} - s.3) = \min(114, 420 - 80\times 3) = 114\,\text{mm}$$

$$L_4 = \min(l_{0.004}, d_{\text{net}} - s.4) = \min(114, 420 - 80\times 4) = 100\,\text{mm}$$

$$L_5 = \min(l_{0.004}, d_{\text{net}} - s.5) = \min(114, 420 - 80\times 5) = 20\,\text{mm}$$

$$L_{\text{tot}} = L_1 + L_2 + L_3 + L_4 + L_5 = 80 + 114 + 114 + 100 + 20 = 428\,\text{mm}$$

FRP contribution to shear capacity can now be expressed by:

$$V_f = 2\pi d_b \tau_b L_{\text{tot}} = 2\times\pi\times 6.35\times 6.9\times 428 = 117.6\,\text{kN}$$

To prevent the concrete crushing, the condition

$V_s + V_f < 0.66\sqrt{f_c'}b\,d$ should be verified.

$$V_s = \frac{A_s f_y d}{s} = \frac{142\,\text{mm}^2 \times 276\,\text{MPa} \times 610\,\text{mm}}{150} = 159.4\,\text{kN}$$

$$V_s + V_f = 159.4 + 117.6 = 277\,\text{kN}$$

$$0.66\sqrt{f_c'}b\,d = 0.66\sqrt{27.6} \times 300 \times 610 = 634.5\,\text{kN}$$

Hence, $V_s + V_f < 0.66\sqrt{f_c'}bd$ is satisfied.

$$V_c = \frac{\sqrt{f_c'}}{6}bd = \frac{\sqrt{27.6}}{6} \times 300 \times 610 = 160.2\,\text{kN}$$

Design Shear Capacity,

$$\phi V_n = \phi(V_c + V_s + \psi_f V_f) = 0.85[160.2 + 159.4 + 0.85 \times 117.6] = 356.6\,\text{kN} > V_u = 324.7\,\text{kN, O.K.}$$

E5.11. Design Example 11

An existing column in building is 550 × 550 mm cross-section and is reinforced with 4–16 mm diameter steel bars and 9.5 mm diameter ties at 400 mm center to center. The column length is 5 m and is subjected to double bending. If displacement ductility requirement is 5, design suitable strengthening pattern to avoid the failure of column due to shear and plastic hinge formation. Neglect the contribution of axial compression force to the shear strength of column. Find the axial load carrying capacity of strengthened column. Assume 50 mm clear concrete cover to the main axial reinforcements.

Yield strength of steel = 414 MPa, Modulus of elasticity of steel = 200 GPa, Cylinder strength of concrete = 36 MPa, and Modulus of elasticity of concrete = 28 400 MPa.

GFRP/EPOXY system properties

Thickness per ply = 1.29 mm

Guaranteed tensile strength = 550 N/mm^2

Guaranteed rupture strain = 0.020

Modulus of elasticity of GFRP = 25 000 N/mm^2

Environmental factor, $C_E = 0.75$

Solutions

Step 1: Compute the design material properties

Design strength of GFRP bars, $f_{fu} = C_E f_{fu}^* = 0.75 \times 550 = 412.5\,\text{MPa}$

Design rupture strain of GFRP bars, $\varepsilon_{fu} = C_E \varepsilon_{fu}^* = 0.75 \times 0.020 = 0.015$

Step 2: Calculate effective strain (ε_{fe}) level in GFRP shear reinforcements

Assuming GFRP is continuously wrapped along the length of column.

$$\varepsilon_{fe} = 0.004 \leq 0.75\varepsilon_{fu} = 0.75 \times 0.015 = 0.013$$

Hence, take $\varepsilon_{fe} = 0.0004$.

Step 3: Calculate the ideal moment capacity of column section and thickness of GFRP for shear strengthening

Cross-section area of steel bars, $A_s = \frac{\pi}{4} \times 16^2 \times 4 = 804.3\,\text{mm}^2$

Cross-section area of steel ties, $A_{sh} = \frac{\pi}{4} \times 9.5^2 \times 2 = 141.8\,\text{mm}^2$

Center to center distance between tie bars, $D' = 459.5\,\text{mm}$

Shear strength contribution of steel ties,

$$V_s = \frac{A_{sh} f_{yh} D'}{s} \cot\theta = \frac{141.8 \times 414 \times 459.5}{400} \times \cot(45) = 67.4\,\text{kN}$$

Gross area of column section, $A_g = 550 \times 550 = 302\,500\ \text{mm}^2$

Effective shear area, $A_e = 0.8A_g = 0.8 \times 302\,500 = 242\,000\ \text{mm}^2$

Since the column is in double bending,

The length of the plastic hinge region, $L_P = 0.08L + 0.022 f_{yl} d_{bl} \geq 0.044 f_{yl} d_{bl}$

$$L = \frac{5}{2} = 2.5\,\text{m} = 2500\,\text{mm}$$

$L_p = 0.08 \times 2500 + 0.022 \times 414 \times 16 = 346.08\ \text{mm} \geq 0.044 f_{yl} d_{bl} = 0.044 \times 415 \times 16 = 292.16\ \text{mm}$

So, $L_p = 346.08$ mm

Using the relationship between displacement ductility and curvature ductility, we have

$$\mu_\Delta = 1 + 3(\mu_\phi - 1)\frac{L_P}{L}\left(1 - 0.5\frac{L_P}{L}\right) \Rightarrow 5 = 1 + 3(\mu_\phi - 1) \times \frac{346.08}{2500} \times \left(1 - 0.5 \times \frac{346.08}{2500}\right)$$

From the above relationship, we have, $\mu_\phi = 11.35$

For $\mu_\phi = 11.35$,

$$k = 0.042 + \frac{(0.083 - 0.042)}{14 - 5} \times (11.35 - 5) = 0.071 \quad \text{(inside the plastic hinge region)}$$

$k = 0.25$ (outside the plastic hinge region)

Step 4: Compute the shear strength contribution of concrete, V_c

i. Inside the plastic hinge region

$$V_c = \frac{0.071 \times \sqrt{36}}{1000} \times 242\,000 = 103\,\text{kN}$$

ii. Outside the plastic hinge region

$$V_c = \frac{0.25 \times \sqrt{36}}{1000} \times 242\,000 = 363\,\text{kN}$$

Step 5: Calculate the ideal moment capacity

Assuming neutral axis lies above the top compression steel, the equilibrium equation,

$$A_s f_y = 0.85 f'_c b \beta_1 c, \quad \text{where, } c \text{ is the depth of neutral axis.}$$

The stress block factor, β_1 is given by,

$$\beta_1 = 0.85 - (0.145 \times 36 - 4) \times 0.05 = 0.7$$

Thus, $804.3 \times 414 = 0.85 \times 36 \times 550 \times 0.79c \Rightarrow c = 25.1\,\text{mm} < 50\,\text{mm}$ (concrete cover)

Hence, $c = 25.1$ mm lies above the top compression steel. Check for strain in steel using strain compatibility condition. Strain in top layer of steel bars,

$$\varepsilon_{s1} = \frac{0.003}{25.1} \times (58 - 25.1) = 3.93 \times 10^{-3} > \varepsilon_y = 0.002$$

Similarly, strain in bottom layer of steel,

$$\varepsilon_{s2} = \frac{0.003}{25.1} \times (492 - 25.1) = 0.0558 \times 10^{-3} > \varepsilon_y = 0.002$$

Above strain levels in steel bars show that bars have yielded before crushing of concrete at failure, so $f_{s1} = f_{s2} = f_y = 414$ MPa. Moment Capacity of Section,

$$M_i = \frac{804.3 \times 414 \times \left[\left(58 + \frac{492 - 58}{2} \right) - \frac{0.79 \times 25.1}{2} \right]}{10^6} = 88.5\,\text{kN-m}$$

Step 6: Compute the column shear demand based on full flexural strength

$$V_o = 1.5 \frac{M_i}{L} = 1.5 \times \frac{88.50}{2.5} = 53.1\,\text{kN-m}$$

Step 7: Compute total shear strength

i. *Outside the plastic hinge region:*

$V_n = 0$ (due to axial load contribution given as zero)

$V_c = 363\,\text{kN}$

$V_s = 67.6\,\text{kN}$

$$V_n + V_c + V_s = 0 + 363 + 67.6 = 430.6\,\text{kN} > \frac{V_o}{\psi_f \phi_v} = \frac{53.1}{0.95 \times 0.85} = 65.76\,\text{kN}$$

Hence, no shear strengthening is required outside the plastic hinge region.

ii. *Inside the plastic hinge region:*

$$V_n + V_c + V_s = 0 + 103 + 67.6 = 170.6\,\text{kN} > \frac{V_o}{\psi_f \phi_v} = \frac{53.1}{0.95 \times 0.85} = 65.76\,\text{kN}$$

Hence, no shear strengthening is required inside the plastic hinge region also. However, for the sake of extra safety, we may provide a layer of ply throughout the length of column. Thickness provided =1.29 mm

Step 9: Flexural hinge confinement requirement based on plastic hinge condition

Thickness required, $t_{\text{frp}} = \dfrac{0.09\,h(\varepsilon_{cc} - 0.004)\,f'_{cc}}{\phi_F\, f_{fu}\, \varepsilon_{fu}}$

where, $f'_{cc} = 1.5 f'_c$, ϕ_F = Flexural capacity reduction factor = 0.9

$$\varepsilon_{cc} = 0.004 + \frac{2.8\,\rho_v f_{fu} \varepsilon_{fu}}{f'_{cc}}$$

Let the length of plastic hinge region, $L_{vi} = 1.5h = 1.5 \times 550 = 825\,\text{mm}$

Providing 10 layers of FRP, volumetric reinforcement ratio,

$$\rho_{fv} = \frac{A_{fv} L_{vi}}{A_g L} = \frac{\left[(550 + 2 \times 10 \times 1.29)^2 - 550^2\right]}{550^2 \times 2500} \times 825 = 0.0317$$

Required ultimate confined compressive strain of concrete,

$$\varepsilon_{cc} = 0.004 + \frac{2.8 \times 0.0317 \times 412.5 \times 0.015}{1.5 \times 36} = 0.0142$$

Required thickness of FRP,

$$t_{\text{frp}} = \frac{0.09 \times 550 \times (0.0142 - 0.004) \times 1.5 \times 36}{0.9 \times 412.5 \times 0.015} = 27.3\text{ mm} > 12.9\text{ mm, N.G.}$$

Provide 25 layers,

$$\rho_{fv} = \frac{A_{fv}L_{vi}}{A_g L} = \frac{\left[(550+2\times 25\times 1.29)^2 - 550^2\right]}{550^2\times 2500}\times 825 = 0.0819$$

$$\varepsilon_{cc} = 0.004 + \frac{2.8\times 0.0819\times 412.5\times 0.015}{1.5\times 36} = 0.0303$$

$$t_{\text{frp}} = \frac{0.09\times 550\times (0.0303-0.004)\times 1.5\times 36}{0.9\times 412.5\times 0.015} = 12.6\,\text{mm} < 25\times 1.29\text{ mm (Provided), O.K.}$$

So provide 25 layers of ply with total thickness of FRP = 25 × 1.29 = 32.25 mm
Length of primary confinement region,

So provide primary plastic hinge region length, $L_{c1} = 350$ mm with FRP thickness of 32.25 mm.
Secondary plastic hinge region length,

$$L_{c2} = L_{vi} - L_{c1} = 825 - 350 = 475\,\text{mm} \geq \begin{cases} L_p = 346.08\text{ mm} \\ 0.5h = 0.5\times 550 = 275\text{ mm} \\ \dfrac{L}{8} = \dfrac{2500}{8} = 312.5\text{ mm} \end{cases}$$

So provide secondary plastic hinge confinement region of length, $L_{c2} = 475$ mm of FRP thickness equal to 32.25/2 = 16.125 mm giving rise to 13 layers with total thickness of 16.77 mm.

Step 10: Calculate the axial load carrying capacity of confined column

$$f'_{cc} = f'_c\left[2.25\sqrt{1+7.9\frac{f_l}{f'_c}} - 2\frac{f_l}{f'_c} - 1.25\right]$$

$$\rho_f = \frac{2nt_f(b+h)}{bh} = \frac{2\times 1\times 1.29(550+550)}{550\times 550} = 9.38\times 10^{-3}$$

Efficiency factor, $k_a = 1 - \dfrac{(b-2r)^2 + (h-2r)^2}{3bh(1-\rho_g)}$

where, $\rho_g = \dfrac{A_{st}}{A_g} = \dfrac{804.3}{302500} = 2.6588\times 10^{-3}$

Taking, $r = 15$ mm (corner radius)

$$k_a = 1 - \frac{(550-2\times 15)^2 + (550-2\times 15)^2}{3\times 550\times 550\times\left(1-2.6588\times 10^{-3}\right)} = 0.4$$

$$f_l = \frac{k_a\rho_f f_{fe}}{2} = \frac{k_a\rho_f E_f\varepsilon_{fe}}{2} = \frac{0.4\times 9.38\times 10^{-3}\times 0.004\times 25000}{2} = 0.1876\,\text{N/mm}^2$$

$$f'_{cc} = 36\left[2.25\sqrt{1+7.9\times\frac{0.1876}{36}} - 2\times\frac{0.1876}{36} - 1.25\right] = 37.3\,\text{N/mm}^2$$

$$(\phi P_n)_{\text{confined column}} = 0.80\times 0.70\left[0.85\times 0.95\times 37.3(302500-804.3)+414\times 804.3\right] = 5275.6\,\text{kN}$$

E5.12. Design Example 12

A deficient simply supported RC beam is to be strengthened to carry a nominal ultimate midspan load of 58 kN using two-point loading system in addition to its self-weight on a long-term basis. The beam is an interior beam. The total span of beam is 2.74 m while the effective span is 2.54 m. The effective span of central two-point loading spreader beam is 0.81 m. The beam is initially reinforced with two grade 60 (f_y = 414 MPa) steel bars of 16 mm diameter. Use of CFRP Leadline bar is recommended for NSM strengthening in the tension zone. Design a suitable CFRP Leadline/ Epoxy system for NSM strengthening. The specified strength of CFRP leadline bar is 2860 MPa while rupture strain is 1.9%. The cylindrical concrete strength is 31 MPa while cube compressive strength is 39 MPa. Take bond strength of FRP bars as 6.9 MPa, and modulus of elasticity, E_f=147 GPa.

Solutions

Step 1: Compute the design material properties

Design strength of CFRP leadline bars, $f_{fd} = C_E f_{fu} = 0.95\times 2860 = 2717\,\text{MPa}$

Design rupture strain of FRP bars, $\varepsilon_{fd} = C_E \varepsilon_{fu} = 0.95\times 0.019 = 0.018$

Step 2: Compute the existing substrate strain,

Dead load of the beam, $w_D = 25(\text{kN/m}^3)\times(0.254\ \text{m}\times 0.178\ \text{m}) = 1.13\,\text{kN/m}$

Unfactored dead load moment (sustained load before strengthening),

$$M_D = \frac{1.13\times 2.54^2}{8} = 0.91\,\text{kN-m}$$

Modulus of elasticity of concrete, $E_c = 26.4\,\text{GPa}$

Modular ratio of steel and concrete,

$$m_s = \frac{E_s}{E_c} = \frac{200}{26.4} = 7.6$$

With 9.5 mm diameter stirrups and 25.4 mm clear concrete cover, the effective concrete cover to main steel reinforcement (Fig. E5.13) is given by,

$$d' = 25.4 + 9.5 + \frac{16}{2} = 42.9\,\text{mm}$$

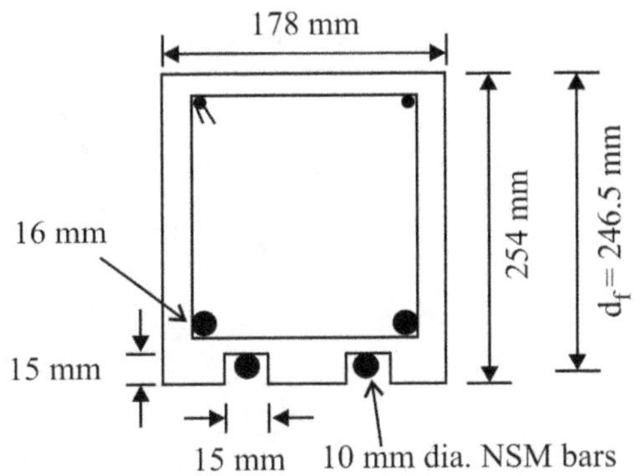

Figure E5.13. NSM FRP strengthened RC beam with two NSM bars.

Effective depth, $d = 254 - 42.9 = 211.1$ mm.

Distance of centroid of NSM bars from extreme compression fibers,

$$d_f = 254 - \frac{15}{2} = 246.5\,\text{mm}$$

$$A_s = 2 \times \frac{\pi}{4} \times 16^2 = 402\,\text{mm}^2$$

Let the neutral axis depth of the unstrengthened beam is c_o. By equating the moment of compression and tension area about the neutral axis, we have,

$$\frac{bc_o^2}{2} = m_s A_s \left(d - c_o\right) \Rightarrow \frac{178 \times c_o^2}{2} = 7.6 \times 402 \times \left(211.1 - c_o\right) \Rightarrow c_o = 69.68\,\text{mm}$$

Transformed moment of inertia of unstrengthened beam section,

$$I_{\text{tro}} = \frac{bc_o^3}{3} + m_s A_s \left(d - c_o\right)^2 = \frac{178 \times 69.68^3}{3} + 7.6 \times 402 \times \left(211.1 - 69.68\right)^2 = 8.1176 \times 10^7\ \text{mm}^4$$

Existing substrate strain,

$$\varepsilon_{bi} = \frac{M_o(d_f - c_o)}{E_c I_{\text{tro}}} = \frac{0.91 \times 10^6 \times (246.5 - 69.68)}{26.4 \times 10^3 \times 8.1176 \times 10^7} = 7.51 \times 10^{-5}$$

Step 3: Compute minimum and maximum crack spacing

Effective area of tension reinforcement,

$$A_e = \frac{2 d_c b}{\text{No. of steel bars}} = \frac{2 \times 42.9 \times 178}{2} = 7636.2\ \text{mm}^2$$

The concrete tensile strength, $f_{ct} = 0.36\sqrt{f_{cu}} = 0.36\sqrt{39} = 2.25$ MPa

Bond strength of steel, $u_s = 1.2 \times 6.9 = 8.28\,\text{MPa}$

Bond strength of FRP bars, $u_f = 6.9\,\text{MPa}$

$$\sum O_s = 2\pi d_{bs} = 2 \times \pi \times 16 = 100.5\,\text{mm}$$

$$\sum O_f = 2\pi d_{bf} = 2 \times \pi \times 10 = 62.8\,\text{mm}$$

$$l_{\min} = \frac{A_e f_{ct}}{u_s \sum O_s + u_f \sum O_f} = \frac{7636.2 \times 2.25}{8.28 \times 100.5 + 6.9 \times 62.8} = 13.6\,\text{mm}$$

$$l_{\max} = 2 l_{\min} = 2 \times 13.6 = 27.2\,\text{mm}$$

$$L_{P1} = \frac{2.54}{2} - \frac{0.81}{2} = 0.865\,\text{m} = 865\text{mm}$$

Here, $l_{\min} = 13.6$ mm < 50 mm

Hence, $L_{P2} = 1.86 \times 13.6^2 - 127 \times 13.6 + 2436 = 1052.8\,\text{mm}$

$$L_P = \min(L_{P1}, L_{P2}) = \min(865, 1052.8) = 865\,\text{mm}$$

$$h' = 42.9 - \frac{16}{2} - \frac{15}{2} = 27.4\,\text{mm}$$

$$\sigma_{\text{fudel,max}} = \frac{2b\,L_P l_{\max}}{3n\pi d_b^2 h'} f_{ct} = \frac{2 \times 178 \times 865 \times 27.2}{3 \times 2 \times \pi \times 10^2 \times 27.4} \times 2.25 = 364.9\text{ MPa}$$

Step 4: Compute the depth of neutral axis at failure

Let, $c = 80$ mm

Based on crushing of concrete,

$$\varepsilon_{fr} = 0.003\left(\frac{d_f - c}{c}\right) - \varepsilon_{bi} = 0.003\left(\frac{246.5 - 80}{80}\right) - 7.51 \times 10^{-5} = 6.1687 \times 10^{-3}$$

$f_{fec} = 147 \times 10^3 \times 6.1687 \times 10^{-3} = 906.8\,\text{MPa} > 364.4$ MPa ($\sigma_{fudel,\max}$). Delamination governs design

Thus, $\varepsilon_{fe} = \dfrac{364.9}{147 \times 10^3} = 2.4823 \times 10^{-3}$

Strain in concrete, $\varepsilon_c = \dfrac{2.4833 \times 10^{-3} + 7.51 \times 10^{-5}}{246.5 - 80} = 1.23 \times 10^{-3} < 0.003$

$$\varepsilon_o = \frac{2 \times 31}{26.4 \times 1000} = 2.348 \times 10^{-3} > \varepsilon_c$$

$$\alpha = \frac{1.23 \times 10^{-3}}{2.348 \times 10^{-3}} - \frac{1}{3} \times \left(\frac{1.23 \times 10^{-3}}{2.348 \times 10^{-3}}\right)^2 = 0.43$$

$$\varepsilon_s = \frac{2.5606 \times 10^{-3} \times (211.1 - 80)}{(246.5 - 80)} = 2.016 \times 10^{-3}$$

$$f_s = 200 \times 2.016 \times 10^{-3} \times 1000 = 403\,\text{MPa} < f_y \qquad \text{O.K.}$$

$$A_s = 2 \times \frac{\pi}{4} \times (16)^2 = 402\,\text{mm}^2$$

$$f_{fe} = \sigma_{fudel,\max} = 364.9\,\text{MPa}$$

$$A_f = 2 \times \frac{\pi}{4} \times (10)^2 = 157\,\text{mm}^2$$

$$c = \frac{402 \times 403 + 364.9 \times 157}{0.43 \times 31 \times 178} = 92.4\,\text{mm} > 80\text{ mm, N.G.}$$

Let $c = 90$ mm

$$\varepsilon_{fe} = 0.003\left(\frac{d_f - c}{c}\right) - \varepsilon_{bi} = 0.003\left(\frac{246.5-90}{90}\right) - 7.51\times10^{-5} = 5.138\times10^{-3}$$

$f_{fec} = 147\times10^3 \times 5.138\times10^{-3} = 755\,\text{MPa}$ >364.4 MPa ($\sigma_{fudel,max}$). Delamination governs design

Thus,

$$\varepsilon_{fe} = \frac{364.9}{147\times10^3} = 2.4823\times10^{-3}$$

Strain in concrete,

$$\varepsilon_c = \frac{2.4833\times10^{-3} + 7.51\times10^{-5}}{246.5-90} = 1.4725\times10^{-3}$$

$$\varepsilon_o = \frac{2\times31}{26.4\times1000} = 2.348\times10^{-3} > \varepsilon_c$$

$$\alpha = \frac{1.4725\times10^{-3}}{2.348\times10^{-3}} - \frac{1}{3}\times\left(\frac{1.4725\times10^{-3}}{2.348\times10^{-3}}\right)^2 = 0.42$$

$$\varepsilon_s = \frac{2.5606\times10^{-3}\times(211.1-90)}{(246.5-90)} = 1.98\times10^{-3}$$

$$f_s = 200\times1.98\times10^{-3}\times1000 = 396\,\text{MPa} < f_y \qquad \text{O.K.}$$

$$A_s = 2\times\frac{\pi}{4}\times(16)^2 = 402\,\text{mm}^2$$

$$f_{fe} = \sigma_{fudel,\max} = 364.9\,\text{MPa}$$

$$A_f = 2\times\frac{\pi}{4}\times(10)^2 = 157\,\text{mm}^2$$

$$c = \frac{396\times402 + 364.9\times157}{0.42\times31\times178} = 93.0\,\text{mm} \cong 90\,\text{mm}$$

So take, $c = 90$ mm.

Step 5: Compute the depth of neutral axis at failure

$$\beta_1 = \frac{\frac{1}{3} - \frac{\varepsilon_c}{12\varepsilon_o}}{1 - \frac{\varepsilon_c}{3\varepsilon_o}} = \frac{\frac{1}{3} - \frac{1.4725\times10^{-3}}{12\times2.348\times10^{-3}}}{1 - \frac{1.4756\times10^{-3}}{3\times2.348\times10^{-3}}} = 0.36$$

$$\gamma_c = \beta_1 c = 0.36\times90 = 32.4\,\text{mm}$$

$$M_n = A_s f_s (d - \gamma_c) + \psi_f A_f f_{fe} (d_f - \gamma_c)$$
$$= 402 \times 396 \times (211.1 - 32.4) + 0.85 \times 157 \times 364.9 \times (246.5 - 32.4)$$
$$= 38.87 \text{ kN-m}$$

$$\phi M_n = 0.7 \times 38.87 = 27 \text{ kN-m}$$

Required Bending Strength, $M_u = 1.2M_D + 1.6M_L = 26.18 \text{ kN-m} < \varphi M_n = 27 \text{ kN-m}$, O.K.

Step 6: Check for Serviceability

a. *Check for stress in steel bars at service load:*

Unfactored dead load moment = 0.91 kN-m

$$\text{Unfactored live load moment} = \frac{58}{1.6 \times 2}\left(\frac{2.54 - 0.81}{2}\right) = 15.67 \text{ kN-m}$$

$$\text{Modular ratio, } m_f = \frac{E_f}{E_c} = \frac{147}{26.4} = 5.6$$

$$\text{Modular ratio, } m_s = \frac{E_s}{E_c} = \frac{200}{26.4} = 7.6$$

Let the neutral axis depth is, $c = kd$

Equating first moment of transformed area of tensile reinforcement and that of compression zone about the neutral axis, we have:

$$\frac{bc^2}{2} = m_s A_s (d - c) + m_f A_f (d_f - c) \Rightarrow \frac{178c^2}{2} = 7.6 \times 402 \times (211.1 - c) + 5.6 \times 157 \times (246.5 - c)$$
$$\Rightarrow c^2 + 44.21c - 9682.48 = 0$$
$$\Rightarrow c = 78.7 \text{ mm} \Rightarrow kd = 78.7 \text{ mm}$$

Stress in steel at service load,

$$f_{s,s} = \frac{\left[M_s + \varepsilon_{bi} A_f E_f \left(d_f - \frac{kd}{3}\right)\right](d - kd)E_s}{A_s E_s \left(d - \frac{kd}{3}\right)(d - kd) + A_f E_f \left(d_f - \frac{kd}{3}\right)(d_f - kd)}$$

$$= \frac{\left[16.59 \times 10^6 + 7.51 \times 10^{-5} \times 157 \times 147 \times 10^3 \left(246.5 - \frac{78.7}{3}\right)\right](211.1 - 78.7) \times 200 \times 10^3}{402 \times 200 \times 10^3 \left(211.1 - \frac{78.7}{3}\right)(211.1 - 78.7) + 157 \times 147 \times 10^3 \times \left(246.5 - \frac{78.7}{3}\right)(246.5 - 78.7)}$$

$$= \frac{4.2905 \times 10^{14}}{2.820915344 \times 10^{12}} = 159.3 \text{ MPa}$$

Thus, $f_{s,s}$ = 159.3 MPa < 0.8 f_y = 0.8 × 414 = 331.2 MPa, O.K.

This condition ensures that steel has not experienced inelastic deformation.

b. *Check for creep-rupture:*

Assuming that the full dead load and live load are sustained loads.

$$M_s = 16.59 \times 10^6 \text{ N-mm}$$

Stress in FRP bars at sustained load,

$$f_{f,s} = \frac{E_f f_{s,s}(d_f - kd)}{E_s(d - kd)} - \varepsilon_{bi} E_f = \frac{147 \times 10^3 \times 159.3 \times (246.5 - 78.7)}{200 \times 10^3 \times (211.1 - 78.7)} - 7.51 \times 10^{-5} \times 147 \times 10^3 = 137.4 \text{MPa}$$

Design strength of FRP bars, $f_{fd} = 2717$ MPa

Creep-rupture stress limits, $F_{f,s} = 0.55 \times 2717 = 1494.4$ MPa

Thus, $f_{f,s} = 128.96 \text{MPa} < F_{f,s} = 1494.4 \text{MPa}$, O.K.

c. *Check for crack width:*

Effective concrete cover, $d_c = h - d = 254 - 211.1 = 42.9$ mm

Effective tension area, $A_e = \dfrac{2 d_c b}{2} = \dfrac{2 \times 42.9 \times 178}{2} = 7636.2 \text{mm}^2$

Stress in steel reinforcement at service load, $f_{s,s} = 159.3$ MPa

Strain gradient, $\beta = \dfrac{h - kd}{d - kd} = \dfrac{254 - 78.7}{211.1 - 78.7} = 1.32$

Crack spacing,

$$w = 1.1 \times 10^{-5} \beta f_{s,s} \sqrt[3]{d_c A_e} = 1.1 \times 10^{-5} \times 1.32 \times 159.3 \times \sqrt[3]{42.9 \times 7636.2} = 0.16 \text{mm} < 0.4 \text{mm}, \text{ O.K.}$$

d. *Check for Long-term Deflection of Beam:*

Groove size = $1.5 d_b = 1.5 \times 10 = 15$ mm

Taking 15 mm by 15 mm groove.

$$I_g = \frac{bh^3}{12} = \frac{178 \times 254^3}{12} = 243074782.7 \text{ mm}^4$$

$$f_r = 0.62\sqrt{f_c'} = 0.62\sqrt{31} = 3.45 \text{MPa}$$

$$M_{cr} = \frac{2 f_r I_g}{h} = \frac{2 \times 3.45 \times 243074782.7}{254} = 6603212.6 \text{ N-mm} \cong 6.6 \text{kN-m}$$

$$I_{cr} = \frac{bd^3k^3}{3} + m_f A_f (d_f - kd)^2 + m_s A_s (d - kd)^2$$

$$= \frac{bc^3}{3} + m_f A_f (d_f - c)^2 + m_s A_s (d - c)^2$$

$$= \frac{178 \times 78.7^3}{3} + 5.6 \times 157 \times (246.5 - 78.7)^2 + 7.6 \times 402 \times (211.1 - 78.7)^2$$

$$= 107\,234\,058.4\,\text{mm}^4$$

$$I_e = \left[\left(\frac{M_{cr}}{M_a}\right)^3 I_g + \left[1 - \left(\frac{M_{cr}}{M_a}\right)^3 I_{cr}\right]\right] \le I_g$$

$$= \left[\left(\frac{6.6}{15.67}\right)^3 \times 243\,074\,782.7 + \left[1 - \left(\frac{6.6}{15.67}\right)^3\right] \times 107\,234\,058.4\right]$$

$$= 117\,383\,795.8\,\text{mm}^4 < I_g$$

$$\Delta_{LT} = 0.6\xi(\Delta)_s = 0.6 \times 2(\Delta)_s = 1.2(\Delta)_s$$

Deflection due to dead load,

$$\Delta_{iDL} = \frac{5w_D l^4}{384E_c I_e} = \frac{5}{384}\frac{1.13(\text{N/mm}) \times (2540\,\text{mm})^4}{26.4 \times 10^3 (\text{N/mm}^2) \times 117\,383\,795.8\,\text{mm}^4} = 0.198\,\text{mm}$$

$$E_c I_e \Delta_{iL} = -\frac{P}{12}x^3 + \left(\frac{PL_1^2}{4} + \frac{PL_1L_2}{4}\right)x + \frac{P}{12}(x - L_1)^3$$

where,

$$x = 0.865 + \frac{0.81}{2} = 1.27\,\text{m} = 1270\,\text{mm}$$

$$P = 36.3\,\text{kN} = 36\,300\,\text{N}$$

$$y_{\text{midspan}} = -\frac{-\frac{36\,300}{12} \times (1270)^3 + \frac{36\,300 \times 865}{4}(865 + 810) \times 1270 + \frac{36\,300}{12}(1270 - 865)^3}{26.4 \times 10^3 \times 117\,383\,795.8} = 3.45\,\text{mm}$$

$$\Delta_s = (\Delta_i)_{DL} + (\Delta_i)_{LL} = 0.198 + 3.45\,\text{mm} = 3.65\,\text{mm}$$

$$\Delta_{LT} = 1.2 \times 3.65 = 4.38\,\text{mm}$$

$$\frac{L}{240} = \frac{2540}{240} = 10.6\,\text{mm}$$

Thus, $\Delta_{LT} = 4.38\,\text{mm} < \dfrac{L}{240}$ O.K.

Required development length,

$$L_d = \frac{d_b f_{fd}}{4\tau_b} = \frac{10 \times 2717}{4 \times 6.895} = 985\,\text{mm}$$

Step 7: Design for shear

$$l = 2.54\,\text{m}$$

Factored dead load, $w_D = 1.2 \times 1.13 = 1.356\,\text{kN/m}$

$$V_u = \frac{w_D l}{2} - w_D d + \frac{W_L}{2 \times 1.6} = \frac{1.356 \times 2.54}{2} - 1.356 \times 0.211 + \frac{58}{2 \times 1.6} = 30.43\,\text{kN}$$

Shear strength contribution by concrete,

$$V_c = \frac{\sqrt{f_c'}}{6} bd = \frac{\sqrt{31}}{6} \times \frac{178 \times 211.1}{1000} = 34.87\,\text{kN}$$

Assuming 2-legged 9.5 mm diameter steel stirrups @ 1000 mm c/c, the shear contribution of steel stirrups is given by,

Effective stress in steel stirrups, $f_{sv} = 0.002E_s = 0.002 \times 200 \times 10^3 = 400\,\text{MPa}$

$$A_{sv} = 2 \times \frac{\pi}{4} \times 9.5^2 = 141.76\,\text{mm}^2$$

$$V_s = \frac{A_{sv} f_{sv} d}{s} = \frac{141.76 \times 400 \times 211.1}{1000 \times 1000} = 11.97\,\text{kN}$$

Let spacing of CFRP NSM shear reinforcement,

$$s = 50\text{ mm} < \frac{d_{\text{net}}}{2} = \frac{239 - 2 \times 34.9}{2} = 169.2\,\text{mm}$$

$$\tau_b = 6.895\,\text{MPa}$$

$$d_b = 10\,\text{mm}$$

$$l_{0.004} = \frac{0.001 d_b E_b}{\tau_b} = \frac{0.001 \times 10 \times 147 \times 1000}{6.895} = 213.2\,\text{mm}$$

$$n = \frac{169.2}{50} = 3.38 \approx 3 \quad \text{(rounded down)}$$

$$\frac{n}{2} = \frac{3}{2} = 1.5 \approx 1 \quad \text{(rounded down)}$$

$$L_1 = \min(l_{0.004}, s.1) = \min(213.2, 50 \times 1) = 50\,\text{mm}$$

$$L_2 = \min(l_{0.004}, d_{\text{net}} - 2s) = \min(213.2,\ 169.2 - 2 \times 50) = 69.2\,\text{mm}$$

$$L_3 = \min(l_{0.004}, d_{\text{net}} - 3s) = \min(213.2, 169.2 - 3 \times 50) = 19.2\,\text{mm}$$

$$L_{\text{tot}} = 50 + 69.2 + 19.2 = 138.4\,\text{mm}$$

The shear contribution of CFRP reinforcement is given by,

$$V_f = \frac{2\pi d_b \tau_b L_{tot}}{1000} = \frac{2\pi \times 10 \times 6.895 \times 138.4}{1000} = 60\,\text{kN}$$

$$V_n = V_c + V_s + \psi V_f = (34.87 + 11.97 + 0.85 \times 60) = 97.84\,\text{kN}$$

Design shear strength, $\phi V_n = \phi_s V_n = 0.85 \times 97.84 = 83.164\,\text{kN} > V_u = 30.43\,\text{kN}$ O.K.

So shear reinforcement is sufficient.

E5.13. Design Example 13

Problem

A circular reinforced concrete column with a factored axial resistance of 3110 kN is to be strengthened with a carbon FRP wrap. The axial live load requirement for the column, P_L, is 1550 kN, and the axial dead load requirement, P_D, is 1200 kN, leading to a factored axial load, P_f, of 4200 kN. Determine the specifics (# of layers) of the FRP wrap required if the column dimensions, reinforcement details, and material properties are as follows:

Unsupported column length, $l_u = 3000$ mm
Column diameter, $D_g = 500$ mm
Column gross cross-sectional area, $A_g = 196\,350$ mm^2
Area of longitudinal reinforcing steel, $A_{st} = 2500$ mm^2
Steel yield strength, $f_y = 400$ MPa
Concrete compressive strength, $f'_c = 30$ MPa
FRP ultimate strength, $f_{\text{frp}u} = 1200$ MPa
FRP thickness, $t_{\text{frp}} = 0.3$ mm
FRP resistance factor, $f_{\text{frp}} = 0.75$

Solution

1. Check if the column remains short once strengthened. Using the increased factored load on the column, $P_f = 4200$ kN:

$$\frac{\ell_u}{D_g} \le \frac{6.25}{\sqrt{P_f / f'_c A_g}}$$

$$\rightarrow \frac{3000}{500} \leq \frac{6.25}{\sqrt{4200000/(30 \times 196350)}}$$

$$\rightarrow 6 \leq 7.4 \rightarrow \text{O.K.} \quad \text{(so column is short)}$$

2. Compute the confined concrete strength required for the specified load by rearranging the terms in Eq. (5.105) with P_f substituted for $P_{r\max}$:

$$P_{r\max} = k_e[\alpha_1 \phi_c f'_{cc}(A_g - A_s) + \phi_s f_y A_s] \Rightarrow f'_{cc} = \frac{\left(\dfrac{P_f}{k_e} - \phi_s f_y A_s\right)}{\alpha_1 \phi_c \left(A_g - A_s\right)}$$

$$\alpha_1 = 0.85 - 0.0015 f'_c \geq 0.67 \Rightarrow \alpha_1 = 0.85 - 0.0015 \times 30 = 0.81$$

Thus, the required confined concrete strength to achieve the required axial strength increase is:

$$f'_{cc} = \frac{\left(\dfrac{4.2 \times 10^6}{0.85} - 0.85(400)(2500)\right)}{0.81(0.6)(196350 - 2500)}$$

$$= 43.4 \text{ MPa}$$

3. Compute the volumetric strength ratio, ω_w, based on the required strength:

$$\omega_w = \frac{\left(\dfrac{f'_{cc}}{f'_c} - 1\right)}{\alpha_{pc}} = \frac{\left(\dfrac{43.4}{30} - 1\right)}{1} = 0.447$$

4. Compute the required confinement pressure by rearranging (Eq. 5.100):

$$f_{\ell\text{frp}} = \frac{\omega_w \phi_c f'_c}{2} = \frac{0.447(0.6)(30)}{2} = 4.02 \text{ MPa} \quad \rightarrow \quad \text{O.K.}$$

$$f_{\ell\text{frp}} > 4 \text{ MPa} \quad \text{(minimum confinement pressure)}$$

Check the maximum confinement pressure using (Eq. 5.103):

$$f_{\ell\text{frp}} \leq \frac{f'_c}{2\alpha_{pc}}\left(\frac{1}{k_e} - \phi_c\right)$$

$$= \frac{30}{2(1.0)}\left(\frac{1}{0.85} - 0.6\right) = 8.65 \quad \rightarrow \quad \text{O.K.}$$

5. Compute the required number of FRP layers by rearranging (Eq. 5.97):

$$N_b = \frac{f_{\ell \mathrm{frp}} D_g}{2\phi_{\mathrm{frp}} f_{\mathrm{frp}u} t_{\mathrm{frp}}}$$

$$= \frac{4.02(500)}{2(0.75)(1200)(0.3)} = 3.72 \rightarrow \text{Take a 4 layers}$$

6. Compute the factored axial strength of the FRP-wrapped column using Eqs. (5.97), (5.99) (5.100), and 5.104:

$$f_{\ell \mathrm{frp}} = \frac{2N_b \phi_{\mathrm{frp}} f_{\mathrm{frp}u} t_{\mathrm{frp}}}{D_g} = 4.32 \text{ MPa}$$

$$\omega_w = \frac{2 f_{\ell \mathrm{frp}}}{\phi_c f_c'} = 0.48$$

$$f_{cc}' = f_c'\left(1 + \alpha_{pc}\omega_w\right) = 44.4 \text{ MPa}$$

$$P_{r\max} = k_e\left[\alpha_1 \phi_c f_{cc}'\left(A_g - A_c\right) + \phi_s f_y A_s\right] = 4230 \text{ kN}$$

which is greater than $P_f = 4200$ kN → O.K.

Thus, 4 layers of the carbon FRP wrap are adequate to sufficiently increase the axial strength of the column. The reader should note that there are a number of additional checks which must be performed in the actual design of an FRP-wrapped concrete column, including checks for creep and fatigue limits.

E5.14. Design Example 14

Problem

Calculate the factored moment capacity of the beam shown below as per ISIS Canada Design approach. The section is singly-reinforced and is strengthened in flexure with externally-bonded carbon FRP on its soffit. Neglect initial strains in the concrete and steel. Beam dimensions, and reinforcement details are shown in Fig. E5.14, and material properties are as follows:

Concrete strength, $f_c' = 45$ MPa

Internal steel reinforcement: 3 × 10M bars

$A_s = 300 \text{ mm}^2$

Steel yield strength, $f_y = 400$ MPa

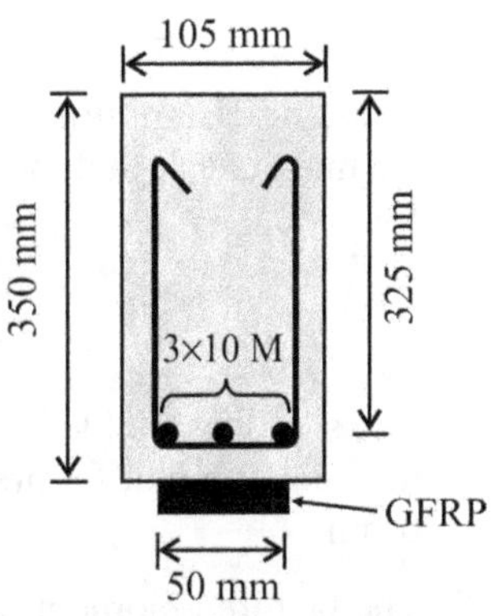

Figure E5.14. Beam dimensions and reinforcement details

Steel elastic modulus, $E_s = 200$ GPa

Carbon FRP: $A_{\text{frp}} = 60$ mm^2; $\varepsilon_{\text{frp}u} = 1.55\%$;

$E_{\text{frp}} = 155$ GPa

Solution

1. Following the ISIS Canada Design procedure for analysis described earlier in this chapter, we assume failure of the beam due to crushing of the concrete in compression, after yielding of the internal steel reinforcement.

2. Determine stress block parameters α_1 and β_1.

$$\alpha_1 = 0.85 - 0.0015 f_c' \geq 0.67 \Rightarrow \alpha_1 = 0.85 - 0.0015 \times 45 = 0.78$$

$$\beta_1 = 0.97 - 0.0025 f_c' \geq 0.67 \Rightarrow \beta_1 = 0.97 - 0.0025 \times 45 = 0.86$$

3. Using equations of equilibrium, find the depth of the neutral axis, c:

$$\phi_c \alpha_1 f_c' \beta_1 bc = \phi_s f_y A_s + \phi_{\text{frp}} E_{\text{frp}} A_{\text{frp}} \varepsilon_{\text{frp}}$$

$$(0.6(0.78)(45)(0.86)(105)c) = (0.85(400)(300)) + \left(0.75(155000)(60)\left(0.0035\left(\frac{350 - c}{c}\right)\right)\right)$$

where, $\varepsilon_{\text{frp}} = \varepsilon_{cu} \dfrac{h - c}{c}$ from strain compatibility. Solving gives $c = 90.5$ mm.

4. Calculate the strain in the FRP, ε_{frp} and compare it to the FRP failure strain, $\varepsilon_{\text{frp}u}$. If $\varepsilon_{\text{frp}} < \varepsilon_{\text{frp}u}$ the assumed failure mode (concrete in compression) is correct and the moment resistance can be determined (step 5). If $\varepsilon_{\text{frp}} > \varepsilon_{\text{frp}u}$, then the assumption was incorrect (and we would have to use step 6 of the procedure for analysis). For this analysis, strain compatibility gives:

$$\varepsilon_{\text{frp}} = \varepsilon_{cu} \frac{h - c}{c} = 0.0035\left(\frac{350 - 90.5}{90.5}\right) = 0.01 < \varepsilon_{\text{frp}u} = 0.0155$$

Thus, the assumption of failure by concrete crushing was correct. To ensure adequate ductility, we must check that the reinforcing steel has yielded. Again, using strain compatibility:

$$\varepsilon_s = \varepsilon_{cu} \frac{d - c}{c} = 0.0035\left(\frac{325 - 90.5}{90.5}\right) = 0.0091 > \varepsilon_y = 0.002$$

The steel has yielded. Thus, the amount of FRP is appropriate. If the steel had not reached its yield strain at failure, the amount of FRP could be reduced to avoid brittle failure of the beam.

5. The factored moment resistance is obtained using the following expression:

$$M_r = \phi_s f_y A_s\left(d - \frac{a}{2}\right) + \phi_{\text{frp}} E_{\text{frp}} A_{\text{frp}} \varepsilon_{\text{frp}}\left(h - \frac{a}{2}\right)$$

$$= 0.85(400)(300)\left(325 - \frac{0.86(90.5)}{2}\right) + 0.75(155000)(60)(0.01)\left(350 - \frac{0.86(90.5)}{2}\right)$$

$$= 50.9 \times 10^6 \text{ N·mm} = 50.9 \text{ kN·m}$$

E5.15. Design Example 15: Shear Strengthening as per ISIS Canada Design Approach

Calculate the factored shear capacity of the beam shown in Fig. E5.15. The section is singly-reinforced and is shear strengthened with externally-bonded glass FRP (GFRP) in the form of vertical U-wraps with a finite width (FRP strips as opposed to a continuous sheet). Beam dimensions, reinforcement details, and material properties are as follows:

Span of the beam, $L = 3.0$ m
Concrete strength, $f_c' = 45$ MPa
Tension steel: 3×10 M bars; $f_y = 400$ MPa; $d = 325$ mm
Stirrup steel: 4.76 mm Ø; $f_y = 400$ MPa; $A_v = 36\text{ mm}^2$
Steel stirrup spacing, $s_s = 225$ mm c/c
GFRP stirrup thickness, $t_{\text{frp}} = 1.3$ mm
GFRP stirrup width, $w_{\text{frp}} = 100$ mm
GFRP stirrup spacing, $s_{\text{frp}} = 200$ mm
GFRP ultimate strain, $\varepsilon_{\text{frp}u} = 2.0\%$
GFRP elastic modulus, $E_{\text{frp}} = 22.7$ GPa
Assume normal density concrete, $\lambda = 1.0$

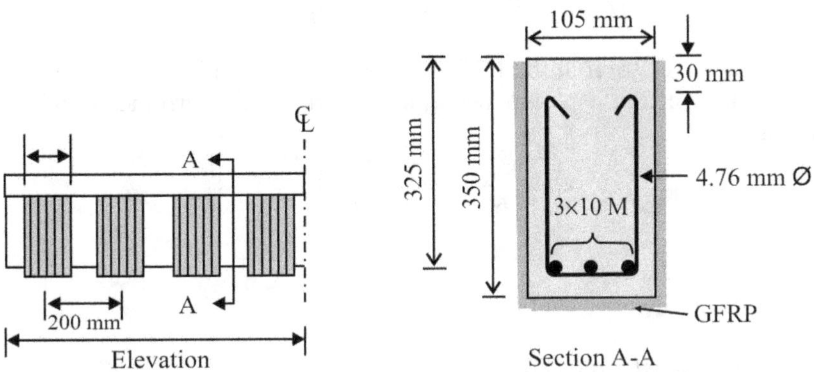

Figure E5.15. Beam dimensions and reinforcement details.

Solution

1. Since the total shear resistance is calculated as the sum of the contributions from the steel, V_s, concrete, V_c, and FRP, V_{frp},

   Shear Contribution of the concrete:

$$V_c = 0.2\phi_c\sqrt{f_c'}b_w d$$
$$= 0.2(0.6)\sqrt{45}(105)(325)$$
$$= 27\,470 \text{ N} = 27.47 \text{ kN}$$

   Shear contribution of steel:

$$V_s = \frac{\phi_s f_y A_v d}{s}$$
$$= \frac{0.85(400)(36)(325)}{225} = 17\,680 \text{ N} = 17.7 \text{ kN}$$

2. To determine the FRP contribution, we must first compute the FRP effective strain. First we find the area of FRP shear reinforcement, A_{frp}:

$$A_{\text{frp}} = 2t_{\text{frp}}w_{\text{frp}} = 2(1.3)(100) = 260 \text{ mm}^2$$

   Next, the FRP shear reinforcement ratio is calculated using the following expression:

$$\rho_{\text{frp}} = \left(\frac{2\cdot t_{\text{frp}}}{b_w}\right)\left(\frac{w_{\text{frp}}}{s_{\text{frp}}}\right)$$
$$= \left(\frac{2(1.3)}{105}\right)\left(\frac{100}{200}\right) = 0.0124 = 1.24\%$$

   and the effective anchorage length is determined as given below:

$$L_e = \frac{25350}{(t_{\text{frp}}E_{\text{frp}})^{0.58}} = \frac{25350}{(1.3\times 22\,700)^{0.58}} = 64.8 \text{ mm}$$

3. Calculate the effective strain in the FRP, $\varepsilon_{\text{frpe}}$. Before calculating the effective strain in the FRP shear reinforcement, the parameters k_1 and k_2 are to be determined to take into account the debonding:

$$k_1 = \left(\frac{f_c'}{27.65}\right)^{2/3} = \left[\frac{45}{27.65}\right]^{2/3} = 1.38$$

$$k_2 = \frac{d_{\text{frp}} - n_e L_e}{d_{\text{frp}}} = \frac{325 - 1(64.8)}{325} = 0.80$$

   Here, $n_e = 1$ as U-shaped wraps are used.

So the effective strain in the FRP is given by,

$$\varepsilon_{\text{frp}e} = \frac{\alpha k_1 k_2 L_e}{9525}$$

$$= \frac{0.8(1.38)(0.80)(64.8)}{9525} = 0.0060$$

4. Determine the second limit on effective strain in the FRP shear reinforcement. The effective-to-ultimate strain ratio, R, is determined as follows:

$$R = \alpha\lambda_1 \left(\frac{f_c'^{2/3}}{\rho_{\text{frp}} E_{\text{frp}}} \right)^{\lambda_2}$$

$$= 0.8(1.23)\left(\frac{45^{2/3}}{0.0124 \times 22\,700} \right)^{0.47} = 0.229$$

where the values of λ_1 and λ_2 have been taken as those of aramid FRP in this example. Thus:

$$\varepsilon_{\text{frp}e} = R\varepsilon_{\text{frp}u} = 0.229(0.02) = 0.0046$$

For design purposes, we use the smallest limiting value of the effective strain in the FRP. In this example, we take the smallest of 0.0040, 0.0046, or 0.0060. Thus, $\varepsilon_{\text{frp}e} = 0.004$.

5. The contribution of the FRP to the shear capacity can be calculated as per the following expression:

$$V_{\text{frp}} = \frac{\phi_{\text{frp}} A_{\text{frp}} E_{\text{frp}} \varepsilon_{\text{frp}e} d_{\text{frp}} (\sin\beta + \cos\beta)}{s_{\text{frp}}}$$

$$= \frac{0.5(260)(22\,700)(0.004)(325(1+0))}{200}$$

$$= 19180 \text{ N} = 19.2 \text{ kN}$$

6. The total shear resistance of the beam is given as a summation of the three contributions:

$$V_r = V_c + V_s + V_{\text{frp}}$$

$$= 27.5 + 17.7 + 19.2 = 64.4 \text{ kN}$$

7. Finally, we must check the limits for maximum shear strengthening,

$$V_r \le V_c + 0.8\lambda\phi_c\sqrt{f_c'}b_w d$$

$$64\,400 \le 27\,500 + 0.8(1.0)(0.6)\sqrt{45}(105)(325)$$

$$= 137\,400 \rightarrow \text{ O.K.}$$

and maximum band spacing:

$$s_{\text{frp}} \le w_{\text{frp}} + \frac{d}{4} \rightarrow 200 \text{ mm} \le 100 + \left(\frac{325}{4}\right) = 181 \text{ mm}$$, N.G. Hence use 180 mm spacing.

Exercise Problems

1. Figure P5.1 shows the cross section details of an interior RC beam located in an office building. The beam is 3 m long and resting on column supports at its both ends with an effective span of 2.8 m. The beam is reinforced with three 16-mm diameter main longitudinal Steel bars and 10 mm diameter steel shear stirrups at c/c spacing of 150 mm. Two 12-mm diameter steel hanger bars are provided at the top of section to support the stirrups. Note these two hanger bars are not provided to increase the strength and stiffness of the RC beam and hence their contributions to structural strength and stiffness are neglected. In addition, the beam is reinforced on its soffit by two layers of CFRP strips as shown in Fig. P5.1. The cross-section dimensions of RC beam are 350 mm height and 250 mm width. The effective concrete cover to the main longitudinal CFRP Leadline bars is 50 mm. The material properties of various constituent materials are as follows:

Concrete

28-day characteristics strength, $f'_c = 31$ MPa

Modulus of elasticity, $E_c = 26352$ MPa

Crushing strain, $\varepsilon_u = 0.003$

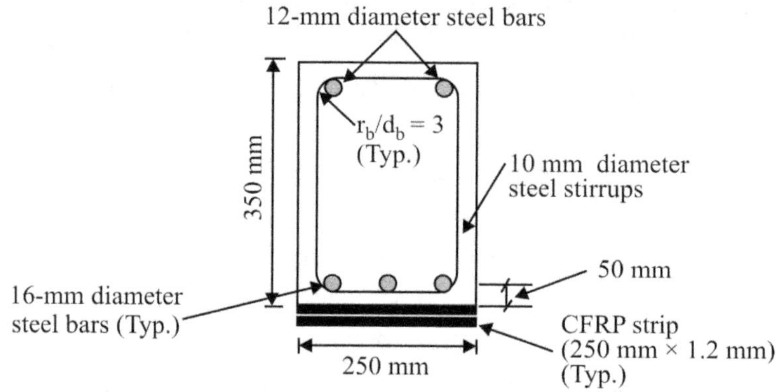

Figure P5.1(a). Cross-section details of rectangular simply supported beam.

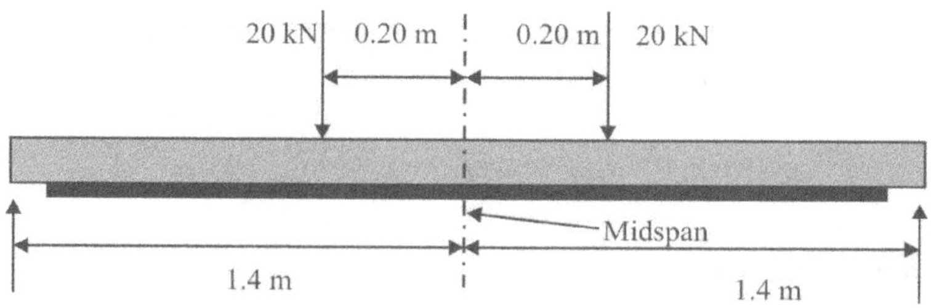

Figure P5.1(b). Longitudinal view of rectangular simply supported beam.

CFRP Strips

Properties specified by manufacturers:

Thickness of each strip= 1.2 mm

Tensile strength, $f^*_{fu} = 2399$ MPa

Rupture strain, $\varepsilon^*_{fu} = 0.014$ mm/mm

Modulus of Elasticity, $E_f = 149$ GPa

Steel main bars and stirrups

Yield strength = 414 MPa

Young's modulus = 200 GPa

Follow the appropriate design standards for FRP reinforced concrete structures such as ACI 440.2R-02 and evaluate the following design parameters for the RC beam based on the above given facts. Assume perfect bond between two layers of CFRP strips. Also, suitably assume the missing data if any.

2. A deficient simply supported RC beam is to be strengthened to carry a uniformly distributed load of 5.8 kN/m (sustained) in addition to its self-weight. The beam is to be used for an interior application. The total span of beam is 2.74 m. The beam is resting on two columns having width of 20 cm along the length of the beam. The beam is initially reinforced with two Grade 60 #5 steel bars (16 mm diameter) longitudinally. The shear reinforcement is provided with #3 (9.5 mm diameter) two-legged steel stirrups of the same material properties as longitudinal reinforcement. Design a suitable near surface mounted GFRP/epoxy system for NSM strengthening. The specified tensile strength of GFRP bar is 620.6 MPa, while specified rupture strain of GFRP bars is 0.014. Modulus of elasticity of GFRP bar is 44 800 MPa. The cubic and cylindrical strengths of concrete are 39 and 31 MPa. Take bond strength of steel and CFRP bars as 13.79 MPa and 6.895 MPa, respectively.

3. As per ISIS Canada design standards, calculate the factored moment capacity of the beam shown below. The section is singly-reinforced and is strengthened in flexure with externally-bonded carbon FRP on its soffit. Beam dimensions, and reinforcement details are shown in Fig. P5.2, and material properties are as follows:

   Concrete strength, $f'_c = 48$ MPa

   Internal steel reinforcement: 3×10 M bars

   $A_s = 300$ mm²

   Steel yield strength, $f_y = 400$ MPa

   Steel elastic modulus, $E_s = 200$ GPa

   Carbon FRP: $A_{\text{frp}} = 60$ mm²; $\varepsilon_{\text{frp}u} = 1.6\%$; $E_{\text{frp}} = 150$ GPa

   Neglect initial strains in the concrete and steel

   Use CSA A23.3 requirements

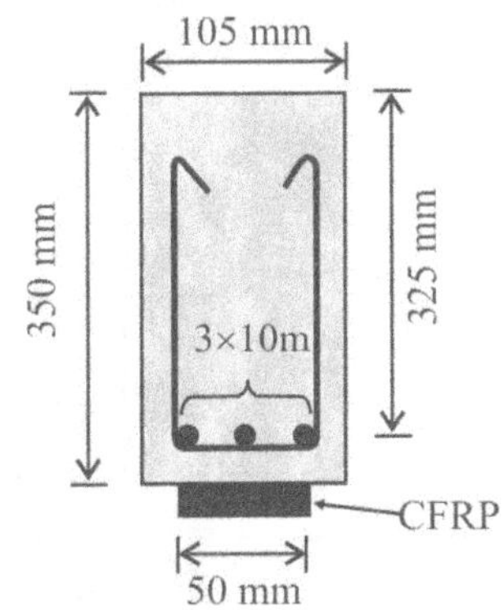

Figure P5.2.

Calculate the factored moment capacity of the beam shown below. The section is singly-reinforced and is strengthened in flexure with externally-bonded carbon FRP on its soffit. Beam dimensions, and reinforcement details are shown in Fig. 4.9, and material properties are as follows:

Concrete strength, $f'_c = 45$ MPa

Internal steel reinforcement: 3 × 10 M bars

$A_s = 300$ mm²; Steel yield strength, $f_y = 400$ MPa

Steel elastic modulus, $E_s = 200$ GPa

Carbon FRP: $A_{frp} = 60$ mm²; $\varepsilon_{frpu} = 1.6\%$; $E_{frp} = 150$ GPa

Neglect initial strains in the concrete and steel.

4. Calculate the factored shear capacity of the beam shown in Fig. P5.3. The section is singly-reinforced and is shear strengthened with externally-bonded glass FRP (GFRP) in the form of vertical U-wraps with a finite width (FRP strips as opposed to a continuous sheet). Beam dimensions, reinforcement details, and material properties are as follows:

Span of the beam, $L = 3.0$ m
Concrete strength, $f'_c = 45$ MPa
Tension steel: 3×10 M bars
$f_y = 400$ MPa; $d = 325$ mm
Stirrup steel: 4.76 mm Ø
$f_y = 400$ MPa; $A_v = 36$ mm^2
Stirrup spacing, $s_s = 225$ mm c/c
GFRP stirrup thickness, $t_{frp} = 1.3$ mm
GFRP stirrup width, $w_{frp} = 100$ mm
GFRP stirrup spacing, $s_{frp} = 150$ mm
GFRP ultimate strain, $\varepsilon_{frpu} = 2.0\%$
GFRP Elastic modulus, $E_{frp} = 30$ GPa

Use CSA A23.3-94 and ISIS Canada Design Standards.

Assume normal density concrete, $\lambda = 1.0$

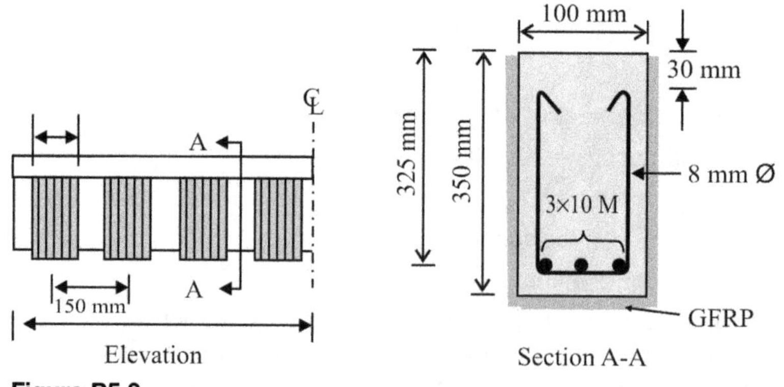

Figure P5.3.

5. A circular reinforced concrete column with a factored axial resistance of 3000 kN is to be strengthened with a carbon FRP wrap. The axial live load requirement for the column, P_L, is 1600 kN, and the axial dead load requirement, P_D, is 1100 kN, leading to a factored axial load, P_f, of 4050 kN. Determine the specifics (# of layers) of the FRP wrap required if the column dimensions, reinforcement details, and material properties are as follows:

   Unsupported column length, $l_u = 3000$ mm
   Column diameter, $D_g = 500$ mm
   Column gross cross-sectional area, $A_g = 196\,350$ mm^2
   Area of longitudinal reinforcing steel, $A_{st} = 2500$ mm^2
   Steel yield strength, $f_y = 400$ MPa
   Concrete compressive strength, $f'_c = 30$ MPa
   FRP ultimate strength, $f_{\text{frp}u} = 1200$ MPa
   FRP thickness, $t_{\text{frp}} = 0.3$ mm
   FRP resistance factor, $\phi_{\text{frp}} = 0.75$

CHAPTER 6

Durability-Based Design Approach for External FRP Strengthening of RC Beams

6.1. Designing of Reinforced Concrete Beams

The following steps can be taken for the process of designing of reinforced concrete (RC) beams to be externally strengthened with fiber reinforced polymer (FRP) plates or fabric sheets for retrofitting and/or repairing applications. Figure 6.1 shows the strain and stress distributions along the depth of cross-section and forces in the steel reinforcements and concrete for an FRP-strengthened beam of rectangular cross-section.

6.1.1. Compute Design Material Properties

Design ultimate tensile strength of FRP plate,

$$f_{fd} = C_E . f_{fu}$$

Design rupture strain of FRP plate,

$$\varepsilon_{fd} = C_E \varepsilon_{fu}$$

where, C_E = environmental tensile strength reduction factor
f_{fu} = specified tensile strength of FRP
ε_{fu} = specified rupture strain of FRP

The values of C_E for carbon fiber reinforced polymer (CFRP)/epoxy system are recommended to be 0.85 and 0.95 for enclosed and unenclosed exposure conditions (ACI 2000), respectively. However, in case the FRP strengthened structural system is exposed to aggressive environmental condition, the nominal FRP-contributed strength of the structure should be further reduced, depending upon the short and long term exposures. It should be noted that the development of design equations, presented in the following sections, is based on the following assumptions: (i) The plane section before bending remains plane after bending; (ii) The ultimate strain capacity of concrete is 0.003 mm/mm (iii) Tensile strength of concrete is ignored; (iv) FRP materials behave linearly elastic up to failure; and (v) Perfect bond exists between the concrete and FRP.

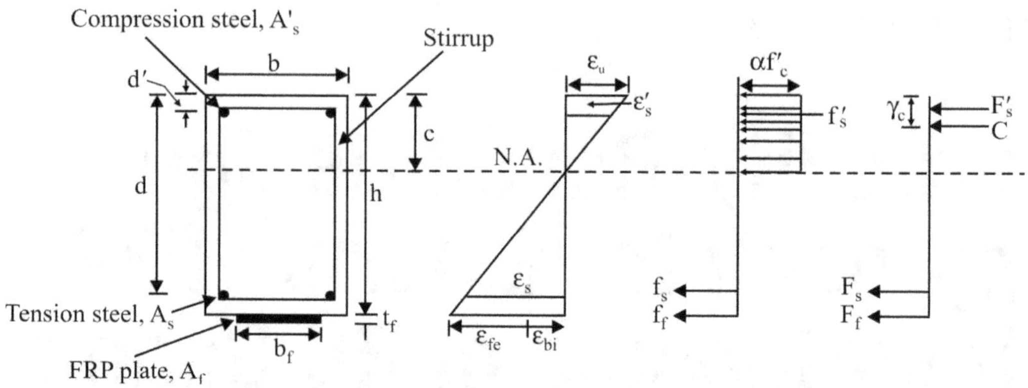

Figure 6.1. Stress, strain, and force diagrams across depth of beam cross-section.

6.1.2. Compute the Existing Substrate Strain

If the unfactored service moment acting on the RC beam before upgrading (strengthening/repairing) is defined as M_o, then strain at the bottom face of the concrete, ε_{bi}, where the composite plate is to be bonded can be obtained by considering linear variation of strain along the depth of the beam.

$$\varepsilon_{bi} = \frac{M_o(h - c_o)}{E_c I_{tro}} \tag{6.1}$$

where, c_o = depth of the neutral axis of unstrengthened beam
E_c = modulus of elasticity of concrete
I_{tro} = moment of inertia of the transformed cracked section of unstrengthened beam

6.1.3. Compute the Balanced Plate Ratio ($\rho_{f,b}$)

Balanced plate ratio gives the maximum cross-section area of the plate to assure yielding of the tensile reinforcement and crushing of the concrete simultaneously. The balanced plate ratio should be calculated depending on whether compression steel yields or not.

$$\text{Critical compression depth, } d_c = \frac{\varepsilon_u - \varepsilon_y}{\varepsilon_u + \varepsilon_y} d$$

Table 6.1. CFRP strength reduction factors (ψ) for different environmental conditionings.

Environmental conditioning	CFRP plate		CFRP fabric	
	Short term	Long term	Short term	Long term
100% Relative humidity	0.82	0.68	0.94	0.90
Dry-heat conditioning	0.87	0.92	1.0	1.0
Alkaline conditioning	1.0	1.0	0.95	0.90
Freeze–thaw conditioning	0.87	0.91	0.94	0.87
Salt-water solution	1.0	0.95	0.95	0.93

where, ε_u = ultimate compression strain of concrete
ε_y = yield strain of steel
d' = distance from the extreme compression fiber to the centroid of compression steel

(a) Compression steel yields ($d' \leq d_c$): The balanced plate ratio corresponding to yielding of compression steel at balanced condition can be computed using Eq. (6.2). This equation is obtained using equilibrium and compatibility conditions, which takes into account the Hognestad's parabola of idealized stress–strain curve (Park and Paulay, 1975) for concrete.

$$\rho_{f,b} = \frac{\alpha f_c' k_1 + (\rho' - \rho) f_y}{\varepsilon_{fe} E_f} \tag{6.2}$$

where, k_1 = neutral axis depth ratio (c/d)

$$= \frac{\varepsilon_u}{\varepsilon_u + \varepsilon_y}$$

$$\varepsilon_{fe} = \frac{(h - k_1 d)}{k_1 d}\varepsilon_u - \varepsilon_{bi} \leq k_m \varepsilon_{fd} \tag{6.3}$$

c = depth to the neutral axis from the extreme compression fiber
d = distance from the extreme compression fiber to the centroid of tension steel
E_f = modulus of elasticity of FRP plate
f'_c = compressive strength of concrete
α = Mean stress factor (Ann *et al.*, 1991)

$$= 1 + \frac{\varepsilon_u}{\varepsilon_o}\left(1 - \frac{\varepsilon_u}{3\varepsilon_o} - \frac{\varepsilon_o^2}{\varepsilon_u^2}\right) - \left(\frac{0.15}{0.004 - \varepsilon_o}\right)\left(\frac{\varepsilon_u}{2} - \varepsilon_o\right) - \left(\frac{0.075}{0.004 - \varepsilon_o}\right)\left(\frac{\varepsilon_o^2}{\varepsilon_u}\right) \quad \text{for } \varepsilon_o \leq \varepsilon_u$$

$$= \frac{\varepsilon_u}{\varepsilon_o} - \frac{\varepsilon_u^2}{3\varepsilon_o^2} \quad \text{for } \varepsilon_o \geq \varepsilon_u$$

where $\varepsilon_o = \dfrac{2f_c'}{E_c}$

$$\rho = \frac{A_s}{bd} \tag{6.4}$$

$$\rho' = \frac{A_s'}{bd} \tag{6.5}$$

$$\rho_f = \frac{A_f}{bd} \tag{6.6}$$

A_s = cross-section area of tensile steel
A'_s = cross-section area of compression steel
A_f = cross-section area of FRP plate
k_m = delamination factor (REPLARK SYSTEM, 2000)

The value of factor k_m can be evaluated as per Eq. (6.7), based on SI units:

$$k_m = 1 - \frac{nE_f t_f}{428\,000} \leq 0.90 \quad \text{for } nE_f t_f \text{ (N/mm)} \leq 214\,000 \tag{6.7a}$$

$$= \frac{107\,000}{nE_f t_f} \leq 0.90 \quad \text{for } nE_f t_f \text{ (N/mm)} \geq 214\,000 \tag{6.7b}$$

(b) Compression steel does not yield ($d_c < d'$): In case the yielding of compression steel at balanced condition does not occur, Eq. (6.8) can be used to compute balanced plate ratio ($\rho_{f,b}$).

$$\rho_{f,b} = \frac{\dfrac{(k_1 d - d')}{k_1 d}\varepsilon_u E_s \rho' + \alpha f_c' k_1 - \rho f_y}{\varepsilon_{fe} E_f} \tag{6.8}$$

6.1.4. Compute the Maximum Allowable Plate Ratio ($\rho_{f,\max}$)

Based on the factor (ACI 318), the allowable plate ratio, $\rho_{f,\max}$, can be computed using Eq. (6.9), as follows:

$$\rho_{f,\max} = 0.75\rho_{f,b} \tag{6.9}$$

6.1.5. Proportion of the FRP Plate

Choose a plate with appropriate dimensions and specified material properties such that plate ratio remains below allowable plate ratio. Here, specified material properties refer to the property of material provided by manufacturers.

6.1.6. Compute the Balanced Plate Ratio ($\rho_{f,bb}$) to Determine Failure Modes

The balanced plate ratio, $\rho_{f,bb}$ refers to the condition at which the maximum compressive stress in the concrete and maximum effective tensile stress in the composite plate reach simultaneously. This can be used to characterize the two primary modes of failure such as crushing of concrete and plate failure. Here, plate failure refers to the failure of the beam by the onset of delamination and/or plate rupture. It must be noted that the occurrence of these two primary beam failure modes are based on the assumption that the premature beam failure due to the onset of delamination or concrete cover, delamination may reduce the maximum effective stress in the FRP plate at the ultimate failure of beam. The balanced plate ratio, $\rho_{f,bb}$, is expressed by Eqs. (6.10) and (6.11).

$$\rho_{f,bb} = \frac{\alpha f_c' k_2 \dfrac{h}{d} + \varepsilon_s' E_s \rho' - \rho f_y}{k_m f_{fd}} \quad \text{for } \rho_f \leq \rho_{f,\max} \tag{6.10}$$

$$\rho_{f,bb} = \frac{\alpha f_c' k_2 \dfrac{h}{d} + \varepsilon_s' E_s \rho' - \varepsilon_s E_s \rho}{k_m f_{fd}} \quad \text{for } \rho_f > \rho_{f,\max} \tag{6.11}$$

$$k_2 = \frac{\varepsilon_u}{\varepsilon_u + k_m \varepsilon_{fd} + \varepsilon_{bi}} \tag{6.12}$$

$$\varepsilon_s' = \left(1 - \frac{d'}{k_2 d}\right) \varepsilon_u \leq \varepsilon_y \tag{6.13}$$

$$\varepsilon_s = \left(\frac{1 - k_2}{k_2}\right) \varepsilon_u \leq \varepsilon_y \tag{6.14}$$

6.1.7. Determine the Critical Plate Ratio ($\rho_{f,c}$)

This ratio is determined to ensure yielding of compression steel at ultimate beam failure by the rupture of plate or by the concrete crushing.

(a) Rupture of plate: The critical plate ratio corresponding to the rupture of the plate is defined as, $\rho_{f,cf}$ and is expressed by Eq. (6.15) as follows:

$$\rho_{f,cf} = \frac{\alpha f_c' \frac{c}{d} + (\rho' - \rho) f_y}{k_m f_{fd}} \tag{6.15}$$

where,

$$c = \frac{\varepsilon_y h + k_m \varepsilon_{fd} d'}{k_m \varepsilon_{fd} + \varepsilon_y} \tag{6.16}$$

(b) Crushing of concrete: The critical plate ratio corresponding to the crushing of concrete is defined as $\rho_{f,cc}$ and is expressed by Eq. (6.17), as follows:

$$\rho_{f,cc} = \frac{\alpha f_c' k_3 \frac{d'}{d} + (\rho' - \rho) f_y}{E_f \varepsilon_{fe}} \tag{6.17}$$

where,

$$k_3 = \frac{\varepsilon_u}{\varepsilon_u - \varepsilon_y} \tag{6.18}$$

$$\varepsilon_{fe} = \frac{\varepsilon_u (h - k_3 d')}{k_3 d'} - \varepsilon_{bi} \leq k_m \varepsilon_{fd} \tag{6.19}$$

6.1.8. Determine the Mode of Failure

The following criteria are used to determine the failure modes of strengthened beams to compute their nominal moment capacity. These failure criteria are based on the assumption that premature failure due to the onset of delamination and concrete cover, delamination is prevented by taking into account the factor, k_m in design equations:

a. Plate failure and yielding of tension and compression steel for $\rho_f \leq \rho_{f,bb}$ and $\rho_f \geq \rho_{f,cf}$
b. Plate failure and yielding of tension steel but no yielding of compression steel for $\rho_f \leq \rho_{f,bb}$ and $\rho_f \leq \rho_{f,cf}$
c. Crushing of concrete and yielding of tension and compression steel for $\rho_f \geq \rho_{f,bb}$ and $\rho_f \geq \rho_{f,cc}$
d. Crushing of concrete and yielding of tension steel but no yielding of compression steel for $\rho_f \geq \rho_{f,bb}$ and $\rho_f \leq \rho_{f,cc}$

6.1.9. Nominal Moment Capacity of Strengthened Beams

Based on the aforementioned failure modes, the nominal flexural capacity of the strengthened beam, M_n is calculated using one of the following equations:

(a) Plate failure and yielding of tension and compression steels:

$$A'_s f_y(\gamma_c - d') + A_s f_y(d - \gamma_c) + \psi k_m A_f f_{fd}(h - \gamma_c) \tag{6.20}$$

where,

$$\gamma_c = \beta_1 c \tag{6.21}$$

$$\beta_1 = \frac{\frac{1}{3} - \frac{\varepsilon_c}{12\varepsilon_o}}{1 - \frac{\varepsilon_c}{3\varepsilon_o}} \quad \text{if } 0 \leq \varepsilon_c \leq \varepsilon_o \quad \text{(An } et\ al.\text{, 1991)} \tag{6.22a}$$

$$= 1 - \frac{\varepsilon_c^3 - 5.1\varepsilon_o\varepsilon_c^2 - 0.004\varepsilon_o^2 + 0.024\varepsilon_c^2}{\varepsilon_c(3.925\varepsilon_o^2 - 10.2\varepsilon_o\varepsilon_c - 0.9\varepsilon_c^2 - 0.016\varepsilon_o + 0.048\varepsilon_c)} \quad \text{if } \varepsilon_o \leq \varepsilon_c \leq \varepsilon_u \tag{6.22b}$$

$$\varepsilon_c = \frac{(k_m\varepsilon_{fd} + \varepsilon_{bi})c}{h - c} \tag{6.23}$$

ψ = FRP strength reduction factor to take into account the short and long term aggressive environmental exposure

$$c = \frac{k_m A_{fd} f_{fd} + A_s f_y - A_s' f_y}{\alpha f_c' b} \tag{6.24}$$

$$\alpha = \frac{\varepsilon_c}{\varepsilon_o} - \frac{\varepsilon_c^2}{3\varepsilon_o^2} \quad \text{for } 0 \leq \varepsilon_c \leq \varepsilon_o \tag{6.25a}$$

$$= 1 + \frac{\varepsilon_c}{\varepsilon_o}\left(1 - \frac{\varepsilon_c}{3\varepsilon_o} - \frac{\varepsilon_o^2}{\varepsilon_c^2}\right) - \left(\frac{0.15}{0.004 - \varepsilon_o}\right)\left(\frac{\varepsilon_c}{2} - \varepsilon_o\right) - \left(\frac{0.075}{0.004 - \varepsilon_o}\right)\left(\frac{\varepsilon_o^2}{\varepsilon_c}\right) \quad \text{for } \varepsilon_o \leq \varepsilon_c \leq \varepsilon_u \tag{6.25b}$$

The value of c has to be obtained using iterative procedure to ensure equilibrium of forces in steel, FRP plate and concrete.

(b) Plate failure and yielding of tensile steel but no yielding of compression steel: Using Eq. (6.23), obtain the value of concrete strain, ε_c, at the extreme compression fiber. For a particular value of, ε_c, the value of α can be obtained using Eq. (6.25). Recalculate the value of c using Eq. (6.26).

$$c = \frac{k_m A_{fd} f_{fd} + A_s f_y - \varepsilon_s' E_s A_s'}{\alpha f_c' b} \tag{6.26}$$

where,

$$\varepsilon_s' = \frac{(k_m \varepsilon_{fd} + \varepsilon_{bi})(c - d')}{h - c} \leq \varepsilon_y \tag{6.27}$$

If assumed value of c and that obtained by Eq. (6.26) are the same, the nominal moment capacity can be obtained using Eq. (6.28).

$$M_n = A'_s f'_s(\gamma_c - d') + A_s f_y(d - \gamma_c) + \psi k_m A_f f_{fd}(h - \gamma_c) \tag{6.28}$$

$$f'_s = E_s \varepsilon'_s \leq f_y \tag{6.29}$$

(c) Crushing of concrete and yielding of tensile and compression steels: For this mode of failure, nominal moment capacity of strengthened beam can be determined by Eq. (6.30). The location of the centroid of concrete compression force, γ_c and α can be determined using Eq. (6.21) and Eq. (6.25), respectively, for ε_c being equal to ε_u.

$$M_n = A'_s f_y(\gamma_c - d') + A_s f_y(d - \gamma_c) + \psi A_f \varepsilon_{fe} E_f(h - \gamma_c) \tag{6.30}$$

where,

$$\varepsilon_{fe} = \frac{(h - c)\varepsilon_u}{c} - \varepsilon_{bi} \leq k_m \varepsilon_{fd} \tag{6.31}$$

$$c = \frac{A_f E_f \varepsilon_{fe} + (A_s - A_s') f_y}{\alpha f_c' b} \tag{6.32}$$

(d) Crushing of concrete and yielding of tensile steel but no yielding of compression steel: For this mode of failure, nominal moment capacity of the strengthened beam can be obtained using Eq. (6.33). The value of α and γ_c can be determined by substituting the value of ε_c, which is equal to crushing strain of concrete, ε_u, in Eqs. (6.21) and (6.25), respectively.

$$M_n = A'_s f'_s(\gamma_c - d') + A_s f_y(d - \gamma_c) + \psi A_f \varepsilon_{fe} E_f(h - \gamma_c) \tag{6.33}$$

The neutral axis depth, c can be expressed by Eq. (6.34) using equation of equilibrium.

$$c = \frac{A_f \varepsilon_{fe} E_f + A_s f_y - A_s' f_s'}{\alpha f_c' b} \tag{6.34}$$

$$\varepsilon_s' = \frac{(c - d')\varepsilon_u}{c} \leq \varepsilon_y \tag{6.35}$$

$$f_s' = E_s \varepsilon_s' \tag{6.36}$$

The value of effective plate strain, ε_{fe}, can be computed by Eq. (6.31).

6.1.10. Compute Design Moment Capacity

The design moment capacity, M_d, of an FRP strengthened beam can be determined using strength reduction factor, ϕ, which takes into account the loss of ductility, which may be caused due to externally bonding the FRP plates to the concrete surface. The strength reduction factor, ϕ, can be determined as per Eq. (6.38) (ACI 318).

$$M_d = \phi M_n \tag{6.37}$$

$$\phi = 0.90 \qquad \text{for } \varepsilon_s \geq 0.005 \tag{6.38a}$$

$$= 0.70 + 0.20\frac{(\varepsilon_s - \varepsilon_y)}{(0.005 - \varepsilon_y)} \qquad \text{for } \varepsilon_y \leq \varepsilon_s \leq 0.005 \tag{6.38b}$$

$$= 0.70 \qquad \text{for } \varepsilon_s \leq \varepsilon_y \tag{6.38c}$$

It should be noted that the design moment, M_d, should be greater than or equal to the required moment capacity, M_u, of the strengthened beam. The required moment capacity of the strengthened beam is expressed by Eq. (6.39).

$$M_u = 1.2M_D + 1.6M_L \tag{6.39}$$

where, M_D is the dead load moment and M_L is the live load moment.

6.1.11. Allowable Service Stresses

In addition to determining the ultimate strength of the FRP upgraded RC beam, the stresses in the concrete, steel and FRP plate must be determined at service load condition. The service load stresses resulting from all the sustained dead and live loads should be checked against allowable stresses. Table 6.2 lists the allowable service load stresses for the constituent materials. The following expressions can be used to determine the stresses in concrete, steel, and FRP plate under sustained service load condition.

Table 6.2. Allowable service load stresses.

Material	Allowable stress
Tension steel	$0.80f_y$
Compression steel	$0.40f_y$
CFRP plate or fabric	$0.55f_{fd}$

$$f_{s,s} = \frac{\left[M_s + \varepsilon_{bi} A_f E_f \left(h - \frac{kd}{3}\right)\right](d-kd)E_s}{A_s E_s \left(d - \frac{kd}{3}\right)(d-kd) + A_s' E_s \left(\frac{kd}{3} - d'\right)(kd - d') + A_f E_f \left(h - \frac{kd}{3}\right)(h-kd)} \tag{6.40}$$

$$f_{c,s} = f_{s,s}\left(\frac{E_c}{E_s}\right)\frac{kd}{(d-kd)} \tag{6.41}$$

$$f_{fe,s} = f_{s,s}\left(\frac{E_f}{E_s}\right)\left(\frac{h-kd}{d-kd}\right) - \varepsilon_{bi} E_f \tag{6.42}$$

$$f_{s,s}' = f_{s,s}\left(\frac{kd-d'}{h-kd}\right) \tag{6.43}$$

where, k = neutral axis depth coefficient at service load condition.

It should be noted that as the service load stress for the FRP plate under sustained loads (dead load) plus sustained portion of the live load is less than the allowable value, the FRP will have adequate safety against failure due to creep rupture.

E6.1. Design Example 1: A Case Study Problem

Problem statement

A deficient simply supported RC beam is to be strengthened to safely carry a midspan load of 93.5 kN in addition to its self-weight on a long-term basis under aggressive humidity conditions (hot water 40°C). Figure E6.1 shows the cross-section details of the beam. The total span of the beam is 2.743 m, while center-to-center distance between supports is 2.54 m. The beam is internally reinforced with 2 Grade 60 steel bars (yield strength = 415 MPa) of 9.5 mm and 16 mm diameter in compression and tension, respectively. Use of CFRP plate is recommended for external strengthening. Assuming that material being added through the placement in the enclosed space using wet lay-up process and adhesive bonding of pre-fabricated sections with ambient cure. Design a suitable strengthening system using CFRP plates and epoxy. The material properties of CFRP plate and epoxy are given in Table E6.1.

Solution

The design steps for the CFRP strengthened beam are explained as follows.

Step 1: Compute the design material properties

Design ultimate tensile strength of CFRP plate, $f_{fd} = C_E f_{fu}$

$= 0.95 \times 2400 = 2280$ MPa

Design rupture strain of CFRP plate, $\varepsilon_{fd} = 0.95 \times 0.014 = 0.013$

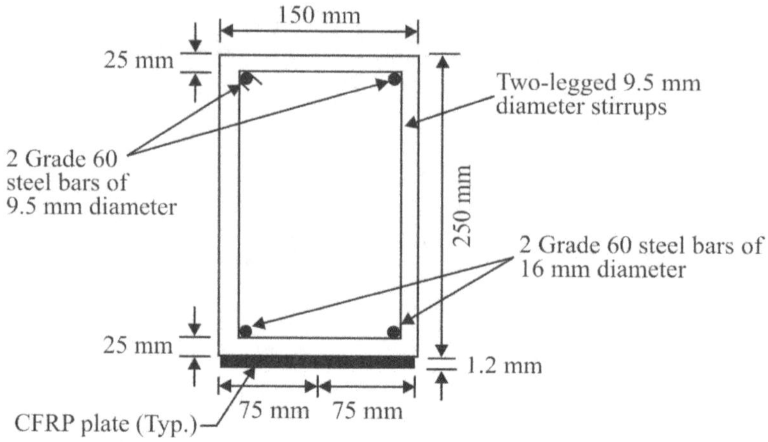

Figure E6.1. Cross-section details of CFRP strengthened beam.

Table E6.1: Mechanical properties of strengthening materials.

Type	Property			
	CFRP plates (Sika CarboDur 50)	CFRP fabric (SikaWrap Hex 103C)	Epoxy for plates (SikaDur30)	Epoxy for fabric (SikaDur Hex 300/306)
Tensile Strength, MPa	2400	960	24.8	72.3
Modulus of Elasticity, GPa	149	73.1	4.46	3.16
Failure Strain (%)	1.4	1.3	1.0	4.8
Shear Strength, MPa	—	—	24.8	123.4
Dimensions, mm	75 wide by 1.2 thick	1.0	—	—

Step 2: Compute the existing substrate strain, ε_{bi}

Dead load of the beam $W_d = 0.150 \times 0.250 \times 25 = 0.94$ kN-m

Unfactored service moment before strengthening $= W_d l^2/8$

$$= \frac{0.94 \times 2.54^2}{8} = 0.76 \text{ kN-m}$$

$f'_c = 31$ MPa

$E_c = 26.4$ GPa

Modular ratio, $m = \dfrac{200}{26.4} = 7.6$

Let the neutral axis depth of unstrengthened beam is c_o. Equating the first moment of tension and compression area (based on concrete) about the neutral axis results in Eq. (6.44).

$$\frac{150\times c_o^2}{2}+7.6\times141.76\times(c_o-25)=7.6\times402.1\times(225-c_o)$$

$$c_o^2+55.11c_o-9527=0 \tag{6.44}$$

From the solution of Eq. (6.44),

$$c_o = 73.9 \text{ mm}$$

Transformed moment of inertia of elastic cracked unstrengthened beam section,

$$I_{tro} = 92.53 \times 10^6 \text{ mm}^4$$

Existing substrate strain,

$$\begin{aligned}\varepsilon_{bi} &= \frac{M_o\,(h-c_o)}{E_c I_{tro}}\\ &= \frac{0.76\times10^6\times(250-73.9)}{26.4\times10^3\times92.53\times10^6}\\ &= 5.5\times10^{-5}\end{aligned}$$

Step 3: Compute the balanced plate ratio, $\rho_{f,b}$

$$\varepsilon_u = 0.003$$

$$\varepsilon_y = \frac{415}{2\times10^5} = 2.1\times10^{-3}$$

Critical compression depth,

$$d_c = \frac{\varepsilon_u-\varepsilon_y}{\varepsilon_u+\varepsilon_y}$$

$$\begin{aligned}d &= \frac{0.003-2.1\times10^{-3}}{0.003+2.1\times10^{-3}}\times225\\ &= 39.71 \text{ mm}\end{aligned}$$

$d' = 25$ mm $< d_c \Rightarrow$ compression steel yields at the balanced condition.
Thus, balanced plate ratio, $\rho_{f,b}$ is given by Eq. (6.2).

$$nE_f t_f = 1\times149\times10^3\times1.2 = 178\,800 < 214\,000$$

$$k_m = 1-\frac{nE_f t_f}{428000} = 1-\frac{178800}{428000} = 0.582 \le 0.90 \qquad \text{OK}$$

$$k_m = 1-\frac{1019900}{2\,400\,000} = 0.575$$

$$k_m\varepsilon_{fd} = 0.582\times0.013 = 7.57\times10^{-3}$$

$$k_1 = \frac{\varepsilon_u}{\varepsilon_u + \varepsilon_y}$$

$$= \frac{0.003}{0.003 + 2.1\times10^{-3}} = 0.588$$

$$\rho' = \frac{141.94}{150\times225} = 4.21\times10^{-3}$$

$$\rho = \frac{402}{150\times225} = 0.012$$

$$\varepsilon_{fe} = \frac{(h - k_1 d)}{k_1 d}\varepsilon_u - \varepsilon_{bi}$$

$$= \frac{(250 - 0.588\times225)}{0.588\times225}\times0.003 - 5.5\times10^{-5}$$

$$= 2.61\times10^{-3} \le k_m \varepsilon_{fd}$$

$$\varepsilon_o = \frac{2f_c'}{E_c} = \frac{2\times31}{26.4\times10^3} = 2.35\times10^{-3} < \varepsilon_u$$

From Eq. (6.25b), $\alpha = 0.92$

$$\rho_{f,b} = \frac{0.92\times31\times0.588 + (4.21\times10^{-3} - 0.012)\times415}{2.61\times10^{-3}\times149\times10^3}$$

$$= 0.035$$

Step 4: Compute the maximum allowable plate ratio, $\rho_{f,max}$

$$\rho_{f,\max} = 0.75\times0.035$$
$$= 2.625\times10^{-2}$$

Step 5: Proportion the CFRP plate

Choosing two plates of 75 × 1.2 mm dimensions and bonding them side by side along the width of beam cross-section.

Total width of bonded plates, $b_f = 150$ mm

Total thickness of plate, $t_f = 1.2$ mm

Plate ratio, $\rho_r = \dfrac{150\times1.2}{150\times225} = 5.3\times10^{-3} < \rho_{f,max}$ O.K.

Step 6: Compute the balanced plate ratio, $\rho_{f,bb}$ to determine the failure modes

$$k_2 = \frac{\varepsilon_u}{\varepsilon_u + k_m \varepsilon_{fd} + \varepsilon_{bi}} = \frac{0.003}{0.003 + 7.57\times10^{-3} + 5.5\times10^{-5}} = 0.2824$$

$$\varepsilon'_s = \varepsilon_u\left(1 - \frac{d'}{k_2 d}\right) \le \varepsilon_y$$

$$\varepsilon'_s = \left(1 - \frac{25}{0.2824\times225}\right)\times0.003 = 1.82\times10^{-3} < \varepsilon_y$$

$$\rho_{f,bb} = \frac{0.92\times31\times0.2824\times\dfrac{250}{225} + (1.82\times10^{-3}\times4.21\times10^{-3} - 2.1\times10^{-3}\times0.012)\times2\times10^{5}}{0.575\times2280}$$

$$= 4.15\times10^{-3} < \rho_f$$

Since the actual plate area is more than the balanced plate area (at ultimate condition), the crushing of concrete will govern the design.

Step 7: Compute the critical plate ratio, $\rho_{f,cc}$

$$k_3 = \frac{0.003}{(0.003 - 2.1\times10^{-3})} = 3.33$$

$$\varepsilon_{fe} = \frac{0.003(250 - 3.33\times25)}{3.33\times25} - 5.5\times10^{-5}$$

$$= 5.95\times10^{-3} < k_m\varepsilon_{fd} \qquad \text{O.K.}$$

$$\rho_{f,cc} = \frac{0.92\times31\times3.33\times\dfrac{25}{225} + (4.21\times10^{-3} - 0.012)\times415}{149\times10^{3}\times5.95\times10^{-3}}$$

$$= 8.26\times10^{-3} > \rho_f$$

Since $\rho_{f,cc}$ is greater than ρ_f, the condition of yielding of compression steel at ultimate condition is not satisfied. Thus, nominal moment capacity of strengthened beam will be based on crushing of concrete, yielding of tensile steel, and unyielding of compression steel.

Step 8: Compute nominal moment capacity of strengthened beam

The value of α corresponding to the crushing of concrete is 0.92.

$$\varepsilon_o = 2.35\times10^{-3}$$

From Eq. (6.22b), $\beta_1 = 0.2157$

$$\gamma_c = \beta_1 c = 0.2157c$$

where,

$$c = \frac{A_f \varepsilon_{fe} E_f + A_s f_s - A_s' f_s'}{\alpha f_c' b} \tag{6.45}$$

$A_f = 180\ \text{mm}^2$
$A_s = 402\ \text{mm}^2$
$A'_f = 142\ \text{mm}^2$

$$\varepsilon_s' = \left(\frac{c - d'}{c}\right) \times \varepsilon_u = \frac{c - 25}{c} \times 0.003$$

$f_s = f_y = 415$ MPa

$$f_s' = \frac{2 \times 10^5 \times 0.003 \times (c - 25)}{c} = \frac{600(c - 25)}{c}$$

$$\varepsilon_{fe} = \frac{0.003(h - c)}{c} - \varepsilon_{bi} \le k_m \varepsilon_{fd}$$

$$= \frac{0.003(250 - c)}{c} - 5.5 \times 10^{-5} \le k_m \varepsilon_{fd}$$

Assume $c = 71$ mm

$f'_s = 388.7\ \text{MPa} < f_y$ OK

$\varepsilon_{fe} = 7.51 \times 10^{-3} < k_m \varepsilon_{fd}$ OK

From the equilibrium Eq. (6.45), value of $c = 72$ mm ≅ assumed value of c. Thus, take $c = 71$ mm.

$$\gamma_c = 0.2157 \times 71 = 15.3\ \text{mm}$$

Strength reduction factor, ψ (Table 6.2) for hot water conditioning (100% humidity) = 0.68
Nominal moment capacity of strengthened beam, M_n is given by Eq. (6.46).

$$M_n = A'_s f'_s (d' - \gamma_c) + A_s f_s (d - \gamma_c) + \psi A_f \varepsilon_{fe} (h - \gamma_c) \tag{6.46}$$

$$= 55\,195.4(25 - 15.3) + 166\,830(225 - 15.3) + 0.68 \times 201\,418.2(250 - 15.3)$$

$$= 67\,665\,185.43\ \text{N-mm} = 67.7\ \text{kN-m}$$

Required moment capacity of the beam, M_u can be determined using Eq. (6.39).

$$M_u = 1.2 M_D + 1.6 M_L$$

$$= 1.2 \times 0.76 + \frac{93.5 \times 2.54}{4} = 60.3\ \text{kN-m}$$

Step 8: Compute design moment capacity of strengthened beam

Strain in tensile steel at ultimate load of strengthened beam,

$$\varepsilon_s = \frac{0.003(225 - 71)}{71}$$

$$= 0.0065 > 0.005$$

From Eq. (6.37), $\phi = 0.90$

$$M_d = \phi M_n = 0.9 \times M_n = 60.93 \text{ kN-m} > M_u \qquad \text{O.K.}$$

Thus the beam strengthened with 2 CFRP plates is capable of supporting the 93.5 kN of load at ultimate under aggressive hot-water condition with 100% humidity.

Experimental moment carrying capacity of strengthened beam = 60.13 kN-m

% difference in analytical and experimental moment capacity of the strengthened beam

$$= \frac{(60.93 - 60.13)}{60.13} \times 100$$
$$= 1.3\%$$

Step 9: Check for stresses under sustained service load condition

In the present problem, sustained service load is only self-weight of the beam.

Thus, moment due to sustained service load, $M_s = 0.76$ kN-m

The stresses in steel, CFRP plate, and concrete under service load condition can be determined using Eqs. (6.40)–(6.43).

$$\text{Modular ratio, } m_f = \frac{149}{26.4} = 5.64$$

$$\text{Modular ratio, } m = 7.6$$

Let the neutral axis depth of the strengthened beam corresponding to the service load condition is c. Equating the first moment of the compression and tension area (based on concrete) about the neutral axis results in the following quadratic equation:

$$c^2 + 68.64c - 12\,908.73 = 0 \tag{6.47}$$

From the solution of Eq. (6.47),

$$c = 84.4 \text{ mm}$$

$$f_{s,s} = \frac{\left[0.76\times10^6 + 5.5\times10^{-5}\times180\times149\times10^3\times\left(250 - \frac{84.4}{3}\right)\right](250-84.4)\times2\times10^5}{402\times2\times10^5\times\left(225-\frac{84.4}{3}\right)(225-84.4)+142\times2\times10^5\left(\frac{84.4}{3}-25\right)(84.4-25)+180\times149\times10^3\left(250-\frac{84.4}{3}\right)(250-84.4)}$$

$$= 11.20 \text{ MPa} < 0.8f_y \qquad \text{O.K.}$$

$$f_{c,s} = 11.20 \times \left(\frac{1}{7.6}\right)\times\frac{84.4}{(225-84.4)} = 0.9 \text{ MPa} < 0.45f'_c \qquad \text{O.K.}$$

$$f_{fe} = 11.20\times\left(\frac{149}{200}\right)\left(\frac{250-84.4}{225-84.4}\right) - 5.5\times10^{-5}\times149\,000 = 5.0 \text{ MPa} < 0.55f_{fd} \qquad \text{O.K.}$$

$$f'_s = 11.20 \times \frac{84.4 - 25}{(225 - 84.4)} = 4.73 \text{ MPa} < 0.4 f_y \qquad \text{O.K.}$$

Since the service load stresses under the sustained self-weight of the beam are under allowable limits, the CFRP plates will have adequate safety against failure due to creep rupture.

For the sake of completeness, the notations used in the above case study problem are given as follows:

A_f = cross-sectional area of FRP plate, mm^2
A_s = cross-sectional area of tension steel reinforcement, mm^2
A'_s = cross-sectional area of compression reinforcement, mm^2
b = width of the beam, mm
b_f = total width of plates, mm
c = depth of the neutral axis of strengthened beam from the extreme compression fiber, mm
C = resultant compressive force in concrete, kN
c_o = depth of the neutral axis of unstrengthened beam from the extreme compression fiber, mm
C_E = environmental tensile strength reduction factor for FRP materials
d,d' = depths of centroid of tensile and compression steel reinforcements from the extreme compression fiber, respectively, mm
d_c = critical compression depth, mm
E_c = modulus of elasticity of concrete, MPa
E_f = modulus of elasticity of FRP material, MPa
E_s = modulus of elasticity of steel, MPa
f'_c = characteristic strength of concrete, MPa
$f_{c,s}$ = maximum compressive stress in concrete under sustained service loads, MPa
f_f = stress in FRP plate at ultimate condition, MPa
F_f = resultant force in FRP plate at ultimate condition, kN
f_{fe} = effective stress in FRP plate, MPa
f_{fu} = specified tensile strength of FRP plate, MPa
$f_{s,s}$ = maximum stress in tensile steel under sustained service loads, MPa
f_s = stress in tensile steel at ultimate condition, MPa
F_s = resultant force in tensile steel at ultimate condition, kN
$f'_{s,s}$ = maximum stress in compression steel under sustained service loads, MPa
f'_s = stress in compressive steel at ultimate condition, MPa
F'_s = resultant force in compressive steel at ultimate condition, kN
f_y = yield stress of steel, MPa
h = overall depth of the beam, mm
I_{tro} = moment of inertia of the transformed cracked section of unstrengthened beam, mm^4
k = neutral axis depth factor at the service load condition
k_1,k_2 = neutral axis depth factors at balanced and ultimate conditions, respectively.
k_3 = neutral axis depth factor corresponding to yielding of compression steel at the crushing of concrete

k_m = bond-dependent co-efficient for flexure
m = modular ratio of steel and concrete
m_f = modular ratio of FRP plate and concrete
M_d = design moment capacity of strengthened beam, kN-m
M_D = unfactored maximum bending moment due to dead load, kN-m
M_L = unfactored maximum bending moment due to live load, kN-m
M_n = nominal moment capacity of strengthened beam, kN-m
M_o = unfactored service moment acting on RC beam before upgrading, kN-m
M_s = unfactored moment due to sustained portion of service loads, kN-m
M_u = ultimate moment capacity of strengthened beam, kN-m
n = number of layers of FRP plates
t_f = thickness of FRP plate, mm
W_d = Dead load of the beam, kN
α = mean stress factor
ε_{bi} = strain level in the concrete substrate at the time of FRP installation
ε_c = concrete strain at the extreme compression fiber at a particular load
ε_{fd} = design strain for FRP plate
ε_{fe} = effective design strain for FRP plate
ε_{fu} = specified rupture strain for FRP plate
ε_o = concrete strain corresponding to the maximum concrete stress
ε_s = strain in tensile steel
ε'_s = strain in compressive steel
ε_u = ultimate concrete strain
ε_y = yield strain of steel
β_1 = ratio of the depth of the centroid of resultant concrete compression force to the depth of the neutral axis
γ_c = depth of the centroid of resultant concrete compression force from the extreme ompression fiber, mm
ϕ = strength reduction factor
ρ_f = ratio of plate to the beam cross-sectional area
ρ = tensile steel reinforcement ratio
ρ' = compressive steel reinforcement ratio
$\rho_{f,b}$ = balanced plate ratio
$\rho_{f,bb}$ = balanced plate ratio at ultimate
$\rho_{f,bb}$ = critical plate ratio for yielding of compression steel at ultimate condition
$\rho_{f,cf}$ = critical plate ratio for yielding of compression steel at the plate rupture
$\rho_{f,cc}$ = critical plate ratio for yielding of compression steel at crushing of concrete
$\rho_{f,\max}$ = maximum allowable plate ratio
ψ = FRP strength reduction factor corresponding to aggressive environmental condition

Exercise Problems

1. A deficient simply supported RC beam is to be strengthened to safely carry a midspan load of 200 kN in addition to its self-weight on a long-term basis under aggressive humidity conditions (hot water 40°C). Figure E6.1 shows the cross-sectional details of the beam. The total span of the beam is 6 m, while center to center distance between supports is 2.06 m. The beam is internally reinforced with 3 Grade 60 steel bars of 9.5 mm and 15.9 mm diameter in compression and tension, respectively. Use of GFRP plate is recommended for external strengthening. Assuming that material is being added through the placement in the enclosed space using wet lay-up process and adhesive bonding of pre-fabricated sections with ambient cure: design a suitable strengthening system using GFRP plates and epoxy. The material properties of CFRP plate and epoxy are given in Table E6.1.
2. Design the strengthening system mentioned in Question No. 1 using near surface mounted GFRP bars and show the effectiveness of NSM GFRP over GFRP plates in external strengthening of RC beams for a similar aggressive environmental condition.
3. What do you understand by the durability based design of structures? Do you think that FRP rebars and/or FRP plates can be used as a solution for durability based design?
4. Design the strengthening system mentioned in Question No. 1 using CFRP and AFRP rebars as a Near Surface Mounted FRP and compare the cost involved in external strengthening of RC beams using CFRP, GFRP, and AFRP systems. Assume similar conditions for strengtgening using these three FRP systems.

E6.2. Design Example 2: A Case Study Problem

Problem definition

A contractor wants a structural engineer to design a bridge girder to be used in a simply supported slightly skewed (about 15°) two-lane bridge. He puts the condition that the bridge is to be constructed of precast and prestressed double-T bridge girders using pretensioning 10-mm diameter CFRP Leadline tendons and 40-mm diameter external unbonded CFCC post-tensioning strands. In addition, beams could be reinforced with non-presressing 12.5 mm CFCC strands and 10-mm diameter Leadline tendons in the webs and flange section of precast girders, respectively. From the early construction practices, it is known that precast girders are usually prestressed with bonded pretensioning tendons and partially prestressed with unbonded post-tensioning tendons. Assume that 60% of the design post-tensioning force is applied at the precast plant while remaining 40% post-tensioning force is applied at the bridge construction site before casting a deck slab of 75-mm thickness. From the ductility and stiffness consideration of the deck slab, it is desired that available NEFMAC grids be used for deck slab (concrete topping) reinforcement. The properties of the concrete topping, precast concrete, prestressing and non-prestressing tendons and NEFMAC grids are given in the enclosed tables. Plan and cross-sectional details of the bridge are given in Figs. E6.2 and E6.3, respectively. Figures E6.4 and E6.5, are showing typical DT-beam cross-section and elevation with reinforcement details used in early construction, respectively.

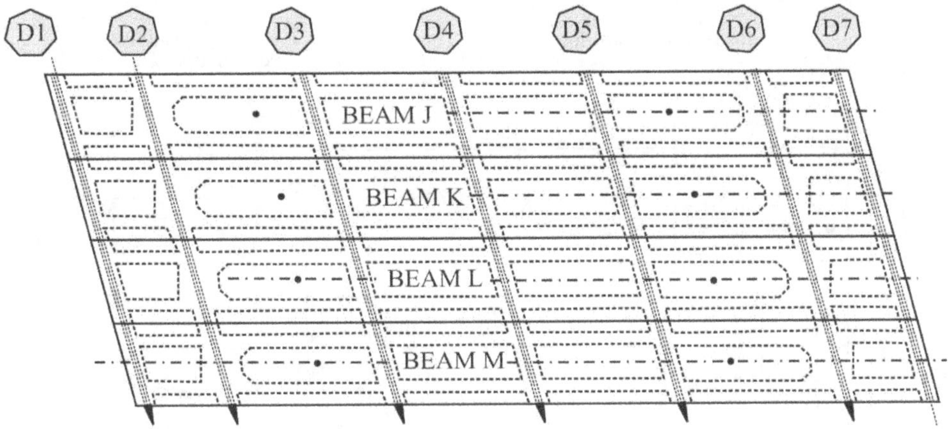

Figure E6.2. Plan view of instrumented bridge.

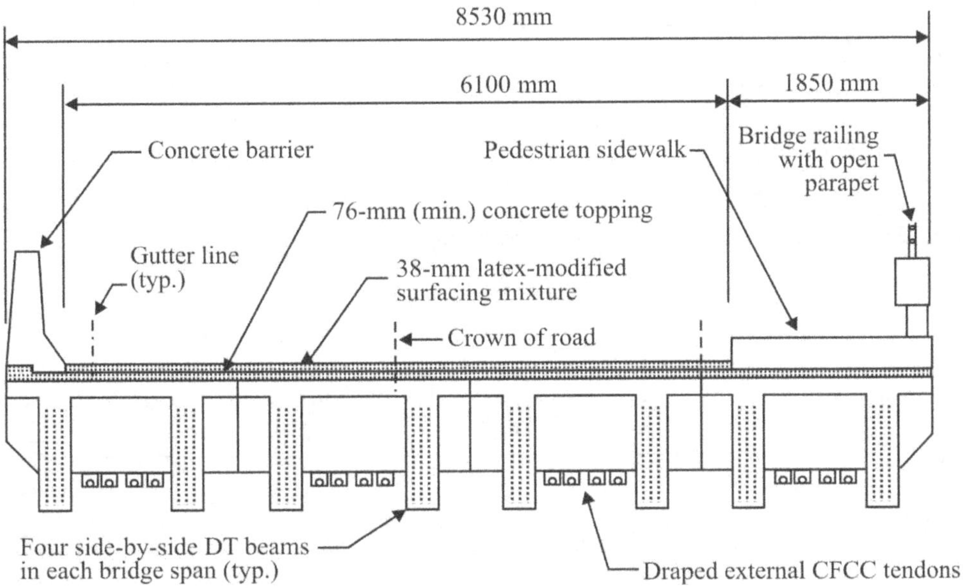

Figure E6.3. Cross-section of bridge reinforced and prestressed using CFRP bars and tendons.

Figure E6.6 shows a DT-beam test setup to predict the response of similar beams to be used in the bridge. Figure E6.7 shows the load distribution factors obtained from AASHTO standard specification, AASHTO LRFD Design specification and from the strains obtained from load test of a similar bridge. Figure E6.8 shows the loading configuration of the two dump trucks (# 649 and 650) used for load test.

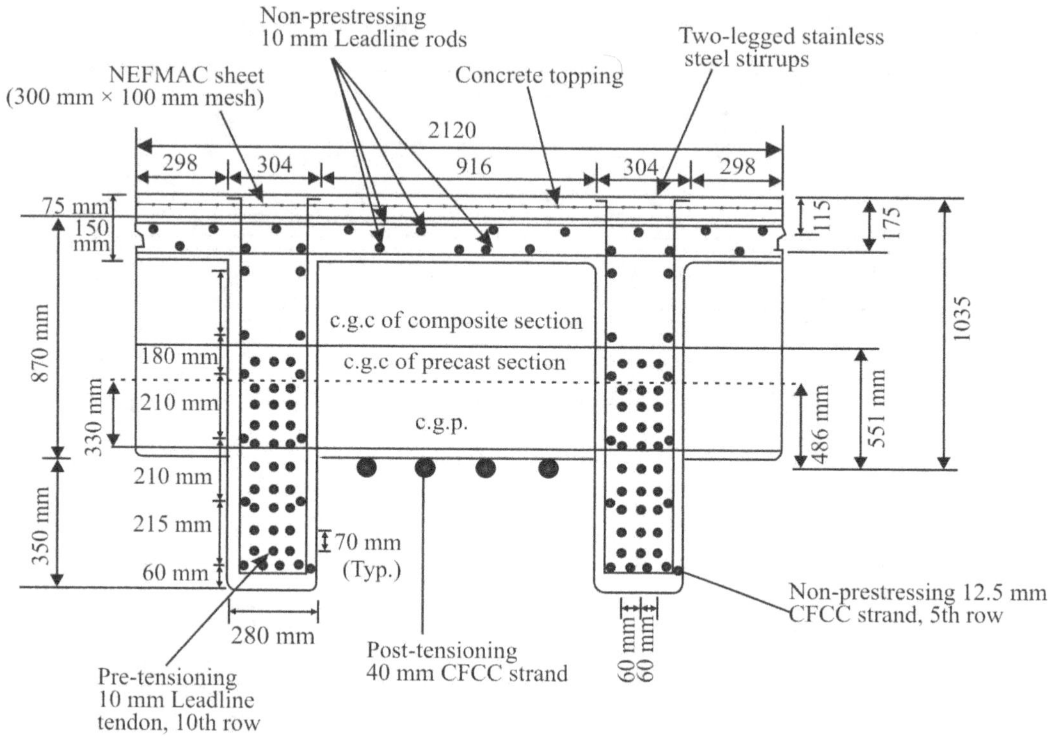

Figure E6.4. Typical cross-section of DT-beam at midspan.

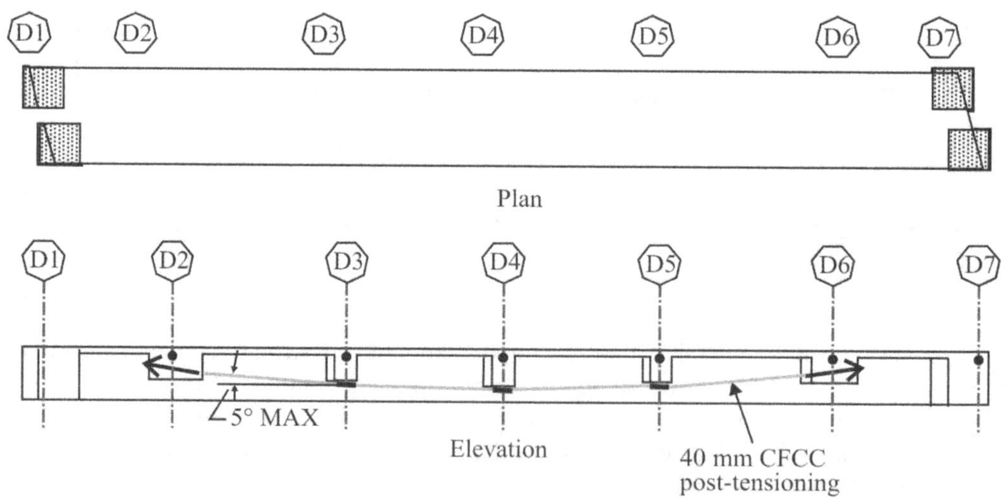

Figure E6.5. External post-tensioning arrangement for CFCC strands.

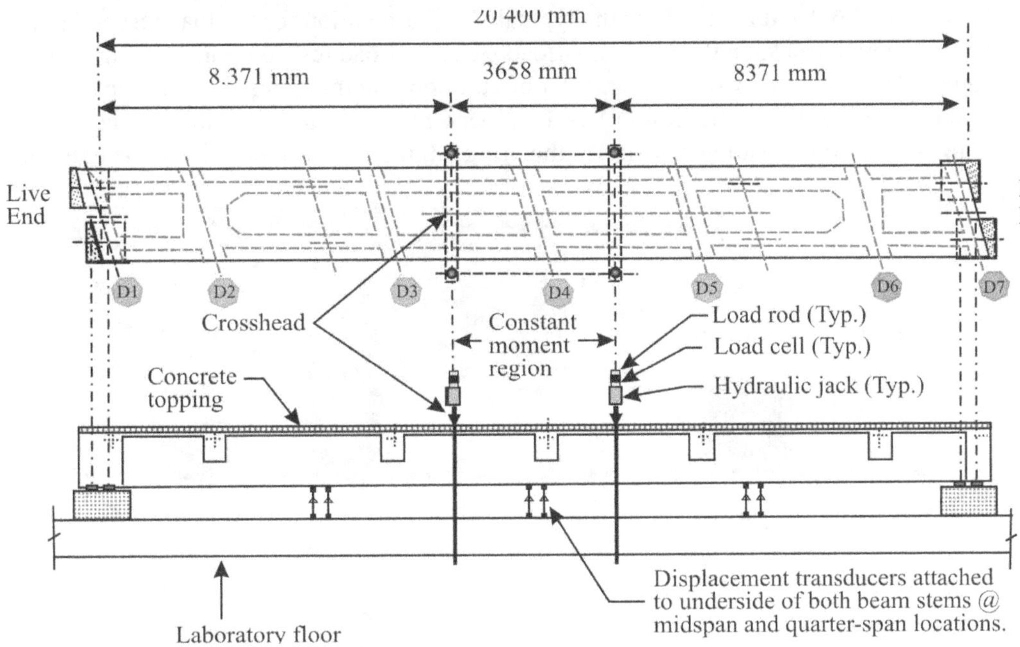

Figure E6.6. Location of 4-point loading system.

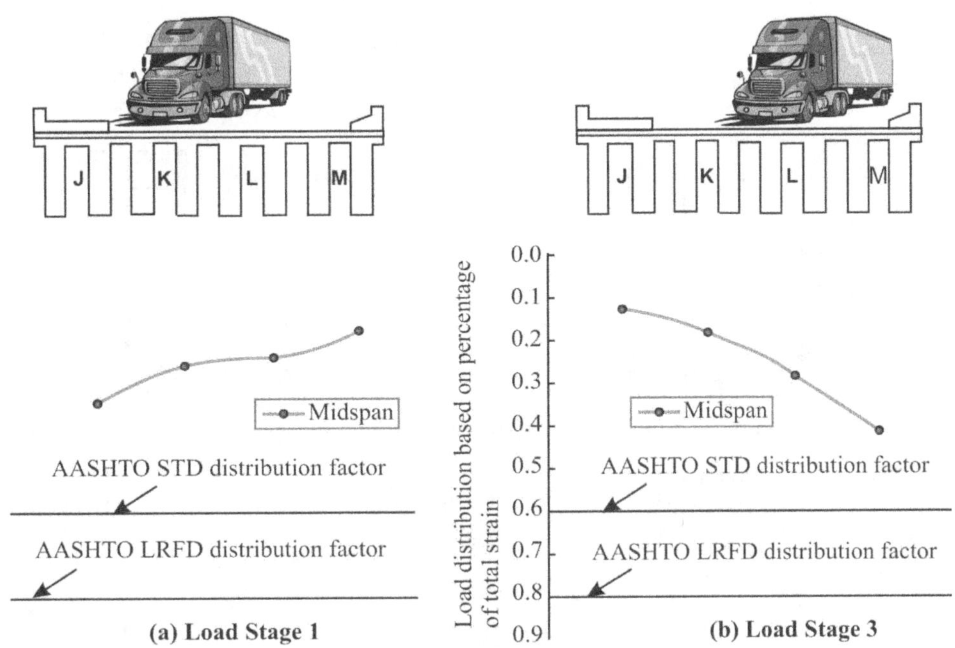

Figure E6.7. Load distribution based on percentage of total strain.

Figures E6.9 and E6.10 show the typical truck orientation for the load test and plan view of the bridge with trucks in the west lane (load stage 1). Load test setup in east lane refers to Load stage 3. Note that the two dump truck configuration loading gives the maximum lane bending moment about 90% of the design live load bending moment including impact obtained from using MS-23 truck loading system for the bridge. It must be also noted that the external strands

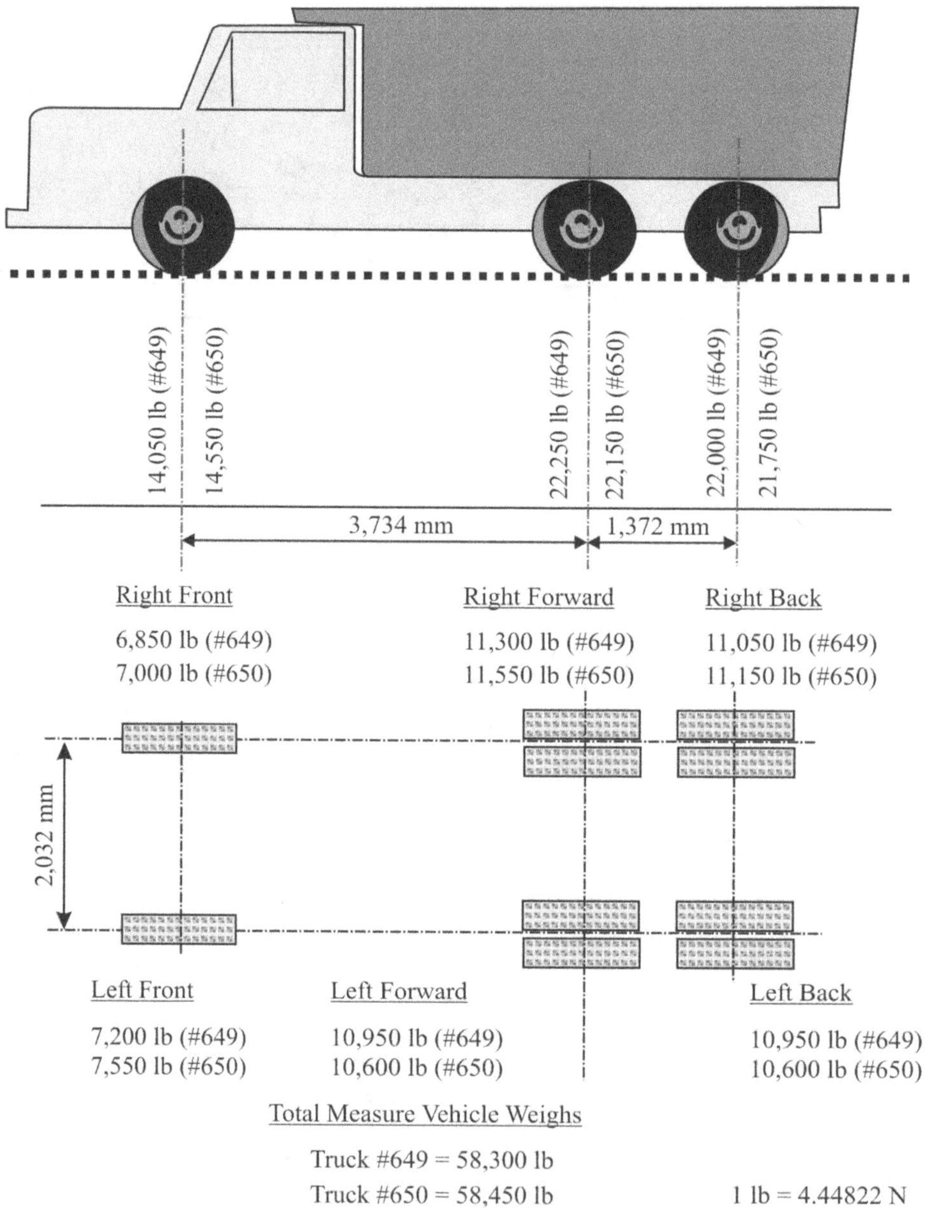

Figure E6.8. Dump truck configuration and measured wheel loads.

are provided between diaphragms D1 and D6 with distance of 15.7 m. Assuming that you are a structural engineer, provide the following design requirements as per the request of contractor:

1. Suggest suitable load distribution factors for the DT-beams with appropriate reasoning.
2. Suggest cross-section dimensions of the DT-beam and degree of prestressing for pretensioning and post-tensioning tendons.

Figure E6.9. Typical truck orientation for each load stage.

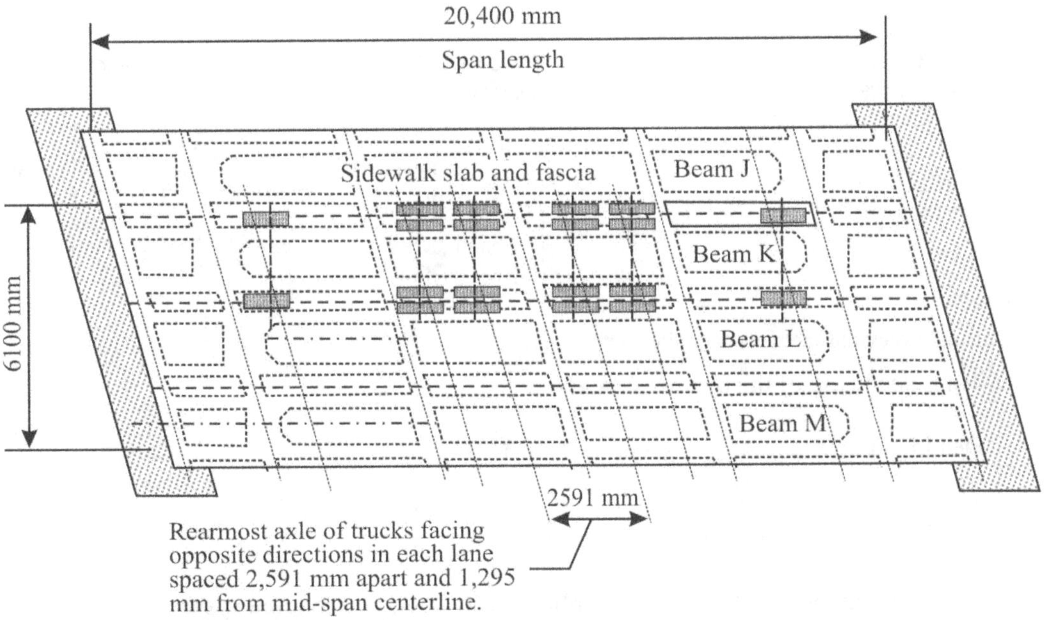

Figure E6.10. Truck positions for load Stage 1 in the west lane of bridge.

3. Anticipate the failure mode of the beam by estimating reinforcement ratio (ρ/ρ_b).
4. Cracking load, nominal moment capacity and design moment capacity of DT-beam.
5. Deflection of the DT-beam at service load, as well as post-tensioning forces in the external unbonded strands.
6. Strains, stresses and forces in pre-tensioning tendons and post-tensioning strands at the ultimate load condition.
7. Using strain controlled approach or any other method you know, determine the complete load versus compressive concrete strain, load-versus deflection and load versus post-tensioning forces in unbonded tendons as per load test setup given in Fig. E6.6. Plot these loads versus strain, post-tensioning forces and deflection using appropriate program.

Table E6.2. Material properties of CFRP tendons/CFCC strands.

Properties	Leadline × (MCC)	CFCC 1 × 7 (Tokyo rope)	CFCC 1 × 37 (Tokyo rope)
Nominal diameter, mm	10	12.5	40
Effective cross-sectional area, mm^2	71.6	76.0	752.6
Guaranteed tensile strength, kN/mm^2	2.26	1.87	1.41
Specified tensile strength, kN/mm^2	2.86	2.10	1.87
Young's modulus of elasticity, kN/mm^2	147	137	127
Elongation, %	1.9	1.5	1.5
Guaranteed breaking load, kN	162	142	1070
Ultimate breaking load, kN	204.7	160	1410

Table E6.3. Material properties of NEFMAC^{TM2} sheets.

Modulus of elasticity, GPa	86.5
Ultimate strength, MPa	1500
Ultimate strain, %	1.8

Table E6.4. Material properties of precast concrete and concrete topping.

Properties	Precast concrete	Concrete topping
Modulus of elasticity, GPa	36.7	31.6
Strength, MPa	53.8	39.3

Solution

1. Calcualte required moment capacity

Total dead load of the beam = 634 kN.

$$\text{Dead load/unit length, } (W_d) = \frac{634}{20400} = 31.08 \text{ kN/m.}$$

$$\text{Maximum dead load moment, } M_D = \frac{w_d \times L^2}{8} = \frac{31.08 \times 20.4^2}{8} = 1616.78 \text{ kN-m}$$

Let W be the total mid-span load applied through 4-point loading system.

The distance between the centres of each support to the nearest load point = 8.371 m.

Design service live load = 464 kN.

Maximum service live load moment, $M_L = \dfrac{464 \times 8.371}{2} = 1942.07$ kN-m.

Required moment capacity of the section (ACI 318)

$$M_{\text{required}} = 1.2M_D + 1.6M_L = 3558.85 \text{ kN-m.}$$

2. Calculation of balanced ratio (ρ_b)

$$\rho_b = 0.85\beta_1 \frac{f_c'}{f_{fu}} \frac{\varepsilon_{cu}}{\varepsilon_{cu} + \varepsilon_{fu} - \varepsilon_{pbmi}}$$

$$\beta_1 = 0.85 - \frac{(7810 - 4000)}{1000} \times 0.05$$

$$\beta_1 = 0.66$$

Prestressing force in each bar = 90 kN
Total prestressing force = 5400 kN
Assuming 25% loss in the prestressing force

$$\varepsilon_{pbmi} = \frac{90 \times 0.75}{71.6 \times 147} = 0.0059$$

$$\varepsilon_{cu} = 0.003$$

The balanced ratio,

$$\rho_b = 0.85(0.66)\left(\frac{53.8}{2860}\right)\left(\frac{0.003}{0.003 + 0.019 - 0.0059}\right)$$

$$\rho_b = 0.0019$$

3. Calculation of cracking moment

Modulus of rupture of concrete, $f_r = 6.0\sqrt{f_c'} = 6.0\sqrt{53.8} = 44$ MPa
Cross-sectional area of flange of DT beam = 318 000 mm^2.
Cross-sectional area of single web of DT beam = 299 600 mm^2.
Cross-sectional area of precast concrete section, A_p = 917 200 mm^2.
Cross-sectional area of concrete topping = 159 000 mm^2.
Cross-sectional area of composite concrete section, A_c = 1 076 200 mm^2

$$\text{C.G. of precast section, } y_p = \frac{(318\,000 \times 75) + 2(299\,600 \times 685)}{917\,200}$$

= 473.5 mm from top of the flange of DT beam

Moment of inertia of precast concrete section (I_p)

$$I_p = \frac{(2120\times150^3)}{12} + 318000(473.5-75)^2 + 2\left[\frac{280\times1070^3}{12} + 299\,600(685-473.5)^2\right]$$

$$=1.35\times10^{11}\,\text{mm}^4.$$

C.G. of composite section (y_c),

$$y_p = \frac{(159\,000\times37.5)+(318\,000\times150)+2(299\,600\times760)}{1076\,200}$$

$$= 473 \text{ mm from top of the concrete topping}$$

Moment of inertia of composite concrete section, (I_c)

$$I_c = \frac{(2120\times75^3)}{12} + 15900(473-37.5)^2 + \frac{(2120\times150^3)}{12}$$

$$+318000(473-150)^2 + 2\left[\frac{280\times1070^3}{12} + 299\,600(760-473)^2\right]$$

$$= 1.7\times10^{11}\,\text{mm}^4$$

Distance of centroid of precast section from the bottom fiber,

$$y_{bp} = 746.5 \text{ mm}$$

Distance of centroid of composite section from the bottom fiber,

$$y_{bc} = 822 \text{ mm}$$

Distance of centroid of precast section from the top fiber,

$$y_{tp} = 548.5 \text{ mm}$$

Distance of centroid of composite section from the top fiber,

$$y_{tc} = 473 \text{ mm}$$

Section modulus corresponding to the bottom fiber of precast section,

$$S_{bp} = \frac{I_p}{y_{bp}} = 180.84 \times 10^6 \text{ mm}^3$$

Section modulus corresponding to the bottom fiber of composite section,

$$S_{bc} = \frac{I_c}{y_{bc}} = 206.81 \times 10^6 \text{ mm}^3$$

Section modulus corresponding to the top fiber of precast section,

$$S_{tp} = \frac{I_p}{y_{tp}} = 246.13 \times 10^6 \text{ mm}^3$$

Section modulus corresponding to the top fiber of composite section,

$$S_{tc} = \frac{I_c}{y_{tc}} = 359.41 \times 10^6 \text{ mm}^3.$$

Total effective pre-tensioning force,

$$F_{\text{pre}} = 5400 \times 0.7 = 3780 \text{ kN}.$$

Total effective post-tensioning force,

$$F_{\text{post}} = 1770 \text{ kN}.$$

Stress at the bottom fiber of the section due to pre-tensioning force

$$= -\frac{F_{\text{pre}}}{A_p} - \frac{F_{\text{pre}} e_b}{S_{bp}}$$

$$= -\frac{3780 \times 10^3}{917\,200} - \frac{3780 \times 330 \times 10^3}{180.84 \times 10^6}$$

$$= -11.02 \text{ N/mm}^2.$$

Since 60% of the total post-tensioning force was used to prestress the precast DT concrete section and remaining 40% of the total post-tensioning force was applied on the composite concrete section (DT-beam plus concrete topping), hence, the stress at the bottom fiber of section due to post-tensioning force

$$= -\frac{0.6 \times F_{\text{post}}}{A_p} - \frac{0.6 \times F_{\text{post}} e_{up}}{S_{bp}} - \frac{0.4 \times F_{\text{post}}}{A_c} - \frac{0.4 \times F_{\text{post}} e_{uc}}{S_{bc}}$$

$$= -\frac{0.6 \times 1770}{917\,200} - \frac{0.6 \times 1770 \times 486}{180.84 \times 10^6} - \frac{0.4 \times 1770}{1076\,200} - \frac{0.4 \times 1770 \times 551}{206.81 \times 10^6}$$

$$= -6.556 \times 10^{-3} \text{ kN/mm}^2$$

$$= -6.556 \text{ N/mm}^2$$

The cracking moment (M_{cr}) can be found from the following expression,

$$= -11.02 - 6.556 + \frac{M_{cr}}{S_{bc}} = f_r$$

$$= -11.02 - 6.556 + \frac{M_{cr}}{206.81 \times 10^6} = 44$$

$$M_{cr} = 12\,734.53 \times 10^6 \text{ N-mm}.$$

$$M_{cr} = 12\,734.53 \text{ kN-m}.$$

Cracking load,

$$P_{cr} = \frac{(M_{cr} - M_D) \times 2}{8.371}$$

$$P_{cr} = \frac{(12734.53 - 1942.07) \times 2}{8.371}$$

$$P_{cr} = 2578.54 \text{ kN} > \text{service load } (1942.07 \text{ kN})$$

4. Compute flexural moment capacity

Ultimate bond reduction co-efficient for external strands,

$$\Omega_u = \frac{5.4}{\frac{L_u}{d_u}}$$

Distance of the external post-tensioning strands from the extreme compression fiber, $d_u = 1035$ mm.

Horizontal distance between the ends of unbonded post-tensioning tendons, $L_u = 15\,670$ mm.

$$\frac{L_u}{d_u} = 15.14$$

$$\Omega_u = 0.36.$$

Distance of the centroid of bottom bonded prestressing tendons from the extreme compression fiber, $d_m = 1193$ mm.

Flange width, $b = 2120$ mm.

Total effective pre-tensioning force, $F_{pi} = 3780$ kN.

Total effective post-tensioning force, $F_{\text{post}} = 1770$ kN.

Flexural stress in the equivalent bonded prestressing tendon at balanced condition, $f_{pbb} = 1300.8$ N/mm^2.

Stress in the equivalent non-prestressing strand of webs at balanced condition, $f_{pnbb} = 1053.56$ N/mm^2.

Stress in the equivalent unbonded strand at balanced condition, $f_{pub} = 498$ N/mm^2.

Stress in the equivalent non-prestressing tendons of flange at balanced condition, $f_{pnf} = 275.45$ N/mm^2.

Total cross-sectional area of bonded prestressing tendons, $A_{pb} = 60 \times 71.6 = 4296$ mm^2.

Total cross-sectional area of non-prestressing strands in the webs, $A_{pn} = 28 \times 76 = 2128$ mm^2.

Total cross-sectional area of unbonded prestressing strands,
$A_{fu} = 4 \times 752.6 = 3010.4$ mm^2.

Total cross-sectional area of non-prestressing tendons in the flange,
$A_{pnf} = 19 \times 71.6 = 1360.4$ mm^2.

5. Reinforcement ratio (ρ)

$$\rho = \frac{\sum_{i=1}^{p} A_{fi}\alpha_i}{bd_m}$$

where, $\alpha_i = \dfrac{f_{bi}}{f_{fu}}$

A_{fi} = cross-sectional area of reinforcement of a particular material
b = flange width of the beam
f_{bi} = total stress in an equivalent tendon of a specific material at the balanced condition.
p = total number of reinforcing materials
d_m = distance of centroid of prestressing tendons at mth row from the extreme compression fiber.

$$\rho = \frac{\sum_{i=1}^{m} F_{pi} + f_{pbb}A_{pb} + f_{pnbb}A_{pn} + F_{pui} + f_{pub}A_{fu} - f_{pnfb}A_{pnf}}{b \times d_m \times f_{fu}}$$

$$\rho = \frac{[3780000 + (1300.8 \times 4296) + (1053.56 \times 2128) + 1770000 + (498 \times 3010.4) - (275.45 \times 1360.4)]}{2120 \times 1193 \times 2860}$$

$$\rho = 0.0015$$

Since, $0.5\rho_b < \rho < \rho_b$

Section is under reinforced.

6. Compute forces in bonded prestressing tendons

Forces in bonded tendons are calculated using strain compatibility as follows:

Strain in bonded prestressing tendons of 10th row,

$$\varepsilon_{pb10} = \left[\frac{0.0131 \times (1193 - n)}{(1193 - n)} + 0.0059\right]$$

Stress in bonded prestressing tendons of 10th row,

$$f_{pb10} = E \times \varepsilon_{pb10}$$

$$f_{pb10} = 147\,000 \times \left[\frac{0.0131 \times (1193 - n)}{(1193 - n)} + 0.0059\right]$$

$$f_{pb10} = 1925.7 + 867.3 = 2793 \text{ N/mm}^2$$

where, n = depth of the neutral axis from the extreme compression fiber

E = modulus of elasticity of bonded tendons = 147 000 N/mm^2.

Resultant force in bonded prestressing tendons of 10th row, $F_{pb10} = f_{pb10} \times A_{fb}$

$$F_{pb10} = 2793 \times 429.6 = 1\,199\,872.8 \text{ N} = 1199.87 \text{ kN}.$$

where A_{fb} = cross-sectional area of bonded prestressing tendons in each row = 6(71.6) = 429.6 mm^2.

Similarly,

$$F_{pb9} = \left[\frac{1925.7(1123 - n)}{(1193 - n)} + 867.3\right] \times 429.6$$

$$F_{pb8} = \left[\frac{1925.7(1053 - n)}{(1193 - n)} + 867.3\right] \times 429.6$$

$$F_{pb7} = \left[\frac{1925.7(983 - n)}{(1193 - n)} + 867.3\right] \times 429.6$$

$$F_{pb6} = \left[\frac{1925.7(913 - n)}{(1193 - n)} + 867.3\right] \times 429.6$$

$$F_{pb5} = \left[\frac{1925.7(843 - n)}{(1193 - n)} + 867.3\right] \times 429.6$$

$$F_{pb4} = \left[\frac{1925.7(773 - n)}{(1193 - n)} + 867.3\right] \times 429.6$$

$$F_{pb3} = \left[\frac{1925.7(703 - n)}{(1193 - n)} + 867.3\right] \times 429.6$$

$$F_{pb2} = \left[\frac{1925.7(633 - n)}{(1193 - n)} + 867.3\right] \times 429.6$$

$$F_{pb1} = \left[\frac{1925.7(563-n)}{(1193-n)} + 867.3\right] \times 429.6$$

Therefore,

$$\sum_{j=1}^{10} F_{pbj} = 429.6 \times \left[\frac{1925.7(8780-10n)}{(1193-n)} + 8673\right]$$

7. Compute forces in bonded non-prestressing tendons

Forces in bonded non-prestressing tendons are calculated using strain compatibility as follows:

Strain in bonded non-prestressing tendons of 6th row,

$$\varepsilon_{pn6} = \frac{0.0131 \times (1235-n)}{1193-n}$$

Stress in bonded non-prestressing tendons of 6th row,

$$f_{pn6} = E \times \varepsilon_{pn6}$$

$$f_{pn6} = 13\,700 \times \frac{0.0131 \times (1235-n)}{(1193-n)}$$

$$M_n = F_R(d - \bar{d})$$

Resultant force in bonded non-prestressing tendons of 6th row,

$$F_{pn6} = f_{pn6} \times A_{fn}$$

$$F_{pn6} = \left[13\,700 \times \frac{0.0131 \times (1235-n)}{(1193-n)}\right] \times 76(8)$$

$$F_{pn6} = \left[\frac{1\,091\,177.6 \times (1235-n)}{(1193-n)}\right]$$

Similarly,

$$F_{pn5} = \left[\frac{545\,588.8 \times (1020-n)}{(1193-n)}\right]$$

$$F_{pn4} = \left[\frac{545\,588.8 \times (810-n)}{(1193-n)}\right]$$

$$F_{pn3} = \left[\frac{545\,588.8 \times (600-n)}{(1193-n)}\right]$$

$$F_{pn2} = \left[\frac{545588.8 \times (420 - n)}{(1193 - n)}\right]$$

$$F_{pn1} = \left[\frac{545588.8 \times (225 - n)}{(1193 - n)}\right]$$

Therefore,

$$\sum_{j=1}^{6} F_{pnj} = \left[\frac{545588.8 \times (5545 - 7n)}{(1193 - n)}\right]$$

8. Compute forces in un-bonded post-tensioning tendons

Forces in un-bonded post-tensioning tendons are calculated using strain compatibility given as follows:

Strain in un-bonded post-tensioning tendons,

$$\varepsilon_{pu} = \frac{0.0131 \times (1035 - n) \times 650}{(1193 - n)} \Omega_u + 0.0046$$

Stress in un-bonded post-tensioning tendons,

$f_{pu} = E \times \varepsilon_{pu}$ Resultant force in un-bonded post-tensioning tendons, $F_{pu} = f_{pu} \times A_{fu}$

where, A_{fu} = cross-sectional area of un-bonded post-tensioning strands 3010.4 mm^2.

$\Omega_u = 0.36$.

$$F_{pu} = \frac{1803024.89(1035 - n)}{(1193 - n)} + 1758675.68$$

9. Calcuation of neutral axis depth (*n*)

From the equation of equilibrium

$$\sum_{j=1}^{10} F_{pbj} + \sum_{j=1}^{6} F_{pnj} + F_{pu} = C$$

$$429.6 \times \left[\frac{1925.7(8780 - 10n)}{(1193 - n)} + 8673\right] + \left[\frac{545588.8(5545 - 7n)}{(1193 - n)}\right]$$

$$+ \left[\frac{1803024.89(1035 - n)}{(1193 - n)}\right] + 1758675.68$$

$$= (0.85 \times 0.86 \times 39.3 \times 2120 \times 75) + 0.85 \times 053.8 \times 2120 \times (0.66n - 75)]$$

$$+ \left[\frac{0.0131(n - 115)}{(1193 - n)} \times 147000 \times 71.6 \times 10\right] + \left[\frac{0.0131(n - 115)}{(1193 - n)} \times 147000 \times 71.6 \times 9\right]$$

$$63\,985.416n^2 - (101.036 \times 10^6)n + (22.297 \times 10^9) = 0.$$

$$n = 265.24 \text{ mm}.$$

10. Compute stresses in bonded prestressing tendons

Using strain compatibility equations from Step 6.

$f_{pb10} = 2793.00$ N/mm^2 < 2860 N/mm^2 O.K.

$f_{pb9} = 2647.70$ N/mm^2 < 2860 N/mm^2 O.K.

$f_{pb8} = 2502.41$ N/mm^2 < 2860 N/mm^2 O.K.

$f_{pb7} = 2357.11$ N/mm^2 < 2860 N/mm^2 O.K.

$f_{pb6} = 2211.82$ N/mm^2 < 2860 N/mm^2 O.K.

$f_{pb5} = 2066.52$ N/mm^2 < 2860 N/mm^2 O.K.

$f_{pb4} = 1921.23$ N/mm^2 < 2860 N/mm^2 O.K.

$f_{pb3} = 1775.93$ N/mm^2 < 2860 N/mm^2 O.K.

$f_{pb2} = 1630.64$ N/mm^2 < 2860 N/mm^2 O.K.

$f_{pb1} = 1485.34$ N/mm^2 < 2860 N/mm^2 O.K.

11. Compute stresses in bonded non-prestressing tendons

Using strain compatibility equations from Step 7,

$f_{pn6} = 1875.95$ N/mm^2 < 2100 N/mm^2 O.K.

$f_{pn5} = 1460.04$ N/mm^2 < 2100 N/mm^2 O.K.

$f_{pn4} = 1053.81$ N/mm^2 < 2100 N/mm^2 O.K.

$f_{pn3} = 647.57$ N/mm^2 < 2100 N/mm^2 O.K.

$f_{pn2} = 299.37$ N/mm^2 < 2100 N/mm^2 O.K.

$f_{pn1} = -77.84$ N/mm^2 < 2100 N/mm^2 O.K.

For tendons in flange of DT beam,

$f_{pnt} = 311.84$ N/mm^2 < 2860 N/mm^2 O.K.

$f_{pnb} = 187.31$ N/mm^2 < 2860 N/mm^2 O.K.

12. Compute stresses in un-bonded post-tensioning tendons

$f_{pu} = 1081.13$ N/mm^2 < 1870 N/mm^2 O.K.

13. Compute resultant forces in bonded prestressing tendons

$$F_{pb10} = 2793.0 \times 429.6 = 1\,199\,872.80 \text{ N} = 1199.87 \text{ kN}.$$

$$F_{pb9} = 2647.70 \times 429.6 = 1\,137\,451.92 \text{ N} = 1137.45 \text{ kN}.$$

$$F_{pb8} = 2502.41 \times 429.6 = 1\,075\,035.36 \text{ N} = 1075.04 \text{ kN}.$$

$$F_{pb7} = 2357.11 \times 429.6 = 1\,012\,614.45 \text{ N} = 1012.61 \text{ kN}.$$

$$F_{pb6} = 2211.82 \times 429.6 = 950\,197.872 \text{ N} = 950.197 \text{ kN}.$$

$$F_{pb5} = 2066.52 \times 429.6 = 887\,776.992 \text{ N} = 887.776 \text{ kN}.$$

$$F_{pb4} = 1921.23 \times 429.6 = 825\,360.408 \text{ N} = 825.360 \text{ kN}.$$

$$F_{pb3} = 1775.93 \times 429.6 = 762\,939.528 \text{ N} = 762.939 \text{ kN}.$$

$$F_{pb2} = 1630.64 \times 429.6 = 700\,522.944 \text{ N} = 700.523 \text{ kN}.$$

$$F_{pb1} = 1485.34 \times 429.6 = 638\,102.064 \text{ N} = 638.102 \text{ kN}.$$

14. Compute resultant forces in bonded non-prestressing tendons

$$F_{pn6} = 1875.95 \times 608 = 1\,140\,577.6 \text{ N} = 1140.577 \text{ kN}.$$

$$F_{pn5} = 1460.04 \times 304 = 443\,852.16 \text{ N} = 443.852 \text{ kN}.$$

$$F_{pn4} = 1053.81 \times 304 = 320\,358.24 \text{ N} = 320.358 \text{ kN}.$$

$$F_{pn3} = 647.57 \times 304 = 196\,861.28 \text{ N} = 196.861 \text{ kN}.$$

$$F_{pn2} = 299.37 \times 304 = 91\,008.48 \text{ N} = 91.008 \text{ kN}.$$

$$F_{pn1} = -77.84 \times 304 = -23\,663.36 \text{ N} = -23.663 \text{ kN}.$$

For tendons in flange of DT beam,

$$F_{pnt} = 311.84 \times 716 = 2\,232\,77.44 \text{ N} = 223.277 \text{ kN}.$$

$$F_{pnb} = 187.31 \times 644.4 = 120\,702.564 \text{ N} = 120.70 \text{ kN}.$$

15. Compute resultant force in un-bonded post-tensioning tendons

$$F_{pu} = 1081.13 \times 3010.4 = 3\,254\,633.75 \text{ N} = 3254.63 \text{ kN}.$$

16. Compute resultant force in concrete

$$C_t = 0.85 \times 0.86 \times 39.3 \times 2120 \times 75 = 4\,567\,799.7 \text{ N} = 4567.80 \text{ kN}.$$

$$C_f = 0.85 \times 53.8 \times 2120 \times [0.66(265.24) - 75] = 9\,700\,421.74 \text{ N} = 9700.42 \text{ kN}.$$

$$C = C_t + C_f + F_{pnt} + F_{pnb} + F_{pn1}$$

$$= 4567.8 + 9700.42 + 223.277 + 120.7 + 23.663 = 14\,635.86 \text{ kN}.$$

$$T = \sum_{j=1}^{10} F_{pbj} + \sum_{j=2}^{6} F_{pnj} + F_{pu} = 9189.867 + 2192.656 + 3254.63 = 14637.153 \text{ kN.}$$

$C \cong T$ O.K.

17. Compute ultimate moment capacity

Centroid of prestressing tendon from the extreme compression fiber = 878 mm.

Centroid of non-prestressing tendon upto 5th layer from the extreme compression fiber = 886.66 mm.

Centroid of post-tensioning tendon from the extreme compression fiber = 1035 mm.

Distance of the resultant tension force from the extreme compression fiber

$$d = \frac{9189.867(878) + 2192.656(886.66) + 3254.63(1035)}{14637.153}$$

$$d = 914.21 \text{ mm}$$

Distance of the center of gravity of resultant compression force from the extreme compression fiber,

$$\bar{d} = \frac{4567.8(37.5) + 9700.42(150) + 223.277(115) + 120.7(175) + 23.663(225)}{14635.86}$$

$$\bar{d} = 114.68$$

Thus, nominal moment capacity,

$$M_n = F_R(d - \bar{d})$$

$$M_n = 14637.153\ (914.21 - 175.06)$$

$$M_n = 10.819 \times 10^6 \text{ kN-mm.}$$

$$M_n = 10819 \text{ kN-m.}$$

Design moment capacity, $M_u = \phi \times M_n$

$$M_u = 0.85 \times 10819$$

$$M_u = 9196.15 \text{ kN-m} > M_{\text{required}} = 3558.85 \text{ kN-m. O.K.}$$

18. Compute stresses due to service load

Moment due to service load, $M = 1616.78 + 1942.07$

$$M = 3558.85 \text{ kN-m} < M_{cr} = 12734.53 \text{ kN-m.}$$

Maximum compressive stress in concrete at the extreme compression fiber,

$$f_{ct} = \frac{F_{\text{pre}}}{A_p} - \frac{F_{\text{pre}} e_b}{S_{bp}} + \frac{0.6 \times F_{\text{post}}}{A_p} - \frac{0.6 \times F_{\text{post}} \times e_{up}}{S_{bp}} + \frac{0.4 \times F_{\text{post}}}{A_c} - \frac{0.4 \times F_{\text{post}} \times e_{uc}}{S_{bc}} + \frac{M}{S_{bc}}$$

$$= \frac{3780 \times 10^3}{917\,200} - \frac{3780 \times 330 \times 10^3}{180.84 \times 10^6} + \frac{0.6 \times 1770 \times 10^3}{917200} - \frac{0.6 \times 1770 \times 10^3 \times 486}{180.84 \times 10^6}$$

$$+ \frac{0.4 \times 1770 \times 10^3}{1076\,200} - \frac{0.4 \times 1770 \times 10^3 \times 551}{206.81 \times 10^6} + \frac{3558.85 \times 10^6}{206.81 \times 10^6}$$

$$= 11.51 \text{ N/mm}^2 < 0.6 f_c' = 32.28 \text{ N/mm}^2$$

Maximum concrete stress due to applied load $= \dfrac{1942.07 \times 10^6}{206.81 \times 10^6} = 9.39 \text{ N/mm}^2$

Stress in bottom prestressing tendons,

$$f_{pb10} = (E_f \times \varepsilon_{pb10}) + \frac{E_f}{E_c} \frac{M(d_{10} - y_{tc})}{I_c}$$

$$= 2793 + \frac{147\,000}{36\,700} \frac{3558.85 \times 10^6 (1193 - 473)}{1.7 \times 10^{11}}$$

$$f_{pb10} = 2853.37 \text{ N/mm}^2 < 2860 \text{ N/mm}^2.$$

Stress in bottom prestressing tendons due to applied load,

$$f_{pb10} = \frac{147\,000}{36\,700} \times \frac{1942.07 \times 10^6 (1193 - 473)}{1.7 \times 10^{11}} = 32.94 \text{ N/mm}^2$$

stress in post-tensioning strands,

$$f_{pu} = (E_f \times \varepsilon_{pu}) \frac{E_f}{E_c} \frac{(M - M_D) e_{yuc}}{I_c} \Omega_u$$

$$f_{pu} = 1081.13 + \frac{147000}{36700} \frac{(3558.85 \times 10^6 - 1616.78 \times 10^6)}{1.7 \times 10^6} \times \frac{2}{3}$$

$$f_{pu} = 1098.27 \text{ N/mm}^2 < 1870 \text{ N/mm}^2.$$

Force in a CFCC strand at service load = 1098.27 × 752.6

= 826 558 N

= 826.558 kN.

19. Maximum deflection under service load

$$M_L = M - M_D = 1942.07 \text{ kN-m}.$$

Distance between the support and nearest load point, $L_1 = 8371$ mm

Longitudinal distance between load points, $L_2 = 3658$ mm.

Deflection due to applied load,

$$\delta_a = \frac{M_L L_1}{8E_C I_C}\left[\frac{8}{3} + 4\left(\frac{L_2}{L_1}\right) + \left(\frac{L_2}{L_1}\right)^2\right]$$

$$\delta_a = \frac{1942.07\times10^6\times(8371)^2}{8\times36700\times1.7\times10^{11}}\times\left[\frac{8}{3} + 4\left(\frac{3658}{8371}\right) + \left(\frac{3658}{8371}\right)^2\right]$$

$$\delta_a = 12.56 \text{ mm}$$

Deflection due to dead load,

$$\delta_d = \frac{1}{E_C I_P}\left[\frac{5W_D L^4}{384}\right]$$

$$\delta_d = \frac{5\times31.08\times(20400^4)}{384\times36700\times1.35\times10^{11}}$$

$$\delta_d = 14.15\,\text{mm}\downarrow$$

Effective pre-tensioning force = 5400 kN.

Deflection due to pre-tensioning force at the instant of initial post-tensioning

$$\delta_{pre} = \frac{E_{\text{pre}} e_b L^2}{8E_C I_P}$$

$$\delta_{pre} = \frac{5400\times10^3\times330\times20400^2}{8\times36700\times1.35\times10^{11}}$$

$$\delta_{pre} = 18.71\,\text{mm}\uparrow$$

Center-to-center distance between diaphragms D_2 and D_6 = 15 670 mm

Deflection due initial post-tensioning,

$$\delta_{po} = \frac{E_{post} e_u L_u^2}{8E_C I_P}$$

$$\delta_{po} = \frac{0.6\times1770\times10^3\times486\times15670^2}{8\times36700\times1.35\times10^{11}}$$

$$\delta_{po} = 3.197\,\text{mm}\uparrow$$

Total deflection of the precast DT-beam after the initial post-tensioning

= 12.56 +14.15 − 18.71 − 3.197

= 4.80 mm ↓

Lose deflection due to transportation and support condition change

$$= 3 \text{ mm. (assumed)} \downarrow$$

Net deflection due to prestressing prior to final post-tensioning

$$= 4.8 + 3$$
$$= 7.8 \text{ mm} \downarrow$$

Increase in deflection due to final post-tensioning

$$= \frac{0.4 \times 1770 \times 10^3 \times 551 \times 15670^2}{8 \times 36700 \times 1.7 \times 10^{11}}$$
$$= 1.92 \text{ mm} \uparrow$$

Self-weight of the concrete topping = 3.975 kN/m.

Deflection due to concrete topping

$$= \frac{5W_D L^4}{384 E_C I_C}$$
$$= \frac{5 \times 3.975 \times (20400^4)}{384 \times 31600 \times 1.7 \times 10^{11}}$$
$$= 1.67 \text{ mm} \downarrow$$

Net deflection at the time of final post-tensioning

$$= 7.8 - 1.92 + 1.67$$
$$= 7.55 \text{ mm} \downarrow$$

E6.3. Design Example 3: A Case Study

Problem statement

Design an intze tank with dimension as shown in Figure E6.11. The tank is supported on eight vertical columns braced at different levels. Use M30 grade of concrete and CFRP reinforcement. Take safe bearing capacity of soil = 250 kN/m^2. Design of staging is not required.

Solution

1. Design constants

For M30 grade of concrete and mild steel reinforcement

$$\sigma_{cbc} = 10 \text{ N/mm}^2$$
$$m = 280/(3\sigma_{cbc}) = 9.33$$

i. For structural members not in contact with water:

For lead line CFRP rods tensile strength = 0.56 × 0.85 × 2860 = 1262.8 MPa

$$k = \frac{m \times \sigma_{cbc}}{(m \times \sigma_{cbc}) + \sigma_{st}} = 0.0688$$

$$J = 1 - (k/3) = 0.977$$

$$R = 0.5\sigma_{cbc} \times k \times J = 0.336$$

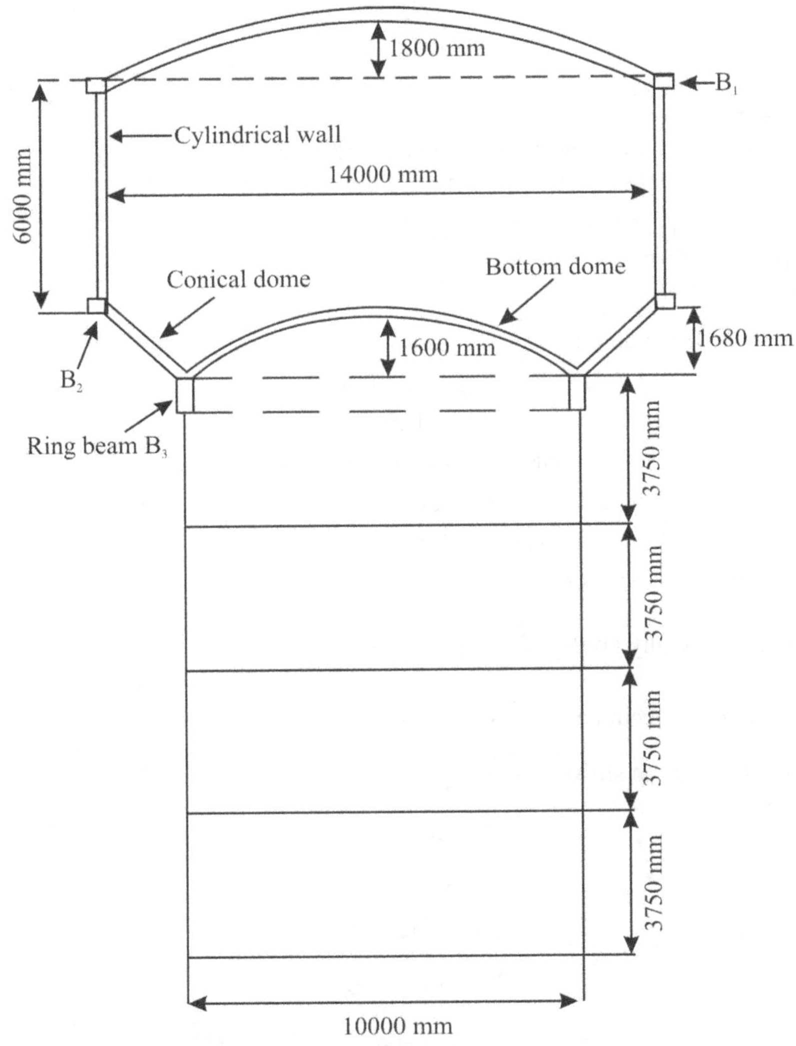

Figure E6.11

ii. Design constants for faces of members in contact with water:

$$(\sigma_{st} = 1118.26 \text{ N/mm}^2)$$
$$k = 0.0825$$
$$J = 0.972$$
$$R = 0.4011$$

iii. Design constants for face of members away from water (for members 225 mm or more thickness)

$$\sigma_{st} = 1215.5 \text{ N/mm}^2$$
$$k = 0.0764$$
$$j = 0.974$$
$$R = 0.3722$$

2. Design of dome

Assume thickness of dome slab = 100 mm and live load = 1.5 kN/m^2
Therefore loads per sq. metre for the design of dome consist of:

i. Self-wt. of dome slab = 0.1 × 25 = 2.5 kN/m^2
ii. Live load = 1.5 kN/m^2
iii. Total load = 4 kN/m^2

Let r be radius of the dome,

$$1.8(2r - 1.8) = (14/2) \times (14/2)$$
$$r = 14.51 \text{ m}$$
$$\sin\phi = 7/14.51 = 0.4824$$
$$\phi = 28^\circ 50'$$
$$\cos\phi = 0.876$$

Checking the max. stresses:

$$\text{Hoop stress at any angle } \theta = \frac{wR}{t}\left[\frac{\cos^2\theta + \cos\theta - 1}{1 + \cos\theta}\right]$$

Max. hoop stress occurs at crown where $\theta = 0°$

$$\text{Hence, max. hoop stress } = \frac{wR}{t} \times \frac{1}{2}$$
$$= \frac{4 \times 1000 \times 14.51}{0.1} \times \frac{1}{2}$$
$$= 2.9 \times 10^5 \text{ N/m}^2$$
$$= 0.29 \text{ N/mm}^2 \quad \text{(safe)}$$

$$\text{Meridional stress at any angle } \theta = \frac{wr}{t}\left[\frac{1 - \cos\theta}{\sin^2\theta}\right] = \frac{wr}{t}\left[\frac{1}{1 + \cos\theta}\right]$$

Max meridional stress occur at the base of dome where, $\theta = \phi = 28^\circ 50'$

Therefore, max value of meridional stress $= \dfrac{4\times1000\times14.51}{0.1}\left[\dfrac{1}{1+\cos 28^\circ 50'}\right]$

$= 0.309\ \text{N/mm}^2$ (safe)

Since, stresses are within limits, provide nominal reinforcement at 0.2% of gross = sectional area of concrete.

$$A_{st} = 0.2\% \times 100 \times 1000 = 200\ \text{mm}^2$$

Provide 8 mm Ø CFRP bars at 200 mm c/c in both directions.

3. Design of top ring beam

The horizontal component of meridional thrust = $T \cos \theta$

Therefore, total tension tending to rupture the ring beam per meter length of its circumference

$= p \times (D/2)$

$= T \cos \theta \times (D/2)$

$= 0.309 \times 100 \times 1000 \times 0.876 \times (14/2)$

$= 189\,450$ N

Area of reinforcement of CFRP required

$A_f = 189\,450/1118.26$

$= 170\ \text{mm}^2$

Number of 8 mm Ø lead line CFRP bars (area of CFRP = 50.5 mm^2)

$= (170/50.5)$

$= 3.36$ (say 4 numbers).

To fix the size of the beam: Equivalent area of the composite section

$= A + (m - 1)A_f$

$= A + (9.33 - 1)\ 4 \times 50.5$

$= (A + 1682.66)\ \text{mm}^2$

Assume allowable stress in the composite section to be 1.5 N/mm^2.

We have, $189\,450/(A + 1682.66) = 1.2$

$A = 124\,650\ \text{mm}^2$

Provide the size of top ring beam = 400 mm × 350 mm

Minimum shear reinforcement: c/c spacing, S_v, of vertical stirrups to meet requirement of minimum shear reinforcement is given by

$$S_v = \frac{A_{sv} \times f_{fv}}{b \times 0.35} \quad \text{(using 2 legged 8 mm Ø CFRP stirrups)}$$

$$A_{sv} = \frac{\pi}{4}(8)^2 = 50.26 \text{ mm}^2$$

$$S_v = \frac{2 \times 50.26 \times 0.004 \times 147\,000}{0.35 \times 350} = 482.5 \text{ mm c/c}$$

Maximum spacing limit = (d/2) = (295/2) = 147 mm or 600 mm

Use 2-legged 8 mm Ø shear stirrups at 145 mm c/c.

4. Design of cylindrical shell

Assuming that cylindrical wall is free to move top and bottom. The wall is subjected to pure tensile stress on account of the horizontal pressure due to the liquid stored.

Maximum hoop tension at the wall base per meter height of wall

$$T = (w \times h \times D)/2$$

Let weight of water (w) be 10 kN/m^3

Now,

$$T = (10 \times 6 \times 14)/2$$
$$= 420 \text{ kN} = 420\,000 \text{ N}$$

Area of Leadline CFRP bars required for Maximum hoop tension

$$A_f = \text{T}/1118.26$$
$$= 420\,000/1118.26 = 380 \text{ mm}^2$$

Provide 380/2 = 190 mm^2 each face.

Spacing, using 8 mm Ø bars (A = 51 mm^2) = 260 mm c/c < 450 mm

Hence, provide 8 mm Ø bars at 250 mm c/c in the form of rings on both faces. The spacing of bars can be increased towards top as the hoop tension is directly proportional to the depth of water.

Therefore, A_f required at 4 m below the top

$$= 4/6 \times 380$$
$$= 254 \text{ mm}^2 \text{ i.e., } 127 \text{ mm}^2 \text{ on each face.}$$

Spacing, using 8 mm Ø bars = 400 mm c/c < 450 mm

Provide 8 mm Ø bars at 400 mm c/c on both faces.

A_f required at 2 m below the top = 2/6 × 380

$$= 127 \text{ mm}^2 \text{ i.e., } 63.5 \text{ mm}^2 \text{ on each face.}$$

Spacing, using 8 mm Ø bars = 800 mm c/c < 450 mm

Provide 8 mm Ø bars at 430 mm c/c on both face.

Thickness of wall

Equivalent area of concrete assuming (t) mm to be the thickness of wall.

$$= [t \times 1000 + (9.33 - 1) \times 380]\ \text{mm}^2$$
$$= (1000t + 3165.4)\ \text{mm}^2$$

Assuming the allowable compressive stress in the composite section to be = 1.2 N/mm^2

We have,

$$\frac{420000}{(1000t + 3165.4)} = 1.2 \quad \text{or } t = 346 \text{ mm}$$

Provide the thickness of wall = 400 mm at its base and taper it to 225 mm at top.

Distribution of steel

Average thickness of the wall slab = (400 + 225)/2 = 312.5 mm

Percentage of FRP to be provided

$$= 0.3 - \left(\frac{312.5 - 100}{500 - 100}\right) \times 0.1$$
$$= 0.25\%$$

Therefore,

$$A_f = \frac{0.25 \times 312.5 \times 1000}{100}$$

= 782 mm^2 or 391 mm^2 on each face.

Spacing using 10 mm Ø CFRP bars = (79 × 1000)/391 = 200 mm c/c

Therefore, provide 10 mm Ø CFRP bars at 195 mm c/c.

5. Design of ring beam (B_2)

This ring beam is to provide at the junction of the cylindrical wall and the conical dome. W_1 (load transmitted through the tank wall at the top of the conical dome) per meter length consists of the following:

i. Load due to top dome:

$$= \text{Area of the slab} \times \text{meridional stress} \times \sin\theta$$
$$= 100 \times 1000 \times 0.309 \times 0.4824$$
$$= 14900 \text{ N/m}$$

ii. Load due to top ring beam:

$$= 0.325 \times (0.4 - 0.2) \times 25000$$
$$= 1625 \text{ N/m}$$

iii. Load due to cylindrical wall

$$= 0.312 \times 6 \times 25\,000$$
$$= 46\,800 \text{ N/m}$$

iv. Load due to ring beam B_2 (assuming the size of beam to be 1 m × 0.6 m)

$$= (1 - 0.325) \times 0.6 \times 25\,000$$
$$= 10\,125 \text{ N/m}$$

Total = 73 450 N/m

Hence, $W_1 = 73\,450$ N/m
The horizontal component of the thrust T, due to W_1,

$$H_1 = W_1 \tan\beta$$

Hoop tension due to H_1

$$= \frac{H_1 \times D}{2}$$
$$= \frac{73450 \times \tan 50^\circ \times 14}{2}$$
$$= 612\,741 \text{ N}$$

Hoop tension on account of the horizontal force due to water pressure on ring beam

$$= \frac{H_2 \times D}{2}$$
$$= whd \times \frac{D}{2}$$
$$= 10\,000 \times 6 \times 0.6 \times 14/2$$
$$= 250\,000 \text{ N}$$

Therefore, total hoop tension for which the beam has to be designed

Or $$H = 612\,741 + 252\,000$$
$$= 864\,741 \text{ N}$$

Area of FRP required

$$A_f = H/1118.26$$
$$= 712 \text{ mm}^2$$

Number of 10 mm Ø CFRP bars required $= \dfrac{712}{25 \times 3.14} = 10$ bars

Provide ten 10 mm Ø bars (A_f provided = 786 mm^2). Check for maximum tensile stress developed in the composite section.

Max. tensile stress developed,

$$= \frac{864\,741}{1000 \times 600 + (9.33 - 1) \times 786}$$
$$= 0.0002 \text{ N/mm}^2 \text{ (safe)}$$

Minimum shear reinforcement:

The c/c spacing, S_v, of vertical stirrups to meet requirement of minimum shear reinforcement is given by,

$$S_v = \frac{A_{sv} \times f_{fv}}{b \times 0.35}$$

Using 2-legged 8 mm Ø CFRP stirrups,

$$A_{sv} = \frac{\pi}{4}(8)^2 = 50.26 \text{ mm}^2$$
$$S_v = \frac{2 \times 50.26 \times 0.004 \times 147\,000}{0.35 \times 350} = 482.5 \text{ mm}^2 \text{ c/c}$$

Maximum spacing limit = $(d/2)$ = (295/2) = 147 mm or 600 mm. Use 2-legged 8 mm Ø shear stirrups at 145 mm c/c.

5. Design of conical dome

Average diameter of the conical dome = (14 + 10)/2 = 12 m

Average depth of water = 6 + (1.68/2) = 6.84 m

Weight of water above the conical dome = $\pi \times 12 \times 6.84 \times 10\,000 \times 2$
$= 5\,157\,240$ N

Assuming 550 mm thick slab. Self-weight of the slab
$= 2.61 \times 0.55 \times 12 \times \pi \times 25\,000$
$= 1\,352\,930$ N

Load from the top dome, cylindrical wall and bottom ring beam (B_2)
$= 76450 \times \pi \times 14$
$= 3\,362\,450$ N

Total load at the base of the conical dome slab
$= 5\,157\,240 + 1\,352\,930 + 3\,362\,450$
$= 9\,872\,320$ N

Total load per metre length at the base of the conical dome slab (w_2)

$$\frac{9872320}{\pi \times 10} = 314290 \text{ N/m}$$

Meridional thrust (T_1) due to the load above,

$$= \frac{w_2}{\cos 50^\circ} = 488950 \text{ N}$$

Hence, Meridional thrust is

$$\frac{488950}{550 \times 1000} = 0.889 \text{ N/mm}^2 \text{ (safe)}$$

Hoop tension in conical dome

To have the general expression for hoop tension, in conical dome, diameter of the conical dome at any height h metre above its base

$$D_k = 10 + 2 \times \frac{2}{1.68} \times h$$
$$= 10 + 2038h$$

Intensity of water pressure at this height:

$$P = [(6 + 1.68) - h] \times 10000 \text{ N/m}^2$$
$$= (7.68 - h) \times 10000 \text{ N/m}^2$$

Weight of the conical dome slab per square metre

$$W_3 = 0.55 \times 25000$$
$$= 13750 \text{ N/m}^2$$

$$\text{Total hoop tension} = \left(\frac{P}{\cos \beta} + w_3 \times \tan \beta\right)\frac{Dh}{2}$$

$$= \left(\frac{(768 - h)10000}{\cos \beta} + 13750 \times \tan 50^\circ\right)\left(\frac{10 + 2038h}{2}\right)$$

$$= (135870 - 15560h)(5 + 1.19h) \qquad \text{(i)}$$

Putting different values for h in the equation (i), we have,

When $h = 0$ hoop tension = 679 350 N

When $h = 1$ hoop tension = 744 720 N

When $h = 1.68$ hoop tension = 768 020 N

From the above result, it is clear that the maximum hoop tension occurs at the top of conical dome slab.

That is at $h = 1.68$

Area of FRP required for max. hoop tension

$$= 768\,020/1118.26$$
$$= 690 \text{ mm}^2$$
$$A_f \text{ on each face} = 345 \text{ mm}^2$$

Spacing of 10 mm Ø CFRP bars (area = 79 mm^2)

$$= (79 \times 1000)/345$$
$$= 220 \text{ mm c/c}$$

Therefore, provide 10 mm Ø CFRP bar at 220 mm c/c on both faces of the slab. (A_f provided is 158 mm^2).

The conical dome also acts as a slab spanning between beam B_2 and B_3 and loaded with the weight of water plus self-weight of slab.

The B.M. for design may be taken = ($WL/12$)

$$W = \frac{5157\,240 + 1352\,930}{\pi \times 10}$$

$$W = 207\,225 \text{ N}$$

$$L = \text{horizontal span} = 2 \text{ m}$$

$$\text{B.M.} = (207\,225 \times 2)/12$$
$$= 34\,537 \text{ N-m}$$

Effective depth of slab (d) = 550 − 40 = 510 mm

Therefore, $$A_f = \frac{34\,537 \times 1000}{0.972 \times 1118.26 \times 510}$$

$$A_f = 63 \text{ mm}^2$$

Spacing of 6 mm Ø FRP bars (Area = 29 mm^2)

$$= (29 \times 1000)/63$$
$$= 450 \text{ mm c/c}$$

Distribution steel,

$$= \frac{0.2}{100} \times 550 \times 1000 = 1100 \text{ mm}^2$$

Steel on each face = 550 mm^2, which is less than reinforcement from B.M. consideration

Hence, provide 8 mm Ø bars at 300 mm c/c at bottom faces of the slab to act as reinforcement for B.M. as well as distribution steel.

Check for max. stress in composite section

$$= \frac{768020}{550 \times 1000 \times (9.33 - 1) \times 158 \times 2} = 1.39 \text{ N/mm}^2 \text{ (safe).}$$

6. Design of bottom spherical dome

Assume the thickness of the dome slab = 300 mm. To find the radius of the dome:

$$(2R - 1.5) = 5 \times 5$$

$$2R = 17.225$$

$$R = 8.61 \text{ m}$$

$$\text{Sin } \phi = 5/8.61 = 0.5807$$

$$\phi = 35^\circ 30'$$

Self-weight of the dome

$$= 2R \times \pi \times 1.6 \times 0.3 \times 25\,000$$

$$= 2 \times 8.61 \times 3.142 \times 1.6 \times 0.3 \times 25\,000$$

$$= 649\,180 \text{ N}$$

Volume of water above the dome

$$= \frac{\pi}{4} \times 10^2 \times (6 + 1.68)$$

$$= \frac{2 \times \pi \times 8.61^8 \times 1.6}{3} - \frac{\pi \times 10^8 \times 6.91}{4 \times 3}$$

$$= 602 - [248.2 - 180.6]$$

$$= 534.4 \text{ m}^3$$

Weight of water = 10000 × 534.4 = 5344000 N

$$\text{Meridional thrust} = \frac{w}{\pi \times D \times \sin\theta}$$

Since, $w = 649\,180 + 5\,344\,000 = 5\,993\,180$ N

$$\text{Meridional trust} = \frac{5993180}{\pi \times 10 \times 0.5807} = 328\,520 \text{ N/m}$$

$$\text{Meridional stress} = \frac{328520}{300 \times 1000} = 1.095 \text{ N/mm}^2$$

$$\text{Hoop stress} = \frac{wR}{t} \times \frac{\cos^2\theta + \cos\theta - 1}{1 + \cos\theta}$$

Hoop stress is maximum, where $\theta = 0 = 0.547$ N/mm^2

Provide minimum reinforcement,

$$= \left[0.3 - \left[\frac{300-200}{450-100}\right]\times 0.1\right]\times\frac{300\times 1000}{100}$$

$$= 816 \text{ mm}^2$$

Spacing, using 10 mm Ø CFRP bars ($A_f = 79$ mm^2)

$$= 96 \text{ mm say } 90 \text{ mm}$$

Provide 10 mm Ø CFRP bars at 90 mm c/c both ways.

7. Design of bottom circular beam (*B*$_3$)

Inward thrust from conical dome or $T_1 \cos\alpha = 488\,950 \sin 50°$

$$= 374\,557$$

$$\cos\alpha = 7.01/8.61 = 0.8142$$

$$\alpha = 35°49'$$

$$\sin\alpha = 0.5806$$

Outward thrust from bottom dome or $T_2 \cos\gamma = 328\,520 \cos\gamma$

$$= 328\,520\times\frac{7}{8.61}$$

$$= 267\,470$$

Net inward pressure $= T_1\cos\alpha - T_2\cos\gamma$

$$= 374\,557 - 267\,470$$

$$= 107\,087 \text{ N/m}$$

Therefore, hoop compression in the beam $= 107\,087\times\frac{10}{2} = 535\,435$ N

Assuming size of the beam = 650 × 1200 mm

Hoop stress $= \frac{535\,435}{650\times 1200} = 0.69$ N/mm^2 (comp.)

This stress being very small, its effect is neglected.

Vertical load on beam per metre run $T_1\sin 40° + T_2\sin\gamma = 488\,950(0.6428) + \left[328\,520\times\frac{5}{8.61}\right]$

$$= 314\,291 + 190\,780$$

$$= 505\,071 \text{ N/m}$$

Self-weight of the beam = 0.65 × 1.2 × 2500 = 19 500 N/m

Total design load for the beam = 505 071 + 19 500

= 524 571 N/m, say 524 570 N/m

It is proposed to support the beam on 8 columns.

Max. -ve B.M. at support $= 0.066 wR^2 \times \frac{\pi}{4}$

$= 0.066 \times 524570 \times 5^2 \times \frac{\pi}{4}$

= 679 796 N-m

Max. +ve B.M. at mid-span $= 0.03 wR^2 \times \frac{\pi}{4}$

$= 0.03 \times 524570 \times 5^2 \times \frac{\pi}{4}$

= 308 998 N-m

Torsional moment $= 0.005 wR^2 \times \frac{\pi}{4}$

$= 0.005 \times 524570 \times 5^2 \times \frac{\pi}{4}$

= 51 500 N-m

Required effective depth of the beam from max. B.M. consideration

$$= \sqrt{\frac{679796 \times 10^3}{650 \times 1.65}}$$

$d = 800$ mm

In order to have reduced shear stress, it is advisable to retain the overall depth of the beam as 0.9 m and effective depth as under.

Assuming 12 mm Ø main bars

d for top face of beam = 900 − 40 – (12/2) = 854 mm

d for bottom face of beam = 900 − 25 – (12/2) = 869 mm

$$A_t \text{ for -ve B.M.} = \frac{679796 \times 10^3}{0.9 \times 854 \times 1118.26}$$

= 790.93 mm²

Using 12 mm Ø bar ($A_Ø = 113.1$ mm²)

No. of bars required = 790.93/113.1 = 7

Provide seven 12 mm Ø CFRP bars (A_t provided = 792 mm^2)

$$A_t \text{ for +ve B.M.} = \frac{308998 \times 10^3}{0.9 \times 869 \times 1118.6}$$

$$= 353.31 \text{ mm}^2$$

Using 12 mm Ø bar ($A_{Ø}$ = 113.1 mm^2)
No. of bars required = 353.31/113.1 = 3.12 say 4
Provide four 12 mm Ø CFRP bars (A_t provided = 340 mm^2)

Design for shear and torsion:

$$\text{Load on column} = 524570 \times 5 \times \frac{\pi}{4} = 2059982 \text{ N}$$

S.F. in the beam at column support V = 2 059 982/2 = 1 029 991 N

$$\tau_v = \frac{V}{bd}$$

$$\tau_v = \frac{1029991}{650 \times 869}$$

$\tau_v = 1.82 \text{ N/mm}^2 \leq \tau_{cmax} = 2.2 \text{ N/mm}^2$, hence, O.K.

$$p = \frac{100 \times A_s}{bd} = \frac{100 \times 7 \times 113.1}{650 \times 869}$$

$$p = 0.14\%$$

$\tau_c = 0.2 \text{ N/mm}^2$ from IS 456 (Table 23)

Since $\tau_v > \tau_c$, shear reinforcement will be required to be designed:

(*a*) *Design of Shear reinforcement*

$$V_c = \tau_c bd$$
$$= 0.2 \times 650 \times 869$$
$$= 112970 \text{ N}$$

$$V_f = V - V_c$$
$$= 1029991 - 112970$$
$$= 917021 \text{ N}$$

Using 2 legged 8 mm Ø stirrups

$$S_v = \frac{\sigma_{fv} \times A_{sv} \times d}{V_f} = \frac{1262.8 \times 100.54 \times 869}{917021} = 120 \text{ mm}$$

Provide two-legged 8 mm Ø stirrups at 120 mm c/c.

(*b*) *Design for torsion:* We know that maximum torsion occurs at the point of contraflexure which is situated at an angle of 9.5° from either support.

Shear force at the point of torsion

$$V_1 = V - V \times \frac{9.5^\circ}{22.5^\circ} = 1029991 - 1029991 \times \frac{9.5^\circ}{22.5^\circ} = 595106 \text{ N}$$

Torsional moment (T) = 51500 Nm = 51 500 000 N-mm

$$\text{Equivalent shear } V_e = V_1 + 1.6 \times \frac{T}{b} = 595106 + 1.6 \times \frac{51500000}{650} = 721875.23 \text{ N}$$

Equivalent nominal shear stress,

$$\tau_{ve} = \frac{V_e}{bd}$$

$$\tau_v = \frac{721875.35}{650 \times 869}$$

$\tau_v = 1.28 \text{ N/mm}^2 \leq \tau_{cmax} = 2.2 \text{ N/mm}^2$, hence, O.K.

Longitudinal reinforcement for torsion

Equivalent B.M.

$$M_{el} = M + M_T$$

$$M_T = \frac{T}{1.7}\left(1 + \frac{D}{b}\right)$$

$$= \frac{51500000}{1.7}\left(1 + \frac{900}{650}\right)$$

$$= 72239819 \text{ N-mm}$$

$$= 72.24 \times 10^6 \text{ N-mm}$$

At point of contraflexure $M = 0$

Therefore, $M_e = M_T = 72.24 \times 10^6$ N-mm

Moment of resistance of a balanced section of the beam of given dimension

$$M_r = R.bd^2$$

$$= 0.336 \times 650 \times 869^2$$

$$= 165 \times 10^6 \text{ N-mm}$$

Since $M_r > M_T$, the section will be designed as a singly reinforced section for torsional moment.

$$A_f = \frac{M_T}{j \times d \times \sigma_{ft}} = \frac{72.24 \times 10^6}{0.977 \times 869 \times 1262.8}; A_f = 67.38 \text{ mm}^2$$

Use 12 mm Ø CFRP bar ($A_f = 113.1$ mm^2)

$$\text{Number of bars} = \frac{67.38}{113.1} = 0.6 \text{ say } 1$$

Number of 12 mm Ø CFRP bar already available at the point of contraflexure are more than 1, hence, no additional bars are to be provided.

Design of transverse reinforcement (stirrups) at the point of maximum torsion

As per code the area of cross-section (A_{fv}) of two legged closed stirrups enclosing the corner longitudinal bars is given by

$$A_f = \frac{T \times s_v}{b_1 \times d_1 \times \sigma_{ft}} + \frac{V_1 \times s_v}{2.5 \times d_1 \times \sigma_{ft}} \quad \text{(ii)}$$

But, the total transverse reinforcement shall not be less than

$$A_{sv} = \frac{(\tau_{ve} - \tau_c) b s_v}{\sigma_{sv}} \quad \text{(iii)}$$

Using two-legged 8 mm Ø stirrups

$$A_{fv} = 100.53 \text{ mm}^2$$
$$b_1 = 650 - 2(25) - 12 = 588 \text{ mm}$$
$$d_1 = 900 - 40 - 12 - 12 = 836 \text{ mm}$$

Substituting the values in equation (ii), we have,

$$100.53 = \frac{51500000 \times s_v}{588 \times 836 \times 1262.8} + \frac{595106 \times s_v}{2.5 \times 836 \times 1262.8} \quad 100.53 = 0.083 s_v + 0.225 s_v$$

$$s_v = 326.4 \text{ mm}$$

Now considering equation (iii)

$$A_{sv} = \frac{(\tau_{ve} - \tau_c) b s_v}{\sigma_{sv}}; 100.53 = \frac{(1.28 - 0.2) \times 650 \times s_v}{1262.8} \quad s_v = 181 \text{ mm}$$

Hence, provide 2-legged 8 mm Ø CFRP stirrups at 180 mm c/c to act as transverse reinforcement for torsion.

Thus, the spacing of 2-legged 8 mm Ø CFRP stirrups at 120 mm c/c near column shall be gradually increased to 180 mm c/c near point of contraflexure from either column supports.

(*C*) *Side face reinforcement:* Since, the web of the beam exceeds 750 mm, side face reinforcement shall be provided, along the two faces.

Total area of side face reinforcement = 0.1% of web area

$$= \frac{0.1}{100} \times 650 \times (900 - 300) = 390 \text{ mm}^2$$

Area of reinforcement on each side = 390/2 = 195 mm^2

Provide Three 10 mm Ø CFRP bars (A_t provided = 235 mm^2)

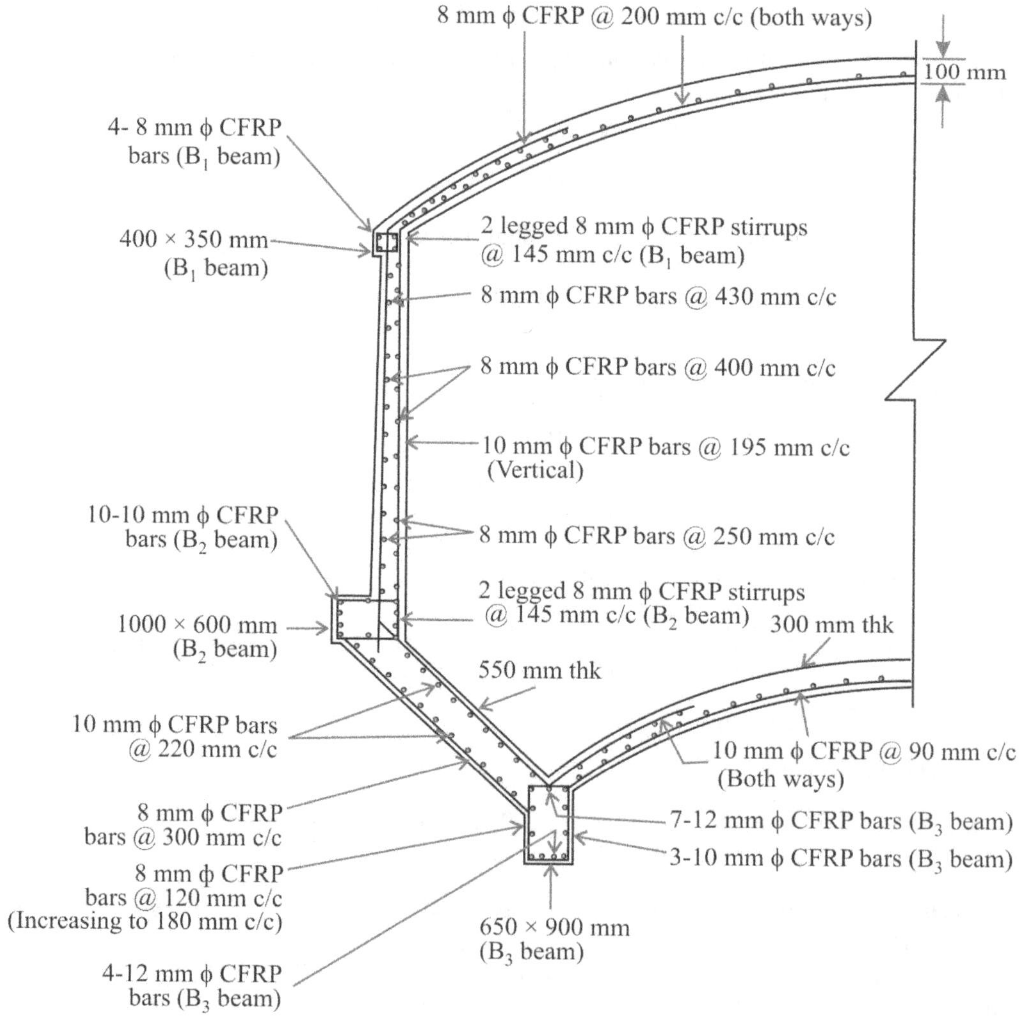

Bibliography

1. ACI 440.1R-06. *Guide for the Design and Construction of Concrete Reinforced with FRP Bars*. Michigan: American Concrete Institute, Farmington Hills, 2006.
2. ACI 440.2R-02. *Guide for the Design and Construction of Externally Bonded FRP Systems for Strengthening Concrete Structures*. Michigan: American Concrete Institute, Farmington Hills, 2002.
3. AASHTO. *LRFD Bridge Design Specifications*. Washington DC: American Association of State Highways and Transportation Officials, 1998.
4. ACI 318. *Building Code Requirements for Structural Concrete and Commentary*. Michigan: American Concrete Institute, Farmington Hills, 2005.
5. ACI 440R-96. *State-of-Art Report on Fiber Reinforced Plastic Reinforcement for Concrete structures*. Michigan: American Concrete Institute, Farmington Hills, 1996.
6. ACI Special Publications SP-215-9. *Field Applications of FRP Reinforcement: Case Studies*. Michigan: American Concrete Institute, Farmington Hills, 2003.
7. An, W., Saadatmanesh, H., and Ehsani, R. "RC Beams Strengthened with FRP Plates—Part II: Analysis and Parametric Study." *J. Structural Engrg., ASCE* 117, no. 11, 3434–3455.
8. Arduini, M., and Nanni, A. "Behavior of Pre-cracked RC Beams Strengthened with Carbon FRP Sheets." *Journal of Composites in Construction* 1, no. 2, 63–70.
9. ASCE. *Journal of Composites for Construction*. American Society of Civil Engineering (1997–2004).
10. Bakht, B. *et al.* "Canadian Bridge Design Code Provisions for Fiber Reinforced Structures." *Journal of Composites for Construction, ASCE* 4, no. 1 (2000): 3–15.
11. Bunsell, A.R., and Renard, J. *Fundamentals of Fibre Reinforced Composite Materials*. Institute of Physics, Series in: *Materials Science and Engineering*. Bristal: Institute of Physics Publishing Ltd., 2005.
12. Burgoyne, C.J. "Ductility and Deformability in Beams Prestressed with FRP Tendons." *FRP Composites in Civil Engineering*, vol. 1, Teng, J.G. ed. Elsevier-Science Ltd., 2001, pp. 15–25.
13. CAN/CSA-S806-02. *Design and Construction of Building Components with Fiber Reinforced Polymers*. Ottawa: Canadian Standrads Association, 2001.
14. CAN/CSA-S6-00. *Canadian Highway Bridge Desin Code*. Ottawa: Canadian Standards Association, 2001.
15. Marshall Industries Composites, Inc. "Material Properties of C-Bar® Reinforcing Bars." *C-Bar Research Bulletin* (97-05) 8 pp.
16. Marshall Industries Composites, Inc. "Material Properties of C-Bar® Reinforcing Bars." *C-Bar Research Bulletin* (97-06), 6 pp.
17. CAN/CSA-S6-00. *Canadian Highway Bridge Design Code*. Toronto, Ontario: Canadian Standards Association International, 2000.
18. Chawla, K.K. *Composite Materials: Science and Engineering*. Springer, 1998.
19. Clements, L.L. "Overview of Composite Materials." *43rd International SAMPE Symposium and Exhibition*, Anaheim, CA, May (1998): 58 pp.
20. DCI. *Product Manual*. USA: Diversified Composites Inc., 2002.
21. FIB. *Externally Bonded FRP Reinforcement for RC Structures*. Lausanne, Switzerland: International Federation for Structural Concrete, 2001, 130 pp.
22. Grace, N.F., and Abdel-Sayed. "Ductility of Prestressed Bridges Using CFRP Strands." *ACI's Concrete International*, Reprint, June (1998): 6 pp.
23. Grace, N.F. *et al.* "Behavior of Prestressed Concrete Box-Beam Bridges Using CFRP Tendons." *PCI Journal* 62, March-April (2006): 26–41.

24. Grace, N.F., and Singh, S.B. "Durability Evaluation of Carbon Fiber-Reinforced Polymer Strengthened Concrete Beams: Experimental Study and Design." *ACI Structural Journal* 102, no. 1, Jan.–Feb. (2005): 40–53.
25. Grace, N.F. *et al.* "Concrete Beams Reinforced with CFRP: Investigating the Shear Performance of Box-Beams Strengthened with Carbon Fiber-Reinforced Polymers." *Concrete International* 27, no. 2, Feb. (2005): 60–64.
26. Grace, N.F. *et al.* "Flexural Response of CFRP Prestressed Highway Bridge Box-Beams." *PCI Journal* 49, no. 1, Jan.–Feb. (2004): 92–104.
27. Grace, N.F. and Singh, S.B. "Design Approach for CFRP Prestressed Concrete Bridge Beams." *ACI Structural Journal* 100, no. 3 (2003): 365–376.
28. Hollaway, L.C., and Head, P.R. *Advanced Polymer Composites and Polymers in the Civil Infrastructures*. Elsevier, 2001.
29. Hollaway, L.C. *Polymers and Polymer Composites in Construction*. London: Thomas Telford Ltd., 1990.
30. Hallaway, L.C., and Leeming, M.B. *Strengthening of Reinforced Concrete Structures using Externally Bonded FRP Composites in Structural and Civil Engineering*. Cambridge: Woodhead Publishing Ltd., 2001, 327 pp.
31. ICE. "FRP Composites: Life Extension and Strengthening of Metallic Structures." *Institution of Civil Engineers, Design and Practice Guides*. London: Thomas Telford Ltd., 2001.
32. ISIS Design Manual No. 3: Reinforcing Concrete Structures with Fiber Reinforced Polymers, Prepared by A Canadian Network of Center of Excellence.
33. "ISIS Design Manual No. 4: Strengthening Reinforced Concrete Structures with Externally Bonded Fiber Reinforced Polymers." Prepared by: A Canadian Network of Center of Excellence.
34. "ISIS Design Manual No. 5: Prestressing Concrete Structures with Fiber Reinforced Polymers." Prepared by: A Canadian Network of Center of Excellence.
35. "ISIS Education Module 1:Mechanics Examples Incorporating FRP materials." Prepared by: A Canadian Network of Center of Excellence.
36. "ISIS Education Module 2:An Introduction to FRP Composites for Construction." Prepared by: A Canadian Network of Center of Excellence.
37. "ISIS Education Module 3:An Introduction to FRP Reinforced Concrete." Prepared by: A Canadian Network of Center of Excellence.
38. Jones, R.M. *Mechanics of Composite Materials*. Second edition (First Indian Reprint 2010). New York: Taylor & Francis Group, 1999.
39. Mitsubishi Chemical Corporation (MCC). "Leadline™ Carbon Fiber Tendon/Bars." *Product Manual* (1994): 14 pp.
40. NEFMAC. *Technical Data Collection*. Canada: Autocon Composites Inc., 1996, pp. 1–10.
41. Nanni, A., Focacci, F., and Cobb, C.A. "Proposed Procedure for the Design of RC Flexural Members Strengthened with FRP Sheets," *Proceedings of ICCI-98*, Tucson, AZ, 1 (1998): 187–201.
42. Open Forums. "Problems and Solutions, Comments and Opinions." *PCI Journal*, Jan.–Feb.(1998): 86–89.
43. Park, R., and Paulay, T. *Reinforced Concrete Structures*. New York: John Wiley and Sons, Inc., 1975.
44. Singh, S.B. "Strength of CFRP Stirrups and Design Approach." *Indian Concrete Journal* 82, no. 8 (2008): 11–17.
45. Singh, S.B. "Relative Strength of Carbon Fibre Reinforced Polymeric Stirrups-an Experimental Study." *ICI Journal* 7, no. 3, Oct.–Dec. (2006): 29–34.
46. Saadatmanesh, H., and Malek, A.M. "Design Guidelines for Flexural Strengthening of RC Beams with FRP Plates." *Journal of Composites for Construction* 2, no. 4, Nov. (1998): 158–164.
47. Teng, J.G. *et al.* "FRP-Strengthened RC Structures." West Sussex: John Wiley and Sons, Ltd., 2002.
48. Tokyo Rope Mfg. Co. Ltd. "Technical Data on CFCC." *Product Manual* (1993): 100 pp.
49. Vinson, J.R., and Sierakowski, R.L. *The Behavior of Structures Composed of Composite Materials*. Second edition, Dordrechit, The Netherlands: Kluwer Academic Publishers, 2002.

INDEX

A

Adhesive, 147
Advantages of NSM FRP Strengthening, 162
Alkaline Conditioning, 264
Allowable Service Stresses, 270
Analysis Procedure, 179
Anchorage Length, 168
Angle-ply Laminate, 27
Aramid Fibers, 15
Antisymmetric Laminate, 29
Axial Load Resistance, 187

B

Balanced Laminate, 27
Balanced Plate Ratio, 264, 266
Balanced Reinforcement Ratio, 70, 91, 287
Balanced Section, 133
Bar Sizes, 46, 47
Bar Identification, 46, 47
Bi-axial Ductility, 191
Bisphenol a Fumerates (BPA), 6
Bond Reduction Coefficient, 290
Bonded Tendons, 96, 291
Bonding Interface, 9
Bond Behavior, 43
Box-Beam, 93
Bridge Street Bridge, 65

C

Cable Stayed, 60
Carbon fibers, 12
CarboShear L, 60
Case Study Problems, 116, 213, 280
CFCC Strands, 282
CFRP Concrete Box-Beam, 132
CFRP Fabric, 264
CFRP Plate, 264
Characteristics of honeycomb core, 23
Circular Columns, 186, 192, 193
Chlorendics, 6
C-glass, 11
Classification of FRP Reinforcements, 54
Cloth, 12
Coefficient of thermal expansion (CTE), 40
Columns, 185
Commercial reinforcing fibers, 14
Completely Wrapped Members, 160
Composite Concrete Section, 288
Composite Mechanics, 26
Compression Failure, 83
Compression Molding, 24
Compressive Behavior, 42
Compression Reinforcement, 181
Concrete Crushing Mode of Failure, 71
Confinement, 186
Confinement Limits, 187
Confinement Pressure, 187
Construction Practices, 48
Concrete Surface Preparation, 147
Constructability, 90
Composite Product Forms, 27
Concrete Cover Requirements, 90
Concrete Tooth Model, 163
Controlling Failure Modes, 152
Cover Delamination, 152
Cracking, 73, 88
Cracking Moment, 87, 92, 134, 287
Crack Width, 74
Crack Spacing, 162
Creep Rupture, 43, 154
Critical Compression Depth, 264
Critical Plate Ratio, 267
Creep Rupture Stress Limits, 78, 154
Cross-linking, 7

Cross-ply Laminate, 27
Crushing of Concrete, 267
Curvature Ductility Factor, 190

D

Debonding Failures, 147
Definition of Material Characteristics, 91
Deflections, 73,75, 88, 102
Deformability, 89
Delamination, 152
Delamination Factor, 265
Density of FRP Bars, 40
Depth of Stress Block, 84
Design Approach, 90, 169
Design Considerations, 10
Design Example, 104, 195, 271
Design for Shear Strength, 171
Design Material Properties, 149, 263
Design Philosophy, 67, 143
Design Guidelines, 61
Design Moment Capacity, 97, 100, 270
Design Principles, 182
Description of fibers, 11
Detailing, 175
Detailing of Shear Stirrups, 80
Development of Design Methods in Japan, 55
Development Length of NSM FRP Bars, 175
Direct Method, 18, 77
Displacement Ductility Factor, 190
Dry-Heat Conditioning, 264
DT-Beam, 93, 280
Ductility, 153
Ductility Oriented Retrofit, 189
Dump Truck Configuration, 284
Durability, 44
Durability Based Design, 263

E

Effects of Fiber Length, 9
Effective Cross Sectional Area, 286
Effective Moment of Inertia, 75, 88
Effective Stress, 193
End Anchorage, 144, 147
E-glass, 11
Environmental Conditioning, 264
Environmental Reduction Factor, 68, 150, 263
Epoxy Resin, 7
Equivalent Moment of Inertia, 174
Equivalent Stress-block Parameter, 86
Evolution of FRP, 2
Exercise Problems, 30, 50, 66, 139, 258, 280
Existing Substrate Strain, 152, 170, 264, 272
External Post-tensioning, 282
External Strengthening, 143, 185

F

Fabrication Scheme, 17
Fabrication of Stirrups, 24
Failure Modes, 82, 144, 171, 178, 188, 266
Fatigue, 44, 154
Fatigue Stress Limits, 79, 154
FHWA Research Program, 64
Fiber placement, 20
Filament winding, 19
Fillers, 5
Fire Endurance, 149
Flanged Section, 104
Flexural Capacity, 92, 132
Flexural Design Approach, 169, 178
Flexural Design Philosophy, 69, 148
Flexural Design Considerations, 148
Flexural Ductility, 189
Flexural Failure, 144
Flexural Failure Modes, 144
Flexural Plastic-Hinge Confinement, 194
Flexural Resistance, 86
Flexural-Shear Cracking, 118
Flexural-Shear Failure, 189
Flexural Strength, 70, 83, 155, 170
Flexural Strengthening, 151, 162, 164, 177
Footing Failure, 189
Forms of Glass Fiber Reinforcements, 12
Freeze-Thaw Conditioning, 264
FRP Composites, 25
FRP Composite Applications, 26
FRP Construction Activities in Europe, 58
FRP for Retrofitting and Repair, 57
FRP Prestressing in the USA, 61
FRP Rupture Mode of Failure, 71
FRP Surface Preparation, 147
Future Uses of FRP, 57

G

Generally Orthotropic Lamina, 27
Generic Shear Crack, 171
Glass Fibers, 11

Guaranteed Breaking Load, 286
Guaranteed Strength, 42, 286
Guaranteed Strain, 42

H

Hand layup process, 18
Handling and Storage of Materials, 48
Hardeners, 8
High temperature effects, 40
History and Uses of FRP Technology, 53
Historical Development of FRP Tendons,
Honeycomb core, 23
Hybrid Reinforcements, 17

I

I-Beam, 94
Ideal Flexural Strength, 193
Importance of Polymer Matrix, 4
Indirect Method, 18
Interfacial Debonding, 146
Intermediate Flexural Crack, 146
Intze Tank, 300
ISIS Canada Design Approach for Flexure, 81, 176
ISIS Canada Design Guidelines for Shear Strengthening, 182
Isophthalic Polyester, 6
Iso Polyester, 6

L

Laminate Terminology, 27
Laminates Types, 28
Lap Splice Clamping, 194
Lap Splice Failure, 189
Leadline Tendons, 280
Limitations of Pultrusion Process, 21
Load Distribution, 283
Long-term Deflection, 77, 174

M

Macromechanics, 26
Mats, 12
Manufacturing of Composites, 17
Manufacturing Methods, 18
Material Characteristics of FRP Bars, 39, 91
Material Properties, 149
Material Resistance Factors, 177
Matrix Polymer, 4
Maximum Allowable Plate Ratio, 266
Maximum Confinement Pressure, 187
Mean Stress Factor, 265
Micromechanics, 26
Minimum Confinement Pressure, 187
Minimum Crack Spacing, 162
Minimum Depth, 75, 89
Minimum Flexural Resistance, 86
Minimum FRP Reinforcement, 73
Minimum Shear Reinforcement, 80
Mode of Failure, 267
Modulus Grades, 46
Modulus of Rupture, 287
Moment Capacity, 91
Moment of Resistance, 179
Multi-Axial Fabric for Structural Components, 24

N

NEFMAC Grids, 280
Nominal Flexural Capacity, 71
Nominal Flexural Strength, 170
Nominal Moment Capacities, 100, 102, 268
Nominal Shear Strength, 119, 159
Nominal Strength, 152
Nonlinear Response, 103
Non-prestressing Tendons, 136
Non-prestressed Soffit Plates, 143
NSM Bars, 161
NSM FRP strengthening, 162

O

Organic fibers, 16
Orthopthalic Polyester, 6
Over-reinforced Beams, 100

P

Physical and Mechanical Properties, 39
Placement and Assembly of Materials, 49
Plastic Curvature, 191
Plastic-Hinge Failure, 188
Plate-end Debonding, 145
Plate-end Interfacial Debonding, 146
Polymer Matrix, 4
Polyster Resin, 6
Post-tensioning Tendons, 96
Potential Failure Modes, 188
Precast Section, 287
Prestressed Concrete Box-Beam, 132
Prestressed Concrete Double-T Beam, 120

Prestressed Piles, 63
Prestressed Soffit Plates, 144
Processing of Composites, 17
Properties of Commercial Reinforcing Fibers, 14
Properties of Spectra Fibers, 16
Proportion of the FRP Plate, 266
Pull Forming Process, 22
Pultrusion Process, 20
Punching Shear Strength, 81

Q

Quality Control and Inspection, 49

R

Rectangular Section, 103
Rectangular Columns, 193
Rehabilitation and Strengthening, 60
Reinforcing Fibers for Structural Composites, 8
Reinforced and Prestressed Concrete, 59
Reinforcement Ratio, 70, 291
Reinforcement Limits, 161
Relative Humidity, 264
Research and Demonstration Projects, 62
Resin Transfer Molding (RTM), 22
Retrofit, 188
Retrofitted Columns, 189
Review of FRP Composites, 3
Ring Beam, 303
Rovings, 12
Rupture of Plate, 267
Rupture Strain, 263

S

Salt-Water Solution, 264
Sandwich Construction, 23
Section Shear Strength, 151, 170
Seismic Retrofit, 188
Serviceability, 73, 88, 154, 174
S-glass, 11
Shear, 79
Shear Behavior, 43
Shear Crack, 171
Shear Demand, 193
Shear Design Philosophy, 79
Shear Failure, 144, 188
Shear Failure Modes, 80
Shear Strength, 79, 151, 159, 169, 171, 191
Shear Strengthening, 158, 166, 182
Shear Strengthening Schemes, 183
Side Face Reinforcement, 316
Significantly Under-reinforced Beams, 92
Sika CarboDur, 50, 272
SikaDur30, 272
SikaWrap Hex 103C, 272
Singly Reinforced Rectangular Section, 155
Slenderness Limits, 186
Soffit Plates, 143, 144
Spacing of Bars, 90
Spacing of FRP Strips, 161
Spectra, 16
Specified Rupture Strain, 263
Specified Tensile Strength, 263, 286
Stacking Sequence, 28
Strains in Bonded Tendons, 96
Strength Grades, 46
Strengthening Limits, 148
Strengthening of Beam and Slabs, 177
Strengthening of Columns, 185
Strength Oriented Retrofit, 189
Strength Reduction Factor, 72, 154
Stresses, 102
Stress-block Parameter, 86
Stress Corrosion, 44
Stress-Ribbon Bridge, 56
Structural Considerations in Processing Polymer Matrix Resins, 8
Structural Fire Endurance, 149
Substrate Strain, 152, 170, 264, 272
Surface Geometry, 46
Surface Preparation, 147
Symmetric Laminate, 27, 28

T

Tail Length, 81
T-Beams, 181
Tensile Behavior, 41
Tension Failure, 85
Tensile Strength, 263
Theoretical Development of Design Equations, 91
Thermosetting Matrix, 5
Thermoplastic Matrix, 5
Time-Dependent Behavior, 43
Tokyo Rope, 286
Tooth Model, 163
Torsion, 314
Transfer of Force in NSM FRP Bars, 176
Two-sided Wrap, 161

Two-way Concrete Slab, 81
Typical FRP Reinforced Concrete Structures in Japan, 55

U

Unbonded Post-tensioning Tendons, 136, 294
Under-reinforced Beams, 98
Ultimate Breaking Load, 286
Ultimate Moment Carrying Capacity, 97
Ultimate Flexural Strength, 155
Ultimate Shear Strength, 171
Ultra-high-molecular-weight-polyethylene fibers, 16
Uni-axial Ductility, 191
U-strip, 147
U-wrap, 161

V

Vinylesters (VE), 6

W

Waterfront Facilities, 63
Woven Roving, 12
Wrapped Members, 160

www.ingramcontent.com/pod-product-compliance
Lightning Source LLC
LaVergne TN
LVHW062158200926
848484LV00006B/1283